With you

THE DENIABLE DARWIN

&

OTHER ESSAYS

THE DENIABLE DARWIN

& OTHER ESSAYS

DAVID BERLINSKI

EDITED BY DAVID KLINGHOFFER

Description

This book collects essays published in journals including *Commentary, The Weekly Standard,* and elsewhere. It centers on three profound mysteries: the existence of the human mind; the existence and diversity of living creatures; and the existence of matter. How they did they come into being?

The author, Dr. David Berlinski, is a senior fellow at the Discovery Institute and formerly a fellow at the Institut des Hautes Études Scientifiques in France. His other books include *The Devil's Delusion: Atheism and Its Scientific Pretensions, Newton's Gift,* and *A Tour of the Calculus.*

Publisher's Note

This book is part of a series published by the Center for Science & Culture at Discovery Institute in Seattle. Previous books include *Darwin's Conservatives: The Misguided Quest,* by John G. West and *Traipsing into Evolution: Intelligent Design and the Kitzmiller vs. Dover Decision,* by David K. DeWolf et al.

Library Cataloging Data

Deniable Darwin & Other Essays by David Berlinski (1942–)
Edited by David Klinghoffer.

557 pages, 6 x 9 x 1.25 inches & 1.8 lb, 229 x 152 x 3.2 cm. & 0.81 kg
Library of Congress Control Number: 2009935347
BISAC Subject: SCI027000   SCIENCE / Life Sciences / Evolution
BISAC Subject: SCI080000   SCIENCE / Essays
BISAC Subject: SCI034000   SCIENCE / History
ISBN-13: 978-0-9790141-2-3   ISBN-10: 0-9790141-2-3 (paperback)
ISBN-13: 978-0-9790141-3-0   ISBN-10: 0-9790141-3-1 (hardback)

Publisher Information

Discovery Institute Press, 208 Columbia Street, Seattle, WA 98101
Internet: http://www. discovery.org/
Published in the United States of America on acid-free paper.
First Edition, First Printing. October 2009.

For Arthur Cody
with gratitude and admiration.

CONTENTS

About the Author

David Berlinski is a senior fellow in the Discovery Institute's Center for Science and Culture. He is the author of numerous books, including *A Tour of the Calculus* (1996), *The Advent of the Algorithm* (2000), *Newton's Gift* (2000), *The Secrets of the Vaulted Sky* (2003), *A Short History of Mathematics* (2004), and *The Devil's Delusion: Atheism and Its Scientific Pretensions* (2008). Two of his essays, "On the Origins of the Mind" (November 2004) and "What Brings a World into Being" (March 2001), have been anthologized in *The Best American Science Writing 2005* and *The Best American Science Writing 2002* respectively.

Berlinski received his PhD in philosophy from Princeton and was later a postdoctoral fellow in mathematics and molecular biology at Columbia. He has authored works on systems analysis, differential topology, theoretical biology, analytic philosophy, and the philosophy of mathematics, as well as three novels. He has also taught philosophy, mathematics and English at Stanford, Rutgers, the City University of New York and the Université de Paris. He has held research fellowships at the International Institute for Applied Systems Analysis in Austria and the Institut des Hautes Études Scientifiques. He lives in Paris.

INTRODUCTION

WHEN IN THE THIRTEENTH CENTURY, POPE INNOCENT III REAL-
ized that the Catholic faith had become too complicated an intellectual structure easily to be grasped, he created the Dominican order
and charged it with the double responsibilities of combating error and
inculcating belief. A difficult task was performed with great industry
by men such as Albertus Magnus and Thomas Aquinas. The regnant
position of the Catholic faith came to an end in the early seventeenth
century. After viewing the heavens through a telescope, Cardinal Bellar-
mine acknowledged that in choosing between Copernicus and Ptolemy,
the Church might be forced by the facts to choose Copernicus. The facts
engendered the choice. The great scientific revolution of the West was
set in motion.

It is hardly surprising that from their inception in the seventeenth
century to the present day, the physical sciences, like the Catholic
Church, have required a cadre of dedicated Dominicans, men prepared
to spread the faith while dispelling doubt. Voltaire played this role with
respect to the *Principia Mathematica*, making Newton's noble work
known throughout learned circles in Europe. In the United States and
England, scientists today appear before the public as mandarins to the
manner born. Their books fill every bookstore shelf, and their views have
by a process of Internet multiplication and amplification been conveyed
to an audience that might otherwise have been deprived of the sound of
their voices.

The advent of Newtonian mechanics, many scientists believe, marks
a radical departure in intellectual history. Before Newton, there are all

the old failed systems of thought, and afterwards, there is Science. This is the popular view and so the myth. Like many myths, it contains a portion of the truth. Nothing like Newton's *Principia* appeared before the seventeenth century. Nothing like it occurred afterwards either, circumstances that suggest a singular moment in intellectual history.

THERE ARE NOW FOUR great physical theories: Newton's mechanics, of course, Clerk Maxwell's theory of the electromagnetic field, Einstein's theories of special and general relativity, and quantum mechanics. Each is fundamental. The laws of nature that they express are so compressed and therefore so vatic that they may be written down in no more than half a page. They are extraordinary in their quantitative precision, theory and experiment agreeing in both general relativity and quantum mechanics to something like 13 decimal places. And they are unique. Where the Newtonian revolution has been extended, it has been extended in essentially Newton's terms: A scientific theory must be mathematical; and it must appeal to theoretical structures very far from experience. To the extent that there are sciences that go beyond the great physical theories, they have without exception been parasitic. What we understand of biology we understand in terms of its chemistry; what we do not understand in these terms, we do not understand.

It is odd to imagine that intellectual structures so singular could carry a general burden of belief, one often expressed in religious terms. Nonetheless, as others were once prepared to say that they believed in God, a great many men and women are prepared to affirm that they *believe in* science. It is widely considered inappropriate not to. When the great logician Kurt Gödel remarked to a luncheon companion at the Institute for Advanced Study that he did *not* believe in the physical sciences, his comment, like his paranoia, was considered shocking enough to have become a part of his legend.

Just what follows from a belief in science? It is not at all easy to say. "After the devastating attacks by Wittgenstein and Quine," the philosopher Paul Horwich has argued, "it is now widely believed that the sciences *exhaust* what can be known."

Do they indeed?

If the sciences are identified with the four physical theories, then it is plain that most of what *we* know, from contract law to the history of the Roman empire, has nothing to do with science at all. The claim that "the sciences *exhaust* what can be known," to take the obvious example, is neither an assumption nor a conclusion of any physical theory. If it is known, it is not science, and if it is not known, it has no interest.

If, on the other hand, the sciences are open-ended, then the thesis that the sciences exhaust what can be known comes to nothing more than the observation that what we know we tend to call scientific. This is true enough, but it is not interesting.

What remains of belief in the sciences when belief has precious little content is an attitude. The great virtue of a scientific education, Ernest Nagel once remarked, lies just in its capacity to provoke a liberation from illusion. Nagel did not specify *which* illusions he had in mind, but plainly he believed in getting rid of them all. In this, he was expressing a characteristic deflationary note, a kind of sullen skepticism.

MY OWN VIEW, REPEATED in virtually all of my essays, is that the sense of skepticism engendered *by* the sciences would be far more appropriately directed *toward* the sciences than toward anything else. It is not a view that the scientific community has *ever* encouraged. The sciences require no criticism, many scientists say, because the sciences comprise a uniquely *self*-critical institution, with questionable theories and theoreticians passing constantly before stern appellate review. Judgment is unrelenting. And impartial. Individual scientists may make mistakes, but like the Communist Party under Lenin, science is infallible because its judgments are collective. Critics are not only unwelcome, they are unneeded. The biologist Paul Gross has made himself the master of this attitude and flaunts it on every conceivable occasion.

Now no one doubts that scientists *are* sometimes critical of themselves. Among astrophysicists, for example, backbiting often leads to backstabbing. The bloodletting that ensues is salutary. But the process of peer review by which grants are funded and papers assigned to scientific journals, is, by its very nature, an undertaking in which a court reviews its own decisions and generally finds them good. It serves the useful purpose of settling various scores, but it does not—and it cannot—achieve the ends that criticism is intended to serve.

If the scientific critic finds himself needed wherever he goes, like a hanging judge he finds himself unwelcome wherever he appears, all the more reason, it seems to me, that he really should get around as much as possible.

THE ESSAYS THAT I have collected in this volume were written over roughly a 15-year period, the earliest in 1994, the most recent in 2008. Many were written specifically for *Commentary*. These essays owe much to editor Neal Kozodoy's brilliant efforts on their behalf.

Although they range over a diverse number of subjects, these essays are the expression of concerns that have occupied me for more than forty years. There is a connection in this respect between my first book, *On Systems Analysis: An Essay on the Limitations of Some Mathematical Methods in the Social, Political and Biological Sciences*, which I wrote for the MIT Press in 1976, and my last essay. The essays thus comprise a critical meditation, but one whose focus has been the sciences rather than the arts.

"You must clear your mind of cant," Dr. Johnson advised Boswell. It is not easy; I have been trying. I have named this collection *The Deniable Darwin* because it has been this essay more than any other that has prompted a gratifying commotion. And for obvious reasons. Darwinism has become far more than a narrow and not very interesting nineteenth-century theory of speciation; it is a way of thought, an attitude, and so an ideology. As political correctness is the reigning ideology of social and political life, so Darwinism is the reigning ideology of scientific life.

Both ideologies are forms of cant; they are expressions of attitude and in the case of Darwinism, the attitude runs straight through almost everything that has loosely collected under the name of science itself.

It is most typically an attitude of confident conviction in the scientific enterprise itself, and it has as its correlative an attitude of confident contempt for doubt or even rational uncertainty. There is obviously Darwinism among the Darwinians; but there is Darwinism where physicists get together, and psychologists as well. It is a vile but universal fluid, one that seeps into every interstice, with even journalists splashing happily in the stuff. So I have decided to call this collection *The Deniable Darwin* and I hope that the reader understands that what is being denied is more than that poor drab Darwin ever advanced. I am against the spirit that he engendered.

It remains for me to acknowledge with pleasure my debt to the Discovery Institute. It is not a simple matter. The Institute has since its inception been the subject of a richly conceived campaign of vilification. Paul Gross and Barbara Forrest, who represent the triumph in academic politics of the principle of reflex action, have discovered to their great alarm that the Discovery Institute has a political agenda. Their ensuing indignation has the quality of squid ink. It is black. It sheds no light. And it permits a fast get-a-way.

The Discovery Institute is a think tank and in this respect no different from the Hudson Institute, the Cato Institute, the Hoover Institution, and at least a dozen other institutions that ply the same trade and ply that trade by the same means. The National Center for Science Education, which is devoted entirely to persuading a reluctant public that in all respects Darwin was right and that they are wrong, carries on its affairs untroubled by any accusation that its mission is compromised by its politics. And this is as it should be. How else does anyone in a democratic society make his views known?

Perhaps I should add the obvious.

The Discovery Institute does not propose to inaugurate a form of theocratic kingship in the United States, and far from being a Christian right-wing fundamentalist, I am myself a secular Jew, one who has faithfully maintained since the age of 13 a remarkable indifference to the religious life.

—David Berlinski
Paris, December 2008

BIBLIOGRAPHICAL NOTE

There follows a list of the essays that appear in this volume that previously appeared elsewhere, with biographical information about their original place of publication:

"The Soul of Man Under Physics," *Commentary*, January 1996, 38-46.

"The Deniable Darwin," *Commentary*, June 1996, 19-29.

"Denying Darwin: David Berlinski & Critics," *Commentary*, September 1996, 4-39.

"Keeping an Eye on Evolution: Richard Dawkins, a Relentless Darwinian Spear Carrier, Trips Over Mount Improbable," review of *Climbing Mount Improbable*, by Richard Dawkins (Norton, 1996), *Globe & Mail*, November 2, 1996.

"The End of Materialist Science," *Forbes ASAP*, December 2, 1996, 146-155.

"Full House Follies," review of *Full House: The Spread of Excellence from Plato to Darwin*, by Stephen Jay Gould (Harmony Books, 1996), *Origins & Design*, Winter 1997, 29-31.

"Gödel's Question," in William A. Dembski ed., *Mere Creation: Science, Faith & Intelligent Design* (InterVarsity Press, 1998), 402-426.

"Was There a Big Bang?" *Commentary*, February 1998, 28-38.

"What Brings a World into Being?" *Commentary*, April 2001, 17-23.

"Where Physics and Politics Meet," review of *Memoirs: A Twentieth Century Journey in Science and Politics*, by Edward Teller (Perseus, 2001), *The Weekly Standard*, November 26, 2001, 33-35.

"God, Man and Physics: Does the Universe Aim at Human Life?," review of *The God Hypothesis: Discovering Design in our "Just Right" Goldilocks Universe*, by Michael A. Corey (Rowman & Littlefield, 2001), *The Weekly Standard*, February 18, 2002, 35-38.

"Einstein and Gödel: Friendship Between Equals," *Discover*, March 2002, 38-42.

"Lucky Jim," review of *Genes, Girls, and Gamow: After the Double Helix*, by James D. Watson (Knopf, 2002), *The Weekly Standard*, March 18, 2002, 41–42.

"Has Darwin Met His Match?" *Commentary*, December 2002, 31–41.

"Darwinism Versus Intelligent Design: David Berlinski & Critics," *Commentary*, March 2003, 9–31.

"A Scientific Scandal," *Commentary*, April 2003, 29–36.

"A Scientific Scandal? David Berlinski & Critics," *Commentary*, July–August 2003, 10–26.

"On the Origins of the Mind," *Commentary*, November 2004, 26–35.

"All Those Darwinian Doubts," adapted from a shorter opinion piece in the *Wichita Eagle*, March 9, 2005.

"Academic Extinction: More and More, Evolutionary Theory is Becoming Nothing More than Darwinian Mantra," *Daily Californian* (U.C. Berkeley), April 1, 2005.

"The Strength of Natural Selection in the Wild," Discovery Institute, April 25, 2005.

"An Open Letter to the Amazing Randy," Discovery Institute, June 13, 2005.

"Our Silent Partners," Discovery Institute, August 31, 2005.

"Copernicus Stages a Comeback," *London Gazette*, September 27, 2005.

"On the Origins of Life," *Commentary*, February 2006, 22–33.

"Inside the Mathematical Mind," review of *The Mathematician's Brain*, by David Ruelle (Princeton University Press, 2007), *New York Sun*, August 29, 2007.

"Connecting Hitler and Darwin," *Human Events* (online edition), April 18, 2008.

"The Scientific Embrace of Atheism," Pajamas Media, April 28, 2008.

1996

THE SOUL OF MAN
UNDER PHYSICS

W HAT IS IT? A SENSE OF UNEASE, PERHAPS, SOME PERSISTENT
feeling, as the century slips into the darkness, that the larger
structures of scientific thought and sentiment are disembodied, disorderly somehow. The feeling is familiar, like the taste of tea. A long moment in our collective experience is coming to an end.

The British novelist (and physicist) C. P. Snow argued in the golden
1950s that contemporary culture had acquired two contentious heads,
the one scientific, the other humanistic, each unable to understand the
other and both committed to commandeering the conversation. Snow's
diagnosis exacerbated the disease: intellectual life seemed suddenly to
divide along a fissure separating those who understood the second law
of thermodynamics from those who did not. There ensued a period of
comical soul-searching, as literary critics in particular realized with dis-
may that, just as Snow had suggested, they were incapable of following
a rudimentary scientific argument. Viewed from the perspective of the
present, the whole episode takes on an ineffable air of poignancy, the
1950s comprising perhaps the last years in which educated men retained
the capacity to be embarrassed by their ignorance.

Today the rich fruity preposterous discourse of the academic Left
would seem to suggest that Snow's bifurcated culture has become flamboyantly fractured. Like a glass pane struck a sharp blow, the large project of the physical sciences has been shown at last to be socially regres-
sive, or politically hegemonistic, or hopelessly gendered, or ethnically
contaminated, or simply anachronistic. Quantum mechanics, after all,

was anticipated—so we are told—by proud African warriors, men who in the Sudan or the marshes of the Nile conveyed Schrödinger's wave equation as an oral tradition.

The contemporary American philosopher John Searle has struck a countervailing note. Having heard out Indian philosophers convinced that in past lives they had been Pakistanis, or were destined in future lives to become Pakistanis once again, Searle affirmed magisterially that the "contemporary scientific world view is simply not up for grabs." The Darwinian theory of evolution and the atomic theory of matter are irrefragable.

It is wonderful that Searle should think to uphold two nineteenth-century theories that could both vanish into the void without affecting the contemporary scientific world view in the slightest. Still, he is right in his larger purpose. It is one culture that we now inhabit, and not two or ten thousand. Some part of the ache that today affects the human heart arises, indeed, from a kind of intellectual claustrophobia, a sense that a single and perhaps alien system of thought has somehow preempted the possibilities for the description of the universe, the consolation of the soul.

SOMETIME IN THE SEVENTEENTH century, the dry tinder of discovery, struck profitlessly throughout so many long centuries, blazed suddenly into life. The physical sciences came into creation. That is the common view, and it is completely correct. Before the seventeenth century there was nothing, and afterward, everything.

Myth places the miracle at the moment Isaac Newton conceived the idea that gravity might control both the fall of objects toward the center of the earth and the movement of the planets in the night sky. But the miracle was in fact divided, one half physical, the other mathematical, and it was the mathematical miracle that struck the deeper. Before the laws of nature could be revealed and then recorded, the real world had to be re-created in terms of the real numbers.

The real numbers—not only the natural, or counting, numbers but zero, the negative numbers, the fractions, and the irrational numbers as well—entered the Western imagination in the sixteenth and seventeenth centuries, the creation of ebullient Italian geometers and mathematicians. Creation is the right word, signifying as it does a spontaneous intellectual act, one that brings something into being. But whatever their origins, the real numbers also have a workaday identity, one that is expressible in terms of their infinite decimal expansion. The square root of two may thus be expressed as 1.41421356..., with the dots indicating a continuously evolving identity as ever more numbers are added to the list. With these strange rich numbers in place, the number system is in a certain sense complete. The thing is contained in itself. It is whole. There are no gaps where the numbers simply lapse.

The introduction of the real numbers allowed the landscape of mathematical analysis to be suffused with a thrilling light, one akin in its own way to the light that may be seen or sensed in the great Renaissance paintings. In that lit-up landscape, the infinite was, for the first time in history, charmed into compliance. Men gained the eerie power to ask of certain processes: *suppose they go on forever, what then?*

Within the scheme of thought known as the calculus, discovered almost simultaneously by Gottfried Leibniz and Newton, they found an entirely comprehensible answer. Relationships between and among numbers could be expressed by the flexible and finely geared instrument of a function, an invention that permitted mathematicians to describe numerical patterns as if they were living processes. The concept of a limit made its first appearance on the mathematical stage, denoting the place where certain things tend and then accumulate. (As the fractions get smaller and smaller, for example, they tend inexorably toward a limit at zero.) Sequences were given voice, and strange series contemplated; hidden for centuries from human sight, an array of mathematical operations and processes became for the first time visible.

The creation of the real number system and its perfection in the calculus represented an inward explosion, one that took place against the backdrop of a larger, outward explosion: the realization by the great natural philosophers of the seventeenth century that these same real numbers might be assigned to physical magnitudes such as force, mass, and distance, thus creating an essentially quantitative representation of the world. To be sure, human beings since time immemorial had used the numbers to count and to measure and to reckon. As the animals trooped aboard the ark, Noah, no doubt, ticked them off on his fingertips, two by two. But the representation of the real world in terms of the real numbers was different. It was a richer and more compelling representation, one that for the very first time allowed some ineffable abstract aspect of things to be localized on a computational canvas.

It was change under the aspect of continuity, the very mutability of matter, that was captured on that canvas. Continuity is a manifestation of seamlessness: continuous processes do not break, or snap, or interrupt themselves. And continuity is a physical as well as a mathematical property. It lights up the night sky as the stars crawl solemnly across the heavens. The undulating quantum wave occupying all portions of an infinite dimensional Hilbert space is continuous, and so is the great worm of time that humps and slithers its way through the theory of relativity.

The calculus is humanity's great meditation on the theme of continuity, and the concentration on continuous change is what mysteriously lends to the calculus its diamond-hard edge, its uncanny powers of specificity.

THE REPRESENTATION THAT MATHEMATICS affords of the real world is not complete—no symbolic instrument is ever fully adequate to reality—but it is larger and more spacious and more commanding than any before discovered. Still, a representation can only do so much, namely, reconvey an aspect of reality, the familiar world finding itself peeping from an unfamiliar mirror. The larger promise of the physical sciences has always been that some striking *revelation* lies behind the new, the odd and unfamiliar representation, some way of coordinating appearances and enforcing a sense of order on the vagaries of things.

The world, the physical sciences affirm, is not merely depictable, but *comprehensible*. It has a rational structure. It is animated by a great plan. The catalogue of its facts may be compressed into a few infinitely pregnant laws. There is a form of words adequate to the complexity of experience.

These words mathematics does not in itself provide. They arise when the detritus of experience is sifted by a profound physical imagination—Newton, for example, discovering that all objects in the material universe attract one another in proportion to their mass and in inverse proportion to the square of the distance between them. In the tide of time, there have been only four absolutely fundamental physical theories: Newtonian mechanics; Clerk Maxwell's theory of electromagnetism; Einstein's theory of relativity; and quantum mechanics. They stand in history like the staring stone statues on Easter Island, blank-eyed and monumental.

Each theory is embedded in a continuous mathematical representation of the world; each succeeds in amalgamating far-flung processes and properties into a single, remarkably compressed affirmation, a tight intellectual knot. The supreme expression of each theory is a single mathematical law, one expressed as an equation: a statement in which something that is unknown is specified by contingencies arranged in a certain way. And each of the great theories contains far more than it states, the laws of nature fantastically compressed, as if they were quite literally messages from a timeless intellect.

ENTERTAINING FAINT IMPRESSIONS FROM the far side of the cosmos, where the dust-clouded galaxies wink and shine forlornly, the human eye can convey a human being to the raveled edge of space. But looking anxiously into time and forever asking, *when*—when shall I find love, when happiness, when God?—the soul is shrouded, its basilisk eye unseeing. The past lies fixed and frozen back there somewhere; the future is not only indeterminate and incomplete but opaque, the horizon lit up by only a few large and gloomy certainties: death, taxes, feminism. Human beings are suspended between the unknowable and the inevitable, the place they have always occupied and the place, one suspects, they will always occupy.

The great physical theories provide an exception to this depressing human condition; they are given over to the conquest of time. Under their influence, the universe becomes temporally transparent, at least in part. The conquest of time is written into the symbolic instruments of the physical sciences, their very way of describing things. The laws of nature specify processes in the world, as when (say) a philosopher dropped from a great height accelerates toward the center of the earth, his position changing at every passing moment.

The instruments of discovery within the physical sciences are differential equations. Like equations everywhere, they express a relationship but conceal an unknown. Solving the equation is, just as in elementary algebra, a matter of uncovering the unknown, and the extraordinary feature of the calculus is that it clears a way in which such equations may in general be solved.

The result is an instrument of remarkable power. Galileo's law of falling bodies, for example, gives direction not only to the philosopher flying downward but to objects in free fall everywhere. That philosopher having commenced his descent by slipping (so I imagine) on a copy of Carl Sagan's *Cosmos*, one left carelessly on a narrow mountain ledge, Galileo's law controls events thereafter, predicting precisely where the philosopher will be at every subsequent moment, the law unraveling forward in time like film thread moving over sprockets, the philosopher flash-frozen frame by frame, from the moment of his initial spastic efforts to regain his balance to his comic cart wheeling in space and thence to the final ignominious wet plop some seconds later when he becomes one again with the Platonic forms which in life afforded him such satisfaction.

Nice, eh? The law of nature, I mean; and nice in the way it reveals in miniature how the physical sciences penetrate the future: by the successful combination of local circumstances—*he started out here*—with universal and deterministic processes—Galileo's law. The laws of nature are general. A local flash ties them to place and time, the contingent circumstances of the real world, whereupon they acquire prophetic powers, the specification of the particular acting as a flare illuminating an ancient text.

The *predictable future* is the exuberant manifestation of the Western scientific imagination. Eyebrows waggling and bony fingers pointing, the great scientists stand and control the flow of time, investing their calculations with a retrospective sense of inevitability, if not of moral urgency. So entrenched by now is this notion of the predictable future that it has become a category of consciousness and discourse, a familiar intellectual fixture, one derived from the very nature of description itself.

In contemporary culture, to be sure, prophecy is a debased currency, the prophets bunched up on television where they may be found offering astrological advice on love and work. But the mathematical sciences are entirely different. In physics, prophecies command a degree of accuracy that must be reckoned miraculous. Quantum electrodynamics penetrates the very heart of matter to something like twenty decimal places. It is as if, in determining the distance from New York to Los Angeles, theory and measurement would diverge by no more than the width of a single human hair.

The great theories, singular in so many ways, are singular in this as well: no other intellectual accomplishment exhibits their prophetic powers. In them, pure thought and physical experience coincide to a degree never achieved in any other domain of intellectual or practical life—coincide, that is, to a degree unparalleled in the entire experience of the race, enabling human beings to achieve a specification of points and places in the future utterly at odds with our habitual inability to say where our keys may have been misplaced, or our hearts lost.

BUT NOW IT IS undergoing dissolution, the predictable future. Quantum mechanics served long ago to introduce a note of alien doubt into the deterministic scheme, countenancing a view of the subatomic world in which quanta bounce around for no good reason whatsoever and measurements are bound by an iron collar of uncertainty. But quantum mechanics has long been recognized as unfathomable, its formulas a series of vatic inscriptions no one can read without going blind. Current corrosives are simpler; they have nothing directly to do with the quantum world. They are chaos, randomness, and complexity.

Chaos first. The idea is very simple. Certain systems may be both deterministic *and* unstable. An example is a baseball bat balanced on its end. Left unperturbed, it may remain upright until the end of time. This is one of its destinies. Given a tiny tap, it falls over promptly, thus embracing a quite different destiny. Chaos arises when the bat is embedded in the flow of time. From where it is, there is no real saying where it is going.

More than ninety years ago, the French mathematician Henri Poincaré observed that simple but nonlinear systems might well exhibit chaotic behavior. And he drew the correct conclusion that such systems, even though governed by deterministic equations, were inherently unpredictable. His ideas were left unperturbed for a time but, given a tap in the 1940s and 1950s, became a part of common scientific currency by means of an unusual route.

The weather, Edward Lorenz suggested in 1963, might be *essentially* unpredictable (a conclusion many of us have come to on our own). Certain physical regimes—like certain households—are intrinsically chaotic; small changes in their initial states lead over time to dramatic and unsettling changes in their evolution. Beating its delicate and translucent wings in Borneo, a butterfly might bring about a tornado in Toledo. Like an upright bat, the earth's atmosphere is a system sensitively dependent on its initial conditions.

Having been read and chuckled over by a dozen or so professionals, Lorenz's paper sank promptly from sight. In the 1970s it underwent a spectacular resurrection; it is now widely regarded as prescient. Evidence of chaos has been discerned in phenomena ranging from planetary astronomy to nerve excitation in the giant squid. Outside of science as well, chaos has become an immensely fashionable term of diagnosis, an explanation for every living shambles. And why not? Both ordinary language and ordinary life seem hideously sensitive to small perturbations. An errant hiccup might have induced Gavrilo Princip to miss the Archduke entirely at Sarajevo, with Bosnia-Herzegovina becoming, over the placid decades since 1914, the Switzerland of Southern Europe.

Chaos is a cultural corrosive, one dissolving the tight connection between a deterministic mathematical model and the delivery of a predictable future. In a chaotic system, the very act of measurement induces uncertainty, for no measurement is entirely accurate; imperceptible errors grow exponentially, cascading along the system until the real world and the measured world seem utterly unlike. Chaos introduces something novel in the physical sciences, a pragmatic sense of inescapable error; it compromises the miracle of quantitative prediction, and returns the imagination to an older view of life and experience. The future *is* clouded. We *do* live amid ineradicable uncertainty.

If chaos is one corrosive, randomness is another. A trite and a tired concept, randomness denotes a statistical property of sequences, but one that is difficult to discern and even more difficult to define. The morganatic cousin of meaninglessness in French movies—as when a breathless hoodlum, ending his one chance for happiness, whimsically shoots a policeman—randomness has received a revivifying interpretation in recent years, thanks to the great Russian mathematician Andrey Kolmogorov and to the American mathematician Gregory Chaitin. Influenced no doubt by morphic resonance, Kolmogorov and Chaitin observed a sympathetic current running between randomness and the concept of complexity. The juxtaposition yields an extraordinary idea, one that captures something long sensed but never quite specified.

Here is an illustration. A painting by Jackson Pollock is complex in the sense that nothing short of the painting conveys what the painting itself conveys. As I look at those curiously compelling, variegated, aggressive streaks and slashes, words fail me. In order to describe the painting, I must display it. An Andy Warhol painting, by way of comparison, subordinates itself to a verbal formula: *just run that soup can up and down the page, Andy.*

This, however, is to remain within the realm of rhetoric. The mathematician attends not to paintings but to binary sequences—strings of 0's and 1's. Now a computer program is itself a string of symbols (binary numbers, in fact). A given string is simple, Chaitin and Kolmogorov argued, if the string may be generated by a computer program significantly shorter than the string itself; otherwise, it is complex.

This tight little declaration begins to explain the large irrelevance that now envelops the mathematical sciences. The laws of nature—those compressed and gnomic affirmations—pertain, on the one hand, to the large-scale structures of space and time and, on the other hand, to the jiggling fundaments of the quantum world. Simplicity reigns amid the very large and the very small, and it is there that things may be mathematically described. And yet most strings and thus most things are not simple but complex. They cannot be more simply conveyed; they are what they are, insusceptible of compression. And this is an easily demonstrated, an indubitable, mathematical fact.

It is the world's complexity that is humanly interesting. What do we see when we look elsewhere? Stars blazing glumly in the night sky, the moons of Jupiter hanging like so many testicles, clouds of cosmic dust, an immensity of space, a spare but irritating sound track consisting of the infernal chatter of background radiation. The evidence seems inescapable that the Creator wrought the simple structures in the universe with a few swift strokes of an instrument much like a cosmic trowel. In our own corner of the composition, by contrast, He apparently set to work with a sable-tipped paintbrush, patiently fashioning a blue planet, a lush garden, creatures utterly unlike anything else in the universe, sensuous, moving, alive.

We live within the confines of that canvas. Complexity is everywhere, whether created or contrived, and compression hard to come by—in truth, the human world cannot be much compressed at all. The most we can typically do, a few resolute morals or maxims aside, is to watch the panorama unfold, surprised as always by the turbulent and unsuspected flow of things, the gross but fascinating cascade of life.

THE CALCULUS IS THE great idealization of Western science; from within its crabbed formulas comes the master plan of equation-and-solution that makes the physical sciences possible. Yet the investiture of mathematics in things and processes weakens as one moves along the intellectual chain of command. This is a curious and disabling fact. Material objects on the quantum level may be explained as roiling waves of probability. By contrast, the attempt to discern the outlines of a coherent system of mathematical thought in the structure of biological objects— protozoa, rock stars, human beings—has been a failure.

From one perspective, the conceptual landscape of biology resembles a range of ancient foothills folded against the mathematician's high alpine peaks. The biologist employs a scheme immeasurably simpler than the one adopted by the mathematical physicist. It is, that scheme, discrete, finite, and combinatorial. No mathematics beyond finger counting.

Living systems may be understood in terms of the constituents that make them up, and of these, there are only finitely many. The dissection complete, what remains is a master molecule, DNA, which functions as a code, and the complicated proteins that it organizes and controls. No continuous magnitudes; no real numbers; no rich body of mathematical analysis. No laws, not in the sense in which physics contains laws of nature; no fantastically pregnant, compressed, and quantitative apothegms. Molecular biology should be comprehensible to someone who knows nothing of modern science, continuity, or the calculus, and who can reckon only to powers of ten—a Harvard graduate, say.

Despite the often vulgar language in which they are expressed, the concepts that animate molecular biology are old, familiar, haunting: system, information, code, language, message, organization. They often affirm a message already known: "one generation passeth away and another generation cometh." They are magical in their nature and effect. DNA, in particular, functions as a kind of biochemical demiurge, something that brings an entire organism into existence by a process akin to a casting of spells. They are often inconsistent: the role attributed to DNA is at odds with the obvious fact that the information resident in the genome is inadequate to specify the whole of a complex organism. Like a rubber band under tension, the concepts of molecular biology seem always to snap back to some earlier way of describing life, one in which *purpose* and *design* come prominently into focus.

And they seem, these concepts, often to mark the very margins of our own intellectual inadequacy. Nowhere in nature do we ever observe purely mechanical forces between large molecules giving rise to self-contained, stable, and autonomous structures like a frog or a fern, something able to carry on as a continuous arc from first to last, a physical object changing over time but remaining the *same* object at every stage, some set of forces endowing its identity with permanence so that variations remain bounded and inevitably return the object to the place from which it started. Nothing but a living system exhibits this extraordinary combination of plasticity and stability, a fact we are barely able to describe and entirely unable to explain.

Molecular biology is immune to the great idealization that marks the physical sciences; and what is more, it seems retrograde to the grand metaphysical assumptions on which the physical sciences rest. Those assumptions have passed directly into popular culture. The world, the physical sciences affirm almost with one voice, is physical and not spiritual, numinous, or mental. It is a world of matter. The doctrine of consideration in contract law and the bright bubble of consciousness are illusions. Reality contains only atoms and the void.

But if, by "physical," physical scientists mean concepts like the concepts found in physics, then the conclusion is irresistible that molecular biology is not a physical science at all, but a discipline struggling to express the properties of living systems in a vocabulary and by means of concepts unlike those needed elsewhere. What we see when we *look* at the observable universe is that one god like dark Pluto rules the quantum underworld; quite another, the biological macromolecules.

Physicists reject such a frankly polytheistic view, of course. The laws of physics are controlling, they say, and in the end everything will be made clear. That is what they always say; it is their destiny to say it. But in truth the grand vision of all of human knowledge devolving toward mathematical physics is no longer taken seriously, even by physicists who take it seriously.

"The most extreme hope for science," the physicist Steven Weinberg has written (in his *Dreams of a Final Theory*, 1993), "is that we will be able to trace the explanation of all natural phenomena to final laws *and historical accidents*" (emphasis added). Machiavelli used the word *fortuna* to describe the inexplicable adjurations of fate; it is a word that communicates a certain grave mockery. How has mathematical physics informed the human heart if the explanation for the way things are involves an appeal to fundamental laws *and* to something like a Neapolitan shrug?

AN UNEASY SENSE PREVAILS—IT has long prevailed—that the vision of a purely physical universe is somehow incomplete. We are creatures with rich and various mental experiences. We live in a world of purpose, belief, intention, and meaning. We bring the future into being by the free exercise of our will, a circumstance that mathematical physics is unable to describe, let alone explain. And we are conscious, we have minds.

The great body of continuous mathematics has played no role in the explanation or description of the human mind (however much its very existence may express the powers of that mind). But within living memory a bright new world has been organized to rival the old cunning and continuous world of the physical sciences. Gone is Freud's model of the mind as a haunted house (superego/ego/id); it has been replaced by the powerful image of the mind as a computational device. Careers have been fashioned to accommodate and exploit that image. Unpleasant young people proclaim themselves weird or wired, involve themselves in trendy little magazines, and sprawl over the Internet. The descriptive resources of the English language have been altered, often to risible effect, the term "digital" emerging from the proctologist's vernacular to become a general adjective of choice. "Life is just bytes and bytes of digital information," the biologist Richard Dawkins writes obligingly in *River Out of Eden* (1995). "Pure information," a reviewer adds loyally.

The foundations of the new view were laid more than 60 years ago by a congregation of chalky logicians: the great Kurt Gödel, Alonzo Church, Emil Post, and, of course, the odd and utterly original A. M. Turing, whose lost spirit seems to roam anxiously over the second half of the twentieth century like one of F. Scott Fitzgerald's sad young men. (*Fortuna*, again.)

Turing's simple model of a computing machine is the first of humanity's *intellectual* artifacts. The machine itself is a device for the manipulation of symbols, and since symbols are abstract, a Turing machine may be realized in any medium in which symbols can be inscribed. Given symbols as input, a Turing machine returns symbols as output, reading, writing, and erasing them on an infinitely extended tape. In a sense, of course, that is what lovers and lawyers do as well, the lover using his warm breath, the lawyer foolscap, each making his point by means of an inscription or exchange of symbols.

A Turing machine undertakes its transformations by means of a *program*: a fixed set of rules setting out what it may do and when it may do it. These rules are formal, in the sense that they make no appeal to the machine's emotions or thoughts, but they also reflect the ineliminable purposes of the system's programmer, enabling the machine to realize his aims or ends.

The essential elements of a Turing machine are the symbols it manipulates, the tape on which it writes, the mechanism by which it sees, and the program by which it acts; indeed, these are the essentials of the computational act itself, the process by which intelligence records its thoughts. The extraordinary, spine-chilling, contrary-to-intuition thing is that this imaginary object not only led historically to the construction of the digital computer itself—a striking example of thought bringing matter into existence—but also in some sense exhausts the very concept of rule-governed behavior. *Whatever* may be done by a discrete system moving in steps may be done by a Turing machine.

To THE QUESTION OF how best to describe change, the answer provided by the physical sciences over the course of 300 years has been a system of mathematical equations. Another answer, one new in our experience, is a program. The difference between the two is profound. A program *does* what an equation describes. Equations are indirect, they must be solved. A program is direct, it must be executed. Equations are continuous; programs discontinuous. Equations are infinite; programs finite. The elements of an equation are numbers; the elements of a program are words. An equation penetrates the future; a program does not.

As these distinctions suggest, the vision of the mind as a computational object is—no less than molecular biology—retrograde to the great movements of mathematical physics. In the physical sciences, time and space are represented by the real numbers and (therefore) have a continuous structure. A computer, by contrast, inhabits a world in which time, represented by the ordinary integers, has lost its pliant seamlessness and moves forward in jerky integral steps. A stern series of renunciations is in force. No differential equations. No connection backward to the calculus. No world-defining symmetries of space and time. No analytic continuation, as when the laws of nature conduct the physicist from the present into the future. No quantitative miracles. No miracles at all—provided one excepts the ordinary achingly human ones.

For under this new conceptual order, the prevailing direction of scientific thought has been altered and reversed. Within mathematical physics, things move dissectively, toward the fundamental objects and their fundamental properties and laws. The physical universe itself remains meaningless. The arena controlled by the fundamental laws, though vast, is sterile, the whole thing rather like a fluorescent-lit bowling alley where balls the size of quarks forever ricochet off one another in the hot and soundless night. Down there, no human voices may be heard.

But up here, things are different; they have always been different. Invoking a rich system of meaning and interpretation, human beings explain themselves to themselves in terms of what they wish and what they believe, the immemorial instincts of desire and conviction being sufficient to bring a world into being. That world is suspended in space by the chatter of human voices. A path through the chatter is not dissective, but almost always circular. A man believes that alfalfa sprouts are a cure for shingles; this is reflected in what he says, in what he does, in what he believes, and in what he wishes, each reflection explaining the one that has gone before, the circle beginning to bend back on itself.

There is no way to break the circle to reach a bedrock of physical fact, for there are no physical facts to reach. How could there be? To enter the circle, any purely physical feature of the world must be interpreted and given meaning. Once given meaning, it is no longer purely a physical feature of the world. Under the computational theory of mind, the conceptual circle is not emptied or evacuated, it is enlarged, the formal objects taking their place like guests at a wedding asked to join the dance. The states of a computer carry a significance that goes beyond physics; like words, they play a role in the economy of meaning. And meaning appears only in the reflective and interpretive gaze of human beings.

This point is evident in the simplest of devices. Call twice for the numeral "2" on a calculator and the machine returns a neoned "4." Considered simply as a physical object, the machine is shuttling among shapes, configurations of light, which it realizes by means of the way in which it is constructed: it is capable of nothing else. What makes the charming show of light an *answer* is the fact that someone has been provoked to ask a question. Question and answer belong to the circle of human voices. A purely physical process has been invested with significance, those winking ruby lights given form and content as symbols, representations. Whether the representation is made in terms of light or by the modulation of a woman's voice, the process is always the same. Some feature of the world has been made incandescent.

STREAMING IN FROM SPACE, light reaches the human eye and deposits its information on the stippled surface of the retina. Directly thereafter I see the great lawn of Golden Gate Park; a young woman, nose ring twitching; a panting puppy; a rose bush; and beyond, a file of cars moving sedately toward the western sun. A three-dimensional world has been conveyed to a two-dimensional surface and then reconveyed to a three-dimension image.

This familiar miracle suggests, if anything does, the relevance of algorithms to the actual accomplishments of the mind; indeed, the transformation of dimensions is precisely the kind of activity that might be brought under the control of a formal program, a system of rules cued to the circumstances of vision as it takes place in a creature with two matched but somewhat asymmetrical eyes. David Marr, for example, provides (in *Vision*, 1982) an extraordinary account of the complex transformations undertaken in the mind's cockpit in order to allow the eyes to see things stereoptically.

In a charming book entitled *Descartes' Error* (1994), Antonio R. Damasio writes of the mind as a place where neural representations, or images, arise. Having concluded correctly, say, that a football is heading toward one's nose, the mind signs off on the formal portion of its visual deliberations by means of a vibrant image (and signals the head to duck). This language of representations and images is general throughout the cognitive sciences. The mind, apparently, stores the stuff in various places and then hauls down a representation or two when the need arises.

But wait a minute. Representations? Images? As in, something seen? *In* the mind? But seen by *whom*? And just *who* is doing the representing?

These questions reverberate with a loud, flat, embarrassing bang, their innocence utterly at odds with the sophistication of the various theories they subvert. Is the mind computational? It is. Does it proceed by an application of determinate rules? It does. Very well, consider this: at the conclusion of its computations, the mind bursts into a vivid, light-enraptured awareness of the world. I open *my* eyes and my eyes are filled. There is a panorama to which my eyes may be partial, but it is my eyes that are filled, my experiences possessing both an experiencing subject—me—and the contents of that experience, the scene and its surveyor bound inseparably together as fragments in a figure.

The persistence in theory of a certain embarrassing imbroglio—the mind suddenly opens an arena in which images are thoughtfully examined, or representations are mysteriously made to represent—is evidence of the enormous difficulty entailed in accommodating consciousness within any computational view of the mind's operations.

Although most analytic philosophers have remained materialists, it is consciousness that is now on everyone's lips. Employing an argument prematurely discarded by logicians, the distinguished mathematician Roger Penrose has concluded that consciousness *cannot* be computational: a reformation of quantum theory is required to set the matter right, the transmutation of thought into action taking place in the microtubules of the cell.[1]

Elsewhere, unorthodox quantum physicists have argued for the ubiquity of mind throughout the cosmos, with even the atoms having a say in the scheme of things.[2] An enterprising academic, Colin McGinn, has concluded that the problem of consciousness must forever be insoluble and has made his discovery the foundation of a far-reaching philosophical system. A few philosophers have even been observed administering discreet kicks to the corpse of mind/body dualism: *get up, you fat fool, I need you.*

Do I have anything better? Of course not. "You could not discover the limits of soul," Heraclitus wrote, "not even if you traveled down every road. Such is the depth of its form."

IT IS A FACT. Among the physicists, the old quiet confidence is gone. Men with black burning eyes roam the corridors of thought. They talk of theories that will explain absolutely everything and like barroom drunks fasten on anyone to unburden themselves: *It's strings, that's what it is, I'm*

---

1. *The Emperor's New Mind* (Oxford, 1990) and *Shadows of the Mind* (Oxford, 1994).
2. Nick Herbert, *Elemental Mind* (Plume, 1994).

telling you. There are physicists (like Stephen Hawking or Paul Davies) convinced that they are shortly to know the Mind of God, or that they have seen in the firmament secrets of a cosmic code, or discovered in the dense inaccessible equations of general relativity living proof of the Christian resurrection.[3]

But even as physicists add to their great creation myths, questions follow assertions in a never-ending spiral. Why do the early galaxies show so much structure? How can the universe be younger than its oldest stars? Did space and time have a beginning? A *beginning?* A beginning *in* what? Are you saying that time is relative? Then what is that business about the first three minutes? Just what are *they* relative *to?* At the margins of speculation, strange numerical coincidences haunt the imagination. And there are singularities at the beginning and end of time, places where the laws of physics simply deform themselves and then collapse.

Mathematical physics, it is sometimes said, is the cathedral constructed by our culture. The image is apt. Messy, disorganized, ideologically confused but inescapably compelling, contemporary physics resembles nothing so much as one of those strange structures designed by Antoni Gaudí. There the spooky structure sits in the somber Spanish moonlight, bats flitting about the crenellated belfry, fantastic and odd, with its thousand-and-one idiosyncratic touches, its radically asymmetrical towers, quantum mechanics on the one side, general relativity on the other, its wealth of poorly understood details set amid fearfully difficult and rebarbative mathematics, portions of the great structure incomplete, the workmen having left their tools in stupefaction, the entire glorious edifice bearing in every way the marks of its many creators, the thing deeply moving, intensely human.

It has been the hope of the physical sciences that everything might be explained by an austere, impersonal, abstract, consistent, and complete set of mathematical laws. The hope has acquired the aspect of a

---

3. See F. J. Tipler, *Physics and Immortality* (Doubleday, 1994). The generally favorable critical reception accorded this inadvertently hilarious book is itself a remarkable sign of divine grace.

faith. Within the closed coffin of academic science and analytic philosophy, things are as they always were; but no one who shares a delusion, as Freud memorably remarked, ever recognizes it as such. Elsewhere, confidence is leaking from the most profound and ambitious system of secular thought ever created. Everyone feels that this is so. And everyone is right.

The prevailing world of thought is like some frozen sea, heaving and cracking, with a trickle of shy life rushing beneath its surface, carrying fragrant memories of what has long been forgotten, a world beyond the world of matter. Human beings will always need to interpret themselves in ancient and familiar terms, the intentional circle enlarging but never breaking; for the way things are, they will never find an explanation so complete and so compelling as to make their transcendental urges irrelevant. Something is going and something is gone; some aspect of conviction has been broken. In return, there is something familiar and something recaptured. *For lo, the winter is past. The rain is over and gone; the flowers appear on the earth; The time of singing is come. And the voice of the turtle is heard in our land.*

THE DENIABLE DARWIN

CHARLES DARWIN PRESENTED *ON THE ORIGIN OF SPECIES* TO A DIS-believing world in 1859—three years after Clerk Maxwell had published "On Faraday's Lines of Force," the first of his papers on the electromagnetic field. Maxwell's theory has by a process of absorption become part of quantum field theory, and so a part of the great canonical structure created by mathematical physics. By contrast, the final triumph of Darwinian theory, although vividly imagined by biologists, remains, along with world peace and Esperanto, on the eschatological horizon of contemporary thought.

"It is just a matter of *time*," one biologist wrote recently, reposing his faith in a receding hereafter, "before this fruitful concept comes to be accepted by the public as wholeheartedly as it has accepted the spherical earth and the sun-centered solar system." Time, however, is what evolutionary biologists have long had, and if general acceptance has not come by now, it is hard to know when it ever will.

IN ITS MOST FAMILIAR, textbook form, Darwin's theory subordinates itself to a haunting and fantastic image, one in which life on earth is represented as a tree. So graphic has this image become that some biologists have persuaded themselves they can see the flowering tree standing on a dusty plain, the mammalian twig obliterating itself by anastomosis into a reptilian branch and so backward to the amphibia and then the fish, the sturdy chordate line—our line, *cosa nostra*—moving by slithering stages into the still more primitive trunk of life and so downward to the single irresistible cell that from within its folded chromosomes foretold the living future.

This is nonsense, of course. That densely reticulated tree, with its lavish foliage, is an intellectual construct, one expressing the *hypothesis* of descent with modification. Evolution is a process, one stretching over four billion years. It has not been observed. The past has gone to where the past inevitably goes. The future has not arrived. The present reveals only the detritus of time and chance: the fossil record, and the comparative anatomy, physiology, and biochemistry of different organisms and creatures. Like every other scientific theory, the theory of evolution lies at the end of an inferential trail.

The facts in favor of evolution are often held to be incontrovertible; prominent biologists shake their heads at the obduracy of those who would dispute them. Those facts, however, have been rather less forthcoming than evolutionary biologists might have hoped. If life progressed by an accumulation of small changes, as they say it has, the fossil record should reflect its flow, the dead stacked up in barely separated strata. But for well over 150 years, the dead have been remarkably diffident about confirming Darwin's theory. Their bones lie suspended in the sands of time—theromorphs and therapsids and things that must have gibbered and then squeaked; but there are gaps in the graveyard, places where there should be intermediate forms but where there is nothing whatsoever instead.[4]

Before the Cambrian era, a brief 600 million years ago, very little is inscribed in the fossil record; but then, signaled by what I imagine as a spectral puff of smoke and a deafening *ta-da!*, an astonishing number of novel biological structures come into creation, and they come into creation at once.

Thereafter, the major transitional sequences are incomplete. Important inferences begin auspiciously, but then trail off, the ancestral connection between *Eusthenopteron* and *Ichthyostega*, for example—the great

4. A. S. Romer's *Vertebrate Paleontology* (University of Chicago Press, third edition, 1966) may be consulted with profit.

hinge between the fish and the amphibia—turning on the interpretation of small grooves within *Eusthenopteron's* intercalary bones. Most species enter the evolutionary order fully formed and then depart unchanged. Where there should be evolution, there is *stasis* instead—the term is used by the paleontologists Stephen Jay Gould and Niles Eldredge in developing their theory of "punctuated equilibria"—with the fire alarms of change going off suddenly during a long night in which nothing happens.

The fundamental core of Darwinian doctrine, the philosopher Daniel Dennett has buoyantly affirmed, "is no longer in dispute among scientists." Such is the party line, useful on those occasions when biologists must present a single face to their public. But it was to the dead that Darwin pointed for confirmation of his theory; the fact that paleontology does not entirely support his doctrine has been a secret of long standing among paleontologists. "The known fossil record," Steven Stanley observes, "fails to document a single example of phyletic evolution accomplishing a major morphologic transition and hence offers no evidence that the gradualistic model can be valid."

Small wonder, then, that when the spotlight of publicity is dimmed, evolutionary biologists evince a feral streak, Stephen Jay Gould, Niles Eldredge, Richard Dawkins, and John Maynard Smith abusing one another roundly like wrestlers grappling in the dark.

SWIMMING IN THE SOUNDLESS sea, the shark has survived for millions of years, sleek as a knife blade and twice as dull. *The shark is an organism wonderfully adapted to its environment.* Pause. And then the bright brittle voice of logical folly intrudes: *after all, it has survived for millions of years.*

This exchange should be deeply embarrassing to evolutionary biologists. And yet, time and again, biologists do explain the survival of an organism by reference to its fitness and the fitness of an organism by reference to its survival, the friction between concepts kindling nothing more illuminating than the observation that some creatures have been around for a very long time. "Those individuals that have the most offspring," writes Ernst Mayr, the distinguished zoologist, "are by definition... the fittest ones." And in *Evolution and the Myth of Creationism*, Tim Berra states that "[f]itness in the Darwinian sense means reproductive fitness—leaving at least enough offspring to spread or sustain the species in nature."

This is not a parody of evolutionary thinking; it is evolutionary thinking. *Que sera, sera.* Evolutionary thought is suffused in general with an unwholesome glow. "The belief that an organ so perfect as the eye," Darwin wrote, "could have been formed by natural selection is enough to stagger anyone." It is. The problem is obvious. "What good," Stephen Jay Gould asked dramatically, "is 5 percent of an eye?" He termed this question "excellent."

The question, retorted the Oxford professor Richard Dawkins, the most prominent representative of ultra-Darwinians, "is not excellent at all": "Vision that is 5 percent as good as yours or mine is very much worth having in comparison with no vision at all. And 6 percent is better than 5, 7 percent better than 6, and so on up the gradual, continuous series."

But Dawkins, replied Philip Johnson in turn, had carelessly assumed that 5 percent *of* an eye would see 5 percent as well *as* an eye, and that is an assumption for which there is little evidence. (A professor of law at the University of California at Berkeley, Johnson has a gift for appealing to the evidence when his opponents invoke theory, and vice versa.)

Having been conducted for more than a century, exchanges of this sort may continue for centuries more; but the debate is an exercise in irrelevance. What is at work in sight is a visual *system*, one that involves not only the anatomical structures of the eye and forebrain, but the remarkably detailed and poorly understood algorithms required to make these structures work. "When we examine the visual mechanism closely," Karen K. de Valois remarked recently in *Science*, "although we understand much about its component parts, we fail to fathom the ways in which they fit together to produce the whole of our complex visual perception."

These facts suggest a chastening reformulation of Gould's "excellent" question, one adapted to reality: *could a system we do not completely understand be constructed by means of a process we cannot completely specify?*

The intellectually responsible answer to this question is that we do not know—we have no way of knowing. But that is not the answer evolutionary theorists accept. According to Daniel Dennett (in *Darwin's Dangerous Idea*), Dawkins is "almost certainly right" to uphold the incremental view, because "Darwinism is basically on the right track." In this, he echoes the philosopher Kim Sterelny, who is also persuaded that "something like Dawkins's stories have *got* to be right" (emphasis added). After all, she asserts, "natural selection is the only possible explanation of complex adaptation."

Dawkins himself has maintained that those who do not believe a complex biological structure may be constructed in small steps are expressing merely their own sense of "personal incredulity." But in countering their animadversions he appeals to his own ability to believe almost anything. Commenting on the (very plausible) claim that spiders could not have acquired their web-spinning behavior by a Darwinian mechanism, Dawkins writes: "It is not impossible at all. That is what I firmly believe and I have some experience of spiders and their webs." It is painful to see this advanced as an argument.

DARWIN CONCEIVED OF EVOLUTION in terms of *small* variations among organisms, variations which by a process of accretion allow one species to change continuously into another. This suggests a view in which living creatures are spread out smoothly over the great manifold of biological possibilities, like colors merging imperceptibly in a color chart.

Life, however, is absolutely nothing like this. Wherever one looks there is singularity, quirkiness, oddness, defiant individuality, and just plain weirdness. The male redback spider (*Latrodectus hasselti*), for example, is often consumed during copulation. Such is sexual cannibalism—the result, biologists have long assumed, of "predatory females overcoming the defenses of weaker males." But it now appears that among *Latrodectus hasselti*, the male is complicit in his own consumption. Having achieved intromission, this schnook performs a characteristi somersault, placing his abdomen directly over his partner's mouth. Such is sexual suicide—awfulness taken to a higher power.[5]

It might seem that sexual suicide confers no advantage on the spider, the male passing from ecstasy to extinction in the course of one and the same act. But spiders willing to pay for love are apparently favored by female spiders (no surprise, there); and female spiders with whom they mate, entomologists claim, are less likely to mate again. The male spider perishes; his preposterous line persists.

This explanation resolves one question only at the cost of inviting another: why such bizarre behavior? In no other *Latrodectus* species does the male perform that obliging somersault, offering his partner the oblation of his life as well as his love. Are there general principles that specify sexual suicide among this species, but that forbid sexual suicide elsewhere? If so, what are they?

Once asked, such questions tend to multiply like party guests. If evolutionary theory cannot answer them, what, then, is its use? Why is the Pitcher plant carnivorous, but not the thorn bush, and why does the Pacific salmon require fresh water to spawn, but not the Chilean sea bass? Why has the British thrush learned to hammer snails upon rocks, but not the British blackbird, which often starves to death in the midst of plenty? Why did the firefly discover bioluminescence, but not the wasp or the warrior ant; why do the bees do their dance, but not the spider or the flies; and why are women, but not cats, born without the sleek tails that would make them even more alluring than they already are?

Why? Yes, *why*? The question, simple, clear, intellectually respectable, was put to the Nobel laureate George Wald. "Various organisms try various things," he finally answered, his words functioning as a verbal shrug, "they keep what works and discard the rest."

---

5. The details have been reported in the *New York Times* and in *Science*: evidence that at least some entomologists have a good deal of time on their hands.

But suppose the manifold of life were to be given a good solid yank, so that the Chilean sea bass but not the Pacific salmon required fresh water to spawn, or that ants but not fireflies flickered enticingly at twilight, or that women but not cats were born with lush tails. What then? An inversion of life's fundamental facts would, I suspect, present evolutionary biologists with few difficulties. *Various organisms try various things.* This idea is adapted to any contingency whatsoever, an interesting example of a Darwinian mechanism in the development of Darwinian thought itself.

A comparison with geology is instructive. No geological theory makes it possible to specify precisely a particular mountain's shape; but the underlying process of upthrust and crumbling is well understood, and geologists can specify something like a mountain's *generic* shape. This provides geological theory with a firm connection to reality. A mountain arranging itself in the shape of the letter "A" is not a physically possible object; it is excluded by geological theory.

The theory of evolution, by contrast, is incapable of ruling *anything* out of court. That job must be done by nature. But a theory that can confront any contingency with unflagging success cannot be falsified. Its control of the facts is an illusion.

"CHANCE ALONE," THE NOBEL Prize-winning chemist Jacques Monod once wrote, "is at the source of every innovation, of all creation in the biosphere. Pure chance, absolutely free but blind, is at the very root of the stupendous edifice of creation."

The sentiment expressed by these words has come to vex evolutionary biologists. "This belief," Richard Dawkins writes, "that Darwinian evolution is 'random,' is not merely false. It is the exact opposite of the truth." But Monod is right and Dawkins wrong. Chance lies at the beating heart of evolutionary theory, just as it lies at the beating heart of thermodynamics.

It is the second law of thermodynamics that holds dominion over the temporal organization of the universe, and what the law has to say we find verified by ordinary experience at every turn. Things fall apart. Energy, like talent, tends to squander itself. Liquids go from hot to lukewarm. And so does love. Disorder and despair overwhelm the human enterprise, filling our rooms and our lives with clutter. Decay is unyielding. Things go from bad to worse. And overall, they go *only* from bad to worse.

These grim certainties the second law abbreviates in the solemn and awful declaration that the entropy of the universe is tending toward a maximum. The final state in which entropy is maximized is simply more *likely* than any other state. The disintegration of my face reflects nothing more compelling than the odds. Sheer dumb luck.

But if things fall apart, they also come together. *Life* appears to offer at least a temporary rebuke to the second law of thermodynamics. Although biologists are unanimous in arguing that evolution has no goal, fixed from the first, it remains true nonetheless that living creatures have organized themselves into ever more elaborate and flexible structures. If their complexity is increasing, the entropy that surrounds them is decreasing. Whatever the universe-as-a-whole may be doing—time fusing incomprehensibly with space, the great stars exploding indignantly—*biologically* things have gone from bad to *better*, the show organized, or so it would seem, as a counterexample to the prevailing winds of fate.

How so? The question has historically been the pivot on which the assumption of religious belief has turned. How so? "God said: 'Let the waters swarm with swarms of living creatures, and let fowl fly above the earth in the open firmament of heaven.'" That is how so. And who on the basis of experience would be inclined to disagree? The structures of life are complex, and complex structures get made in this, the purely human world, only by a process of deliberate design. An act of intelligence is required to bring even a thimble into being; why should the artifacts of life be different?

Darwin's theory of evolution rejects this counsel of experience and intuition. Instead, the theory forges, at least in spirit, a perverse connection with the second law itself, arguing that precisely the same force that explains one turn of the cosmic wheel explains another: sheer dumb luck.

If the universe is for reasons of sheer dumb luck committed ultimately to a state of cosmic listlessness, it is *also* by sheer dumb luck that life first emerged on earth, the chemicals in the pre-biotic seas or soup illuminated and then invigorated by a fateful flash of lightning. It is again by sheer dumb luck that the first self-reproducing systems were created. The dense and ropy chains of RNA—*they* were created by sheer dumb luck, and sheer dumb luck drove the primitive chemicals of life to form a living cell. It is sheer dumb luck that alters the genetic message so that, from infernal nonsense, meaning for a moment emerges; and sheer dumb luck again that endows life with its *opportunities*, the space of possibilities over which natural selection plays, sheer dumb luck creating the mammalian eye and the marsupial pouch, sheer dumb luck again endowing the elephant's sensitive nose with nerves and the orchid's translucent petal with blush.

Amazing. *Sheer dumb luck.*

PHYSICISTS ARE PERSUADED THAT things are in the end simple; biologists that they are not. A good deal depends on where one looks. Wherever the biologist looks, there is complexity beyond complexity, the entanglement of things ramifying downward from the organism to the cell. In a superbly elaborated figure, the Australian biologist Michael Denton compares a single cell to an immense automated factory, one the size of a large city:

> On the surface of the cell we would see millions of openings, like the portholes of a vast space ship, opening and closing to allow a continual stream of materials to flow in and out. If we were to enter one of these openings we would find ourselves in a world of supreme technology and bewildering complexity. We would see endless highly organized corridors and conduits branching in every direction away from the perimeter of the cell, some leading to the central memory bank in the nucleus and others to assembly plants and processing units. The nucleus itself would be a vast spherical chamber more than a kilometer in

diameter, resembling a geodesic dome inside of which we would see, all neatly stacked together in ordered arrays, the miles of coiled chains of the DNA molecule.... We would notice that the simplest of the functional components of the cell, the protein molecules, were, astonishingly, complex pieces of molecular machinery.... Yet the life of the cell depends on the integrated activities of thousands, certainly tens, and probably hundreds of thousands of different protein molecules.

And whatever the complexity of the cell, it is insignificant in comparison with the mammalian nervous system; and beyond that, far impossibly ahead, there is the human mind, an instrument like no other in the biological world, conscious, flexible, penetrating, inscrutable, and profound.

It is here that the door of doubt begins to swing. *Chance* and *complexity* are countervailing forces; they work at cross-purposes. This circumstance the English theologian William Paley (1743–1805) made the gravamen of his well-known argument from design:

> Nor would any man in his senses think the existence of the watch, with its various machinery, accounted for, by being told that it was one out of possible combinations of material forms; that whatever he had found in the place where he found the watch, must have contained some internal configuration or other, and that this configuration might be the structure now exhibited, viz., of the works of a watch, as well as a different structure.

It is worth remarking, it is simply a *fact*, that this courtly and old-fashioned argument is entirely compelling. We *never* attribute the existence of a complex artifact to chance. And for obvious reasons: complex objects are useful islands, isolated amid an archipelago of useless possibilities. Of the thousands of ways in which a watch might be assembled from its constituents, only one is liable to work. It is unreasonable to attribute the existence of a watch to chance, if only because it is *unlikely*. An artifact is the overflow in matter of the mental motions of intention, deliberate design, planning, and coordination. The inferential spool runs backward, and it runs irresistibly from a complex object to the contrived, the artificial, circumstances that brought it into being.

Paley allowed the conclusion of his argument to drift from man-made to biological artifacts, a human eye or kidney falling under the same classification as a watch. "Every indication of contrivance," he wrote, "every manifestation of design, exists in the works of nature; with the difference, on the side of nature, of being greater or more, and that in a degree which exceeds all computation."

In this drifting, Darwinists see dangerous signs of a *non sequitur*. There is a tight connection, they acknowledge, between what a watch is and how it is made; but the connection unravels at the human eye—or any other organ, disposition, body plan, or strategy—if only because another and a simpler explanation is available. Among living creatures, say Darwinists, *the design persists even as the designer disappears*.

"Paley's argument," Dawkins writes, "is made with passionate sincerity and is informed by the best biological scholarship of his day, but it is wrong, gloriously and utterly wrong."

The enormous confidence this quotation expresses must be juxtaposed against the weight of intuition it displaces. It is true that intuition is often wrong—quantum theory is intuition's graveyard. But quantum theory is remote from experience; our intuitions in biology lie closer to the bone. We are ourselves such stuff as genes are made of, and while this does not establish that our assessments of time and chance must be correct, it does suggest that they may be pertinent.

THE DISCOVERY OF DNA by James D. Watson and Francis Crick in 1953 revealed that a living creature is an organization of matter orchestrated by a genetic text. Within the bacterial cell, for example, the book of life is written in a distinctive language. The book is read aloud, its message specifying the construction of the cell's constituents, and then the book is copied, passed faithfully into the future.

This striking metaphor introduces a troubling instability, a kind of tremor, into biological thought. With the discovery of the genetic code, every living creature comes to divide itself into alien realms: the alphabetic and the organismic. The realms are conceptually distinct, responding to entirely different imperatives and constraints. An alphabet, on the one hand, belongs to the class of finite combinatorial objects, things that are discrete and that fit together in highly circumscribed ways. An organism, on the other hand, traces a continuous figure in space and in time. How, then, are these realms coordinated?

I ask the question because in similar systems, coordination is crucial. When I use the English language, the rules of grammar act as a constraint on the changes that I might make to the letters or sounds I employ. This is something we take for granted, an ordinary miracle in which I pass from one sentence to the next, almost as if crossing an abyss by means of a series of well-placed stepping stones.

In living creatures, things evidently proceed otherwise. There is *no* obvious coordination between alphabet and organism; the two objects are governed by different conceptual regimes, and that apparently is the end of it. Under the pressures of competition, the orchid *Orphrys apifera* undergoes a statistically adapted drift, some incidental feature in its design becoming over time ever more refined, until, consumed with longing, a misguided bee amorously mounts the orchid's very petals, convinced that he has seen shimmering there a female's fragile genitalia. As this is taking place, the marvelous mimetic design maturing slowly, the orchid's underlying alphabetic system undergoes a series of *random* perturbations, letters in its genetic alphabet winking off or winking on in a way utterly independent of the grand convergent progression toward perfection taking place out there where the action is.

We do not understand, we cannot re-create, a system of this sort. However it may operate in life, randomness in language is the enemy of order, a way of annihilating meaning. And not only in language, but in any language-*like* system—computer programs, for example. The alien influence of randomness in such systems was first noted by the distinguished French mathematician M. P. Schützenberger, who also marked the significance of this circumstance for evolutionary theory. "If we try to simulate such a situation," he wrote, "by making changes randomly...

on computer programs, we find that we have no chance... even to see what the modified program would compute; it just jams."[6]

THIS IS NOT YET an argument, only an expression of intellectual unease: but the unease tends to build as analogies are amplified. The general issue is one of size and space, and the way in which something small may be found amidst something very big.

Linguists in the 1950s, most notably Noam Chomsky and George Miller, asked dramatically how many grammatical English sentences could be constructed with 100 letters. Approximately 10 to the 25th power (10^{25}), they answered. This is a very large number. But a sentence is one thing; a sequence, another. A sentence obeys the laws of English grammar; a sequence is lawless and comprises any concatenation of those 100 letters. If there are roughly (10^{25}) sentences at hand, the number of sequences 100 letters in length is, by way of contrast, 26 to the 100th power (26^{100}). This is an inconceivably greater number. The space of possibilities has blown up, the explosive process being one of *combinatorial inflation*.

Now, the vast majority of sequences drawn on a finite alphabet fail to make a statement: they consist of letters arranged to no point or purpose. It is the contrast between sentences and sequences that carries the full, critical weight of memory and intuition. Organized as a writhing ball, the sequences resemble a planet-sized object, one as large as pale Pluto. Landing almost anywhere on that planet, linguists see nothing but nonsense. Meaning resides with the *grammatical* sequences, but they, those *sentences*, occupy an area no larger than a dime.

How on earth could the sentences be *discovered by chance* amid such an infernal and hyperborean immensity of gibberish? They cannot be

6. Schützenberger's comments were made at a symposium held in 1966. The proceedings were edited by Paul S. Moorhead and Martin Kaplan and published as *Mathematical Challenges to the Neo-Darwinian Interpretation of Evolution* (Wistar Institute Press, 1967). Schützenbergers's remarks, together with those of the physicist Murray Eden at the same symposium, constituted the first significant criticism of evolutionary doctrine in recent decades.

discovered by chance, and, of course, chance plays no role in their discovery. The linguist or the native English-speaker moves around the place or planet with a perfectly secure sense of where he should go, and what he is apt to see.

The eerie and unexpected presence of an alphabet in every living creature might suggest the possibility of a similar argument in biology. It is DNA, of course, that acts as life's primordial text, the code itself organized in nucleic triplets, like messages in Morse code. Each triplet is matched to a particular chemical object, an amino acid. There are twenty such acids in all. They correspond to letters in an alphabet. As the code is read somewhere in life's hidden housing, the linear order of the nucleic acids induces a corresponding linear order in the amino acids. The biological finger writes, and what the cell reads is an ordered presentation of such amino acids—a protein.

Like the nucleic acids, proteins are alphabetic objects, composed of discrete constituents. On average, proteins are roughly 250 amino acid residues in length, so a given protein may be imagined as a long biochemical word, one of many.

The aspects of an analogy are now in place. What is needed is a relevant contrast, something comparable to sentences and sequences in language. Of course nothing completely comparable is at hand: there are *no* sentences in molecular biology. Nonetheless, there is this fact, helpfully recounted by Richard Dawkins: "The actual animals that have ever lived on earth are a tiny subset of the theoretical animals that *could* exist." It follows that over the course of four billion years, life has expressed itself by means of a particular stock of proteins, a certain set of life-like words.

COMBINATORIAL COUNT IS NOW possible. The MIT physicist Murray Eden, to whom I owe this argument, estimates the number of the viable proteins at 10 to the 50th power (10^{50}). Within this set is the raw material of everything that has ever lived: the flowering plants and the alien insects and the seagoing turtles and the sad shambling dinosaurs, the great evolutionary successes and the great evolutionary failures as well.

These creatures are, quite literally, composed of the proteins that over the course of time have performed some useful function, with "usefulness" now standing for the sense of sentencehood in linguistics.

As in the case of language, what has once lived occupies some corner in the space of a larger array of possibilities, the actual residing in the shadow of the possible. The space of all *possible* proteins of a fixed length (250 residues, recall) is computed by multiplying 20 by itself 250 times (20^{250}). It is idle to carry out the calculation. The number is larger by far than seconds in the history of the world since the Big Bang or grains of sand on the shores of every sounding sea. Another planet now looms in the night sky, Pluto-sized or bigger, a conceptual companion to the planet containing every sequence composed by endlessly arranging the 26 English letters into sequences 100 letters in length. This planetary *doppelgänger* is the planet of all possible proteins of fixed length, the planet, in a certain sense, of every *conceivable* form of carbon-based life. And there the two planets lie, spinning on their soundless axes. The contrast between sentences and sequences on Pluto reappears on Pluto's double as the contrast between useful protein forms and all the rest; and it reappears in terms of the same dramatic difference in numbers, the enormous (20^{250}) overawing the merely big (10^{50}), the contrast between the two being quite literally between an immense and swollen planet and a dime's worth of area. That dime-sized corner, which on Pluto contains the English sentences, on Pluto's double contains the living creatures; and there the biologist may be seen tramping, the warm puddle of wet life achingly distinct amid the planet's snow and stray proteins. It is here that living creatures, whatever their ultimate fate, breathed and moaned and carried on, life evidently having discovered the small quiet corner of the space of possibilities in which things *work*.

It would seem that evolution, Murray Eden writes in artfully ambiguous language, "was directed toward the incredibly small proportion of useful protein forms...," the word "directed" conveying, at least to me, the sobering image of a stage-managed search, with evolution bypassing the awful immensity of all that frozen space because in some sense evolution *knew* where it was going.

And yet, from the perspective of Darwinian theory, it is chance that plays the crucial—that plays the *only*—role in generating the proteins. Wandering the surface of a planet, evolution wanders blindly, having forgotten where it has been, unsure of where it is going.

RANDOM MUTATIONS ARE THE great creative demiurge of evolution, throwing up possibilities and bathing life in the bright light of chance. Each living creature is not only what it is but what it might be. What, then, acts to make the possible palpable?

The theory of evolution is a materialistic theory. Various deities need not apply. Any form of mind is out. Yet a force is needed, something adequate to the manifest complexity of the biological world, and something that in the largest arena of all might substitute for the acts of design, anticipation, and memory that are obvious features of such day-to-day activities as fashioning a sentence or a sonnet.

This need is met in evolutionary theory by natural selection, the filter but not the source of change. "It may be said," Darwin wrote,

> that natural selection is daily and hourly scrutinizing, throughout the world, every variation, even the slightest; rejecting that which is bad, preserving and adding up all that is good: silently and insensibly working, whenever and wherever opportunity offers, as the improvement of each organic being in relation to its organic and inorganic conditions of life.

Natural selection emerges from these reflections as a strange force-like concept. It is strange because it is unconnected to any notion of force in physics, and it is force-*like* because natural selection *does* something, it has an effect and so functions as a kind of cause.[7] Creatures, habits, organ systems, body plans, organs, and tissues are *shaped* by natural selec-

---

7. Murray Eden is, as usual, perceptive: "It is as if," he writes, "some pre-Newtonian cosmologist had proposed a theory of planetary motion which supposed that a natural force of unknown origin held the planets in their courses. The supposition is right enough and the idea of a force between two celestial bodies is a very useful one, but it is hardly a theory."

tion. Population geneticists write of selection forces, selection pressures, and coefficients of natural selection; biologists say that natural selection sculpts, shapes, coordinates, transforms, directs, controls, changes, and transfigures living creatures.

It is natural selection, Richard Dawkins believes, that is the artificer of design, a cunning force that mocks human ingenuity even as it mimics it:

> Charles Darwin showed how it is possible for blind physical forces to mimic the effects of conscious design, and, by operating as a cumulative filter of chance variations, to lead eventually to organized and adaptive complexity, to mosquitoes and mammoths, to humans and therefore, indirectly, to books and computers.

In affirming what Darwin showed, these words suggest that Darwin *demonstrated* the power of natural selection in some formal sense, settling the issue once and for all. But that is simply not true. When Darwin wrote, the mechanism of evolution that he proposed had only life itself to commend it. But to refer to the power of natural selection by appealing to the course of evolution is a little like confirming a story in the *New York Times* by reading it twice. The theory of evolution is, after all, a *general* theory of change; if natural selection can sift the debris of chance to fashion an elephant's trunk, should it not be able to work elsewhere—amid computer programs and algorithms, words and sentences? Skeptics require a demonstration of natural selection's cunning, one that does not involve the very phenomenon it is meant to explain.

No sooner said than done. An extensive literature is now devoted to what is optimistically called artificial life. These are schemes in which a variety of programs generate amusing computer objects and by a process said to be similar to evolution show that they are capable of growth and decay and even a phosphorescent simulacrum of death. An algorithm called "Face Prints," for example, has been designed to enable crime victims to identify their attackers. The algorithm runs through hundreds of facial combinations (long hair, short hair, big nose, wide chin, moles, warts, wens, wrinkles) until the indignant victim spots the resemblance

between the long-haired, big-nosed, wide-chinned portrait of the perpetrator and the perpetrator himself.

It is the presence of the *human* victim in this scenario that should
give pause. What is *he* doing there, complaining loudly amid those otherwise blind forces? A mechanism that requires a discerning human
agent cannot be Darwinian. The Darwinian mechanism neither antici-
pates nor remembers. It gives no directions and makes no choices. What
is unacceptable in evolutionary theory, what is strictly forbidden, is the
appearance of a force with the power to survey time, a force that conserves a point or a property because it *will* be useful. Such a force is no
longer Darwinian. How would a blind force know such a thing? And by
what means could future usefulness be transmitted to the present?

If life is, as evolutionary biologists so often say, a matter merely of
blind thrusting and throbbing, any definition of natural selection must
plainly meet what I have elsewhere called a rule against deferred success.[8]
It is a rule that cannot be violated with impunity; if evolutionary theory
is to retain its intellectual integrity, it cannot be violated at all.

But the rule is widely violated, the violations so frequent as to
amount to a formal fallacy.

IT IS RICHARD DAWKINS's grand intention in *The Blind Watchmaker*
to demonstrate, as one reviewer enthusiastically remarked, "how natural selection allows biologists to dispense with such notions as purpose
and design." This he does by exhibiting a process in which the random
exploration of certain possibilities, a *blind* stab here, another there, is
followed by the filtering effects of natural selection, some of those stabs
saved, others discarded. But could a process so conceived—a *Darwinian*
process—discover a simple English sentence: a target, say, chosen from
Shakespeare? The question is by no means academic. If natural selection
cannot discern a simple English sentence, what chance is there that it
might have discovered the mammalian eye or the system by which glu-
cose is regulated by the liver?

---

8. *Black Mischief: Language, Life, Logic & Luck* (1986).

A thought experiment in *The Blind Watchmaker* now follows. Randomness in the experiment is conveyed by the metaphor of the monkeys, perennial favorites in the theory of probability. There they sit, simian hands curved over the keyboards of a thousand typewriters, their long agile fingers striking keys at random. It is an image of some poignancy, those otherwise intelligent apes banging away at a machine they cannot fathom; and what makes the poignancy pointed is the fact that the system of rewards by which the apes have been induced to strike the typewriter's keys is from the first rigged against them. The probability that a monkey will strike a given letter is one in 26. The typewriter has 26 keys: the monkey, one working finger. But a letter is not a word. Should Dawkins demand that the monkey get two English letters right, the odds against success rise with terrible inexorability from one in 26 to one in 676. The Shakespearean target chosen by Dawkins—"Methinks it is like a weasel"—is a six-word sentence containing 28 English letters (including the spaces). It occupies an isolated point in a space of 10,000 million, million, million, million, million, million possibilities.

This is a very large number; combinatorial inflation is at work. And these are very long odds. And a six-word sentence consisting of 28 English letters is a very short, very simple English sentence.

Such are the fatal facts. The problem confronting the monkeys is, of course, a double one: they must, to be sure, find the right letters, but they cannot *lose* the right letters once they have found them. A random search in a space of this size is an exercise in irrelevance. This is something the monkeys appear to know.

What more, then, is expected; what more required? *Cumulative* selection, Dawkins argues—the answer offered as well by Stephen Jay Gould, Manfred Eigen, and Daniel Dennett. The experiment now proceeds in stages. The monkeys type randomly. After a time, they are allowed to survey what they have typed in order to choose the result "which *however slightly* most resembles the target phrase." It is a computer that in Dawkins's experiment performs the crucial assessments, but I prefer to imagine its role assigned to a scrutinizing monkey—the Head Monkey of the experiment. The process under way is one in which stray successes are spotted and then saved. This process is iterated and iterated again. Variations close to the target are conserved *because* they are close to the target, the Head Monkey equably surveying the scene until, with the appearance of a miracle in progress, randomly derived sentences do begin to converge on the target sentence itself.

The contrast between schemes and scenarios is striking. Acting on their own, the monkeys are adrift in fathomless possibilities, any accidental success—a pair of English-like letters—lost at once, those successes seeming like faint untraceable lights flickering over a wine-dark sea. The advent of the Head Monkey changes things entirely. Successes are *conserved* and then conserved again. The light that formerly flickered uncertainly now stays lit, a beacon burning steadily, a point of illumination. By the light of that light, other lights are lit, until the isolated successes converge, bringing order out of nothingness.

The entire exercise is, however, an achievement in self-deception. A *target* phrase? Iterations that *most resemble* the target? A Head Monkey that *measures* the distance between failure and success? If things are sightless, how is the target represented, and how is the distance between randomly generated phrases and the targets assessed? And by whom? And the Head Monkey? What of him? The mechanism of deliberate design, purged by Darwinian theory on the level of the organism, has reappeared in the description of natural selection itself, a vivid example of what Freud meant by the return of the repressed.

This is a point that Dawkins accepts without quite acknowledging, rather like a man adroitly separating his doctor's diagnosis from his own disease.[9] Nature presents life with no targets. Life shambles for-

9. The same pattern of intellectual displacement is especially vivid in Daniel Dennett's description of natural selection as a force subordinate to what he calls "the principle of the accumulation of design." Sifting through the debris of chance, natural selection, he writes, occupies itself by "thriftily conserving the design work... accomplished at each stage." But there is *no* such principle. Dennett has simply assumed that a sequence of conserved advantages will converge to an improvement in design; the assumption expresses a *non sequitur.*

ward, surging here, shuffling there, the small advantages accumulating *on their own* until something novel appears on the broad evolutionary screen—an arch or an eye, an intricate pattern of behavior, the complexity characteristic of life. May we, then, see *this* process at work, by seeing it simulated? "Unfortunately," Dawkins writes, "I think it may be beyond my powers as a programmer to set up such a counterfeit world."[10]

This is the authentic voice of contemporary Darwinian theory. What may be illustrated by the theory does not involve a Darwinian mechanism; what involves a Darwinian mechanism cannot be illustrated by the theory. Darwin Without Darwinism

BIOLOGISTS OFTEN AFFIRM THAT as members of the scientific community they positively welcome criticism. Nonsense. Like everyone else, biologists loathe criticism and arrange their lives so as to avoid it. Criticism has nonetheless seeped into their souls, the process of doubt a curiously Darwinian one in which individual biologists entertain minor reservations about their theory without ever recognizing the degree to which these doubts mount up to a substantial deficit. Creationism, so often the target of their indignation, is the least of their worries.

For many years, biologists have succeeded in keeping skepticism on the circumference of evolutionary thought, where paleontologists, taxonomists, and philosophers linger. But the burning fringe of criticism is now contracting, coming ever closer to the heart of Darwin's doctrine. In a paper of historic importance, Stephen Jay Gould and Richard Lewontin expressed their dissatisfaction with what they termed "just-so" stories in biology.[11] It is by means of a just-so story, for example, that the pop biologist Elaine Morgan explains the presence in human beings of an aquatic diving reflex. An obscure primate ancestral to man, Morgan argues, was actually aquatic, having returned to the sea like the dolphin.

---

10. It is absurdly easy to set up a sentence-searching algorithm obeying purely Darwinian constraints. The result, however, is always the same—gibberish.

11. "The Spandrels of San Marco and the Panglossian Paradigm: A Critique of the Adaptationist Programme," *Proceedings of the Royal Society*, Volume B 205 (1979).

Some time later, that primate, having tired of the water, clambered back to land, his aquatic adaptations intact. Just so.

If stories of this sort are intellectually inadequate—preposterous, in fact—some biologists are prepared to argue that they are unnecessary as well, another matter entirely. "How seriously," H. Allen Orr asked in a superb if savage review of Dennett's *Darwin's Dangerous Idea*,

> should we take these endless adaptive explanations of features whose alleged Design may be illusory? Isn't there a difference between those cases where we recognize Design *before* we understand its precise significance and those cases where we try to make Design manifest *by* concocting a story? And isn't it especially worrisome that we can make up arbitrary traits faster than adaptive stories, and adaptive stories faster than experimental tests?

The camel's lowly hump and the elephant's nose—*these*, Orr suggests, may well be adaptive and so designed by natural selection. But beyond the old familiar cases, life may not be designed at all, the weight of evolution borne by neutral mutations, with genes undergoing a slow but pointless drifting in time's soft currents.

Like Orr, many biologists see an acknowledgment of their doubts as a cagey, a *calculated*, concession; but cagey or not, it is a concession devastating to the larger project of Darwinian biology. Unable to say *what* evolution has accomplished, biologists now find themselves unable to say *whether* evolution has accomplished it. This leaves evolutionary theory in the doubly damned position of having compromised the concepts needed to make sense of life—complexity, adaptation, design—while simultaneously conceding that the theory does little to explain them.

No DOUBT, THE THEORY of evolution will continue to play the singular role in the life of our secular culture that it has always played. The theory is unique among scientific instruments in being cherished not for what it contains, but for what it lacks. There are in Darwin's scheme no biotic laws, no *Bauplan* as in German natural philosophy, no special creation, no *élan vital*, no divine guidance or transcendental forces. The theory functions simply as a description of matter in one of its modes, and living creatures are said to be something that the gods of law indifferently sanction and allow.

"Darwin," Richard Dawkins has remarked with evident gratitude, "made it possible to be an intellectually fulfilled atheist." This is an exaggeration, of course, but one containing a portion of the truth. That Darwin's theory of evolution and biblical accounts of creation play similar roles in the human economy of belief is an irony appreciated by altogether too few biologists.

Postscript: On the Derivation of *Ulysses* from *Don Quixote*

IMAGINE THIS STORY BEING told to me by Jorge Luis Borges one evening in a Buenos Aires café.

His voice dry and infinitely ironic, the aging, nearly blind literary master observes that "the *Ulysses*," mistakenly attributed to the Irishman James Joyce, is in fact derived from "the *Quixote*."

I raise my eyebrows.

Borges pauses to sip discreetly at the bitter coffee our waiter has placed in front of him, guiding his hands to the saucer.

"The details of the remarkable series of events in question may be found at the University of Leiden," he says. "They were conveyed to me by the Freemason Alejandro Ferri in Montevideo."

Borges wipes his thin lips with a linen handkerchief that he has withdrawn from his breast pocket.

"As you know," he continues, "the original handwritten text of the *Quixote* was given to an order of French Cistercians in the autumn of 1576."

I hold up my hand to signify to our waiter that no further service is needed.

"Curiously enough, for none of the brothers could read Spanish, the Order was charged by the Papal Nuncio, Hoyo dos Monterrey (a man of great refinement and implacable will), with the responsibility for copying the *Quixote*, the printing press having then gained no currency in the wilderness of what is now known as the department of Auvergne. Unable to speak or read Spanish, a language they not unreasonably detested, the brothers copied the *Quixote* over and over again, re-creating the text but, of course, compromising it as well, and so inadvertently discovering the true nature of authorship. Thus they created Fernando Lor's *Los Hombres d'Estado* in 1585 by means of a singular series of copying errors, and then in 1654 Juan Luis Samorza's remarkable epistolary novel *Por Favor* by the same means, and then in 1685, the errors having accumulated sufficiently to change Spanish into French, Molière's *Le Bourgeois Gentilhomme*, their copying continuous and indefatigable, the work handed down from generation to generation as a sacred but secret trust, so that in time the brothers of the monastery, known only to members of the Bourbon house and, rumor has it, the Englishman and psychic Conan Doyle, copied into creation Stendhal's *The Red and the Black* and Flaubert's *Madame Bovary*, and then as a result of a particularly significant series of errors, in which French changed into Russian, Tolstoy's *The Death of Ivan Ilyich* and *Anna Karenina*. Late in the last decade of the nineteenth century there suddenly emerged, in English, Oscar Wilde's *The Importance of Being Earnest*, and then the brothers, their numbers reduced by an infectious disease of mysterious origin, finally copied the *Ulysses* into creation in 1902, the manuscript lying neglected for almost thirteen years and then mysteriously making its way to Paris in 1915, just months before the British attack on the Somme, a circumstance whose significance remains to be determined."

I sit there, amazed at what Borges has recounted. "Is it your understanding, then," I ask, "that *every* novel in the West was created in this way?"

"Of course," replies Borges imperturbably. Then he adds: "Although every novel is derived directly from another novel, there is really only one novel, the *Quixote*."

Denying Darwin:
David Berlinski & Critics

In June 1996 Commentary *magazine published David Berlinski's essay "The Deniable Darwin"; page 41 in this volume. The article created quite a controversy and inspired dozens of letters from a number of scientists and philosophers. Below are the letters, and at the end is a reply from Dr. Berlinski to his critics. Originally published in September 1996 in* Commentary.

—Editor

H. Allen Orr

Department of Biology, University of Rochester, Rochester, New York

HAVING THOROUGHLY ENJOYED DAVID Berlinski's recent book, *A Tour of the Calculus,* I am not eager to squabble publicly with him. But I am afraid I must. In his article, "The Deniable Darwin" [June], Mr. Berlinski discusses an essay of mine in which I criticize an extremist style of evolutionary thinking called "adaptationism." He concludes: "Like Orr, many biologists see an acknowledgment of their doubts as a cagey, a *calculated,* concession; but cagey or not, it is a concession devastating to the larger project of Darwinian biology."

I admit I briefly enjoyed the suggestion that my essay is so clever that it manages to undermine Darwinism. But, alas, Mr. Berlinski is exaggerating just a tad. The claim that my criticism of adaptationism, or, … for that matter, any other biologist's criticism of any other "ism," has pulled the rug out from under Darwin is just wrong.

But Mr. Berlinski has his own, fairly novel, criticisms of Darwinism. A couple of these are very clever and, to a nonbiologist, surely seductive. I am writing, therefore, to explain why I think these criticisms are wrong.

But first a few facts. Near the beginning of his essay, Mr. Berlinski states: "Evolution is a process, one stretching over four billion years. It has not been observed." While it is true that no biologist attended the last four billion years of evolution, the claim that evolution has not been observed is simply wrong. Examples are a dime a dozen. When antibiotics were first introduced, most bacteria were susceptible. Antibiotics were handed out like candy and anyone who had read a page of Darwin could have predicted the result: now, many bacteria are resistant. And Mr. Berlinski surely knows what happened when we threw DDT at insects: they evolved insecticide resistance. On a grander scale, botanists have documented the recent evolution of new species and some species have even been recreated from their ancestors in the lab. Though hardly packing the drama of reptiles metamorphosing into mammals (but what do you expect in 100 years of observation?), these are all iron-clad—and witnessed—examples of evolution.

But on to Mr. Berlinski's more novel criticisms of Darwinism. He has two worries. First, can random changes in an "alphabetic" system (like DNA) fuel evolution? And second, if evolution is fueled by such changes, how does it know "where to go?"

Mr. Berlinski's first question is more sophisticated than it might seem. He is not just rehashing the tired argument that Darwinism depends on random mutations and that such changes cannot build something as fancy as an eye. He knows that Darwinism is not just "random mutations" but "random mutations *plus* natural selection," which is a different beast altogether. His worry is more subtle. It is this: DNA is "alphabetic," a discrete language of A's, T's, G's, and C's that somehow encodes all the designs we find in organisms. But how can random perturbations in such a language yield usable material for evolution? In every other language we know of, Mr. Berlinski writes, "randomness... is the enemy of order." Random changes in English yield gibberish. Random changes in computer programs are even worse: we cannot even ask what a "modified program would compute; it just jams." And so, he argues, look what Darwinism really asks of us: it demands we believe that selection uses random changes in DNA, when—by analogy with any other formal language—such changes should yield mere gibberish, hopelessly "jamming" organisms.

Mr. Berlinski's objection is one of those beautiful theories that gets killed by an ugly fact. The fact is: whether or not random DNA changes *should* invariably jam organisms, they do not. While lethal mutations— changes which so derail an organism that it dies—are common, so are mutations of such benign and subtle effect that heroic chemical measures are required just to find them. The existence of subtle, functional, *usable* mutations in DNA is a simple fact that no amount of analogizing with computer programs can make go away. That random changes in computer programs—but not DNA—invariably jam things does not show that there is something wrong with Darwinism but that there is something wrong with the analogy.

What about Mr. Berlinski's second criticism? Even if random mutation could provide usable material for evolution, how does evolution know where to go? Mr. Berlinski retells the story of the proverbial monkeys who—banging away at their typewriters—try to create a phrase from Shakespeare. In Mr. Berlinski's version, a "Head Monkey" preserves any stray letter that matches the target phrase. For the evolutionist who usually tells the story, the point is that it does not take long for this cumulative sifting process to yield the desired phrase: mutation (typing) plus selection (save the matches) works. But Mr. Berlinski is not sold: "A *target* phrase? Iterations that *most* resemble the target? A Head Monkey that *measures* the distance between success and failure? If things are sightless, how is the target represented?" The Designer, allegedly tossed out of a job by Darwinism, has apparently reappeared as the guy who erects the target and measures the distance between here and there.

Mr. Berlinski certainly shows that the monkey analogy is imperfect. But that is true of any analogy... and the monkey analogy captures an important part of Darwinism. It shows that, by saving favorable random changes, evolution can gradually build fancy structures. One need not wait for all the "parts" to appear miraculously at once.

But the analogy completely flubs another part of Darwinism: evolution does not, of course, work toward any "target." So how, then, does evolution know where to go? The answer is the most radical and beautiful part of Darwinism: it does not. The only thing that "guides" evolution is sheer, cold demographics. If a worm with a patch of light-sensitive tissue leaves a few more kids than a worm that cannot tell if the lights are on, that is where evolution will go. And, later, if a worm with light-sensitive tissue and a rough lens escapes a few more predators, that is where evolution will go. Despite all the loose talk (much of it, admittedly, from evolutionary biologists), evolution knows nothing of "design" and "targets."...

In the end, I am afraid that Mr. Berlinski's criticisms do not fare any better than those of other anti-evolutionists. I will be the first to admit, though, that Mr. Berlinski is not a traditional anti-evolutionist. I know him well enough to know that he, unlike them, is neither anti-scientific nor doctrinaire. His criticisms—like those from any good scientist— are, I think, both sincere *and* tentative.

Richard Dawkins
University Museum, Oxford University, Oxford, England

DAVID BERLINSKI'S ARTICLE REMINDS me of the tactics employed by a certain creationist with whom I once shared a platform in Oxford. The great evolutionist John Maynard Smith was also on the bill, and he spoke after this creationist. Maynard Smith was, of course, easily able to destroy the creationist's case, and in his good-natured way he soon had the audience roaring with appreciative laughter at its expense. The creationist had his own peculiar way of dealing with this. He sprang to his feet, palms facing the audience in a gesture eloquent of magnanimous reproof. "No, no!" he cried reproachfully, "Don't laugh. Let Maynard Smith have his say. It's only fair!" This desperate pretense that the audience was laughing at Maynard Smith, when in fact it was laughing with Maynard Smith at the creationist himself, reminds me of Mr. Berlinski's pretending to misunderstand Jacques Monod and me to the extent of thinking we disagree with each other over the issue of chance.

As for the identity of the creationist who tried to pull this little stunt on Maynard Smith, it was none other than David Berlinski. The audience, by the way, saw through his tactic instantly and treated it with hoots of derision.

Daniel C. Dennett

Center for Cognitive Studies, Tufts University, Medford, Massachusetts

I LOVE IT: ANOTHER hilarious demonstration that you can publish bull—t at will—just so long as you say what an editorial board wants to hear in a style it favors. First there was Alan Sokal's delicious unmasking of the editors of *Social Text*, who fell for his fashionably anti-scientific "proof" that according to quantum physics, the world is a social construction. Now David Berlinski has done the same to the editors of *Commentary*, who fell just as hard for his parody of "scientific" creationism. They must really be oppressed by evolutionary theory to publish such inspired silliness without running it by a biologist or two for soundness. Two such similar pranks in a single month make one wonder if this is just the tip of the *Zeitgeist*. What next? A hoax extolling the educational virtues of machine guns for tots in the *American Rifleman?*

I love the rich comic patina of smug miseducation Mr. Berlinski exudes: Latin names for species mixed with elementary falsehoods in about equal measure, the subtle misuse of *"Doppelgänger,"* the "unwitting" creation of a new term, "combinatorial inflation," and the deft touch of "betraying" his cluelessness by referring to Kim Sterelny as "she."

The hints are subtle but conclusive. No serious opponent of evolutionary theory would trot out the ill-considered remarks of the mathematician M. P. Schützenberger—a line of discredited criticism quietly abandoned by others years ago—without so much as a hint about their standing. How could the heroic misunderstanding of Jacques Monod that enables our author to pit Monod against Richard Dawkins be anything but disingenuous? Could any actual professor of mathematics and philosophy "in American and French universities" misrepresent the import of the second law of thermodynamics with such poetic fervor, such blithe overconfidence?

Whoever this David Berlinski is, he is clever enough to fool *Commentary*, and I wouldn't even be surprised if some evolutionists take him seriously enough to rebut him in detail. Even better, some earnest creationists may clasp him to their bosom. That is, one presumes, his larger joke. The only reason I am exposing it now (killjoy that I am) is to make it clear that so far as I know, we evolutionists did not put him up to it. We feel no need to burden our critics with such *agents provocateurs*, but they are welcome to him if they want him.

Arthur M. Shapiro

Division of Biological Sciences, Section of Evolution and Ecology
University of California, Davis, California

. . . WHEN I debate anti-evolutionists, I always warn the audience that a critic can raise more issues in a given amount of time than a defender can answer. The defender is thus vulnerable to the claim that he has failed to address this or that criticism. In replying to David Berlinski's eloquent but deeply flawed article, let me stress that I am not concerned with a point-by-point refutation. I am much more interested in his uses of rhetoric to create unjustified impressions in the mind of the reader....

Most biologists I know are driven by curiosity, not ideology. We recognize that it is pointless to deny that a real world exists, and dangerous to place perfect faith in our own objectivity in interpreting it. That is why science has institutionalized skepticism in the process of criticism and competition.... We are indeed human in becoming enamored of or trapped in our own ideas, but we are surrounded by hungry rivals who work to keep us honest.

That is just as true in evolutionary biology as in other sciences. So how can Mr. Berlinski get away with accusing us of hushing up our disagreements in order to present a united front against our cultural foes? There is no more contentious, pugnacious, querulous bunch of professionals on earth than evolutionary biologists,... and intergenerational and ideological conflicts within academia keep the pot boiling in a very public way. In the process, anti-evolutionists are provided with a mountain of very quotable quotes, usually presented to the public out of context. If we were so concerned with papering over our differences, why would we debate them before the public?...

I was at the Wistar Institute in 1966 when Murray Eden and M. P. Schützenberger presented their arguments, cited approvingly by Mr. Berlinski. If he had read the proceedings of that meeting, he would know that it was organized at the instance of the biologists, who wanted the opportunity to hear and discuss the criticisms of their mathematical colleagues. He would also know that some of the objections he cites now were actually dealt with effectively from the floor. Instead, he disingenuously implies that those arguments still bear as much force now as they did then—that they somehow have been swept under the rug or ignored by biologists.

Mr. Berlinski ignores the fact that mathematics has been the indispensable handmaiden of evolutionary biology for over 70 years. Does he really think nothing interesting has happened in evolutionary biology for 30 years, or that anything interesting that has happened has merely weakened the Darwinian paradigm? Perhaps. After all, with the avalanche of technical, semi-popular, and popular books on paleobiology published in recent years, Mr. Berlinski sends his readers off to consult Alfred S. Romer's classic treatise on vertebrate paleontology, also published in 1966, and showing its age badly.

In another display of disingenuousness, Mr. Berlinski wants evolutionary biologists to explain from first principles why one species of widow spider commits sexual suicide while another does not. He knows perfectly well that the chain of causes is too long, with too many opportunities for historical contingency to operate, for anyone with less than the omniscience of the deity to do that. But just as physics rules out mountains shaped like inverted cones, so, too, does Darwinism rule out certain things. For example, it rules out any organism displaying a structure or behavior that enhances the fitness of some other species while reducing its own: the cartoonist Al Capp's altruistic "Shmoos" were a Darwinian impossibility. That is precisely why "altruism" was such a hot topic among evolutionists in the 1960s and 70s, when numerous biologists tried and failed to demonstrate its existence in nonhuman species.

One final point, which, while seemingly a trivial matter of semantics, illuminates the deepest problem of Mr. Berlinski's essay: our ability to reconstruct the history of life (phylogeny) depends upon the "tree of life" not only *not* being "densely reticulated," but *not reticulated at all.* "Reticulated" means net-like; branches diverge, then anastomose. But in phylogeny, branches, once separate, stay separate. It is this fact that enables us to reconstruct phylogeny through the method of nested, shared resemblances, now usually called "cladistics." That method fails if branches really do anastomose—either through hybridization, which is rare in many groups but frequent in others, or through a more arcane process such as the vectoring of genetic information across taxonomic lines by viruses.

If the tree of life were very reticulate, formal phylogenetic reconstruction would be impossible, and we truly would have a problem—not with the Darwinian paradigm *per se*, but with our ability to know whether it was true or false.

Mr. Berlinski could write a good essay about a living, breathing issue like that—were he not so preoccupied with rehashing the debates of 30 years ago....

Paul R. Gross

Center for Cell Signaling, University of Virginia, Charlottesville, Virginia

BERLINSKI NO. 1 ["THE Soul of Man Under Physics," January] was amusing: especially his silly similes, such as Jupiter's moons in the guise of testicles. Most of the science was OK, although marred by glosses that were themselves errors....

Berlinski No. 2 is much more presumptuous and more seriously wrong. How could *Commentary* not have let *some* biologist or other read it? I do not have the space to deal with all his unsupported or dead-wrong assertions, so I will just make a few counterassertions which perhaps will reduce the encouragement Mr. Berlinski has given creationists and other consumers of anti-science who might be among *Commentary*'s readers.

The claim that intermediate fossil forms are absent is simply false. It is the oldest canard in the creationist handbook. Picking and choosing from among a few of the more florid statements of Stephen Jay Gould and Niles Eldredge does not reveal to the innocent reader... why these two are and have always been convinced of the essential truth of organic evolution as Darwin first described it and of natural selection as at least one mechanism of it.

The "fitness" argument is dealt with, seriously, in every introductory course in biology, including the ones I have taught. Mr. Berlinski's fixation on it was out of date a very long time ago.

"What good is 5 percent of an eye?," is a stupid question if asked in the context in which Mr. Berlinski places it. Moreover, there is new and astonishing evidence of the genetic, hence selective, mechanism underlying the embryology of all eyes—insects, mice, humans.

Mr. Berlinski grossly misrepresents Jacques Monod's argument, as he does the way in which "chance" is used and understood in modern biology. It is possible he really does not know, but I am not sure; maybe he is a postmodernist at heart.

"*Bauplan*," like a few other fancy words, is comically misused.

I could go on and on,… but for a small fee I would be glad to support all these assertions with citations of solid science from the *current* literature of molecular and evolutionary biology. A bit of that literature Mr. Berlinski may have mined; but if so, he has done it the same way that "Intelligent Design Theorists" mine it: to extract a few sentences here and there for the comfort of creationists. This second Berlinski opus amounts to a charge that biologists are in a conspiracy to hide from outsiders the bankruptcy of the central principle of biology.…

As I did once before, I deplore *Commentary's* giving such comfort to Luddites.

Randy M. Wadkins
Silver Spring, Maryland

… IT would take an article longer than David Berlinski's original one just to correct and/or clarify his many misstatements, and I will address only a few of them here. I would strongly suggest that anyone who thinks the points made by Mr. Berlinski seriously detract from evolutionary theory should read the *talk.origins* FAQ's (http://earth. ics.uci.edu:8080/ origins/faqs. html) for a brief synopsis of why virtually everything he says on the matter of evolution is suspect.…

Mr. Berlinski states that "before the Cambrian era… very little is inscribed in the fossil record." Yet the oldest known fossils are of bacteria and stromatolites (containing blue-green algae), both of which are with us today, and are still among the simplest forms of life known. The Cambrian explosion occurred much later, after simple life (and even simple multicellular life) had already appeared. This is one of the many points in the fossil record that support evolution.

Mr. Berlinski fails to mention the specific cases where transitional fossil forms are found in abundance, and demonstrate quite readily how animals we know today have, without doubt, evolved from earlier forms. The exhibit on horse evolution at the Smithsonian Natural History Museum is quite convincing, and the evolution of horses is perhaps the most well-documented of any animal....

Although he points out that gradualism is not considered to be an accurate description of evolution, Mr. Berlinski uses it anyway to argue against evolution. He presents several examples of related species which exhibit traits that, according to him, could not have come about if one came from the other (or from a precursor). But he neglects to point out that a *single* amino-acid mutation in a *single* protein is enough to cause amazing changes in a developing animal, such as eyes that grow on an insects' legs, or chickens that develop webbed duck feet and hair....

Mr. Berlinski further asserts that evolution is random. It is not quite that simple. Evolution is accomplished by random mutations of DNA. However, the process of natural selection is not random at all. Any physician treating a relapsed cancer patient knows full well that the cancer... will be resistant to the original drugs used as therapy (and this resistance may be due to the overexpression of a single protein). This is not random in any sense. Similarly, any physician treating a patient infected with the new strains of antibiotic-resistant bacteria can tell you that there is nothing random about the selection of these bacteria: they grow where their predecessors could not.

More generally, when a red, orange, yellow, green, blue, and purple ball are all placed in a bag, there is a one-in-six chance that I will draw out the green one. If, however, I look into the bag, there is a 100-percent chance I will draw the green one. There is nothing random about this, and it is this nonrandom process (natural selection) that drives evolution. Mr. Berlinski does not see this at all....

Just to show how ineffective Mr. Berlinski's "Advent of the Head Monkey" section is,... while Richard Dawkins was unable to write a computer program that simulates evolution, scientists more adept at programming have indeed been able to do just that. The natural-selection algorithm in Keen and Spain's *Computer Simulations in Biology* generates the phrase basic biological modeling is fun from a string of random letters, based on Dawkins's suggestions. Furthermore, the entire burgeoning field of genetic algorithms is based on similar, but more complex, models of evolution....

The theory of evolution deserves constant criticism, as do all scientific theories. But... it deserves some challenges that are less trivial than the ones Mr. Berlinski proposes....

Karl F. Wessel
Rancho Palos Verdes, California

THE HEART OF DAVID Berlinski's argument has been experimentally refuted, a result that is of more recent vintage than the obsolete mathematical metaphor he employs. In a 1986 experiment performed by Marshall Horwitz and Lawrence Loeb of the University of Washington, 19-base long messenger RNA promoter sequences were deleted from the genomes of E. *coli* bacteria and replaced by randomly synthesized sequences. Of the approximately 10^{11} possible sequences of the type, it turned out that many promoted the function of the deleted natural sequence (which confers resistance to tetracycline) as well as or better than the original. This was true even for a small subset of sequences randomly generated from two bases, i.e., from far fewer than 10^6 of the 10^{11}. Nor did some of the most efficient sequences at all resemble the original.

Given this outcome and others like it, it is clear that something is radically wrong with Mr. Berlinski's analogy of biological genomes to computer programs. Either genomes are nothing like programs, or else at least some programs are far more robust in the face of effects resembling natural selection than he imagines.

In point of fact, for several years people like John Koza of Stanford University have been using analogs of natural selection to evolve computer programs. Many of these evolved programs perform their optimizing tasks better than the best intentionally designed ones, providing the *reductio ad absurdum* of Mr. Berlinski's criticism....

Mr. Berlinski seems to have been misled in his opinions by the unbiological prejudices of mathematicians accustomed to working with formal systems. Koza has addressed several of these biases in print. One of the most characteristic is the belief that the genome must be parsimonious, like the shortest proof of a mathematical theorem or the code of a highly compressed computer program. Maximally compressed programs, in the sense of algorithmic information theory, are necessarily both rare and randomly distributed within the space of all possible programs; in fact, such programs cannot be obtained through evolutionary processes, since genetic distances do not correlate nonrandomly with differences in fitness in this kind of regime....

The fact is that biological genomes are not highly compressed in the information-theoretic sense but are, rather, extravagantly redundant, something that was not appreciated in the era (the 1960s) from which Mr. Berlinski draws his argument. Redundancies in the code help enhance evolvability by reducing the impact of deleterious mutations. At the same time, sexual recombination works to filter out deleterious mutations in the homozygote while helping fixate beneficial ones....

When Mr. Berlinski's anti-Darwinian rhetoric is deflated, essentially nothing of his argument is left. Natural selection is neither more nor less "forcelike" than the Pauli Exclusion Principle; in both cases, probability distributions mimic the effects of forces. Gould and Dawkins are no more argumentative than Mach and Boltzmann, Einstein and Bohr, Glashow and the superstringers. States of punctuated equilibria fall out gracefully from highly plausible computer models of natural selection. William Paley's (and David Berlinski's) intuitions about the Design of the World are neither more nor less reliable than those of flat-earthers, goat-entrail readers, or believers in the Oedipus complex.... Incidentally, DNA was not discovered by James Watson and Francis Crick but by Friedrich Miescher (1869).

Philip H. Smith, Jr.
Chapel Hill, North Carolina

IN HIS ELEGANT, LYRICAL critique of Darwinism, David Berlinski shrewdly derides the smugness of some mainstream evolutionists who consider the Darwinian theory of evolution "fact,"... but the theory is far more robust than his attacks suggest.

Darwinism is a theory not unlike the Indo-European theory in linguistics, which holds that most of the languages of Europe and the Near East descended from a common parent language by a process of evolution. Each theory notices a number of facts—some contemporary, some fossil—and attempts to link them logically. Neither pretends to predict the future, either among biological organisms or among European languages.

In comparing Darwinism unfavorably with the science of geology, Mr. Berlinski indicates that he expects predictive natural laws from evolutionary theory, similar to the laws of physics. But Darwinism is much more a theory of history... than a predictive theory. In this it resembles the Indo-European theory. Although that theory talks in terms of sound-change "laws," these laws are really only observed historical regularities, and claim no predictive power.

Mr. Berlinski says that Darwinism "suggests a view in which living creatures are spread out smoothly over the great manifold of biological possibilities, like colors merging imperceptibly in a color chart," but he finds life in fact to be quite different. Evolution must, of course, proceed by myriad small changes, but not all the resulting life forms will be visible at one time. Indeed, as these numerous small changes occur, the less successful forms will, of necessity, no longer prevail, leaving the field to the latest-evolved forms. It would be unrealistic to expect a continuous rainbow of *all* life forms. In exactly analogous fashion, the linguistic landscape of Europe shows sharp discontinuities between the Romance and the Germanic areas, between Germanic and Slavic, and so on....

Like many another critic of Darwinism, Mr. Berlinski applies the second law of thermodynamics to evolution. He suggests that life "appears to offer at least a temporary rebuke to the second law of thermodynamics." Not so. He has forgotten the phrase "in a closed system," which belongs in the statement of the second law. Life is not a closed system; it draws energy from the increasingly entropic universe to enhance its own order, no less than we do when we forge iron and other materials to make a highly organized automobile (or a watch)....

Both Darwinism and the Indo-European theory have been around for well over 100 years, and each has constructed an interlocked edifice of observed fact (plant, bone, rock; language, document inscription) and consistent explanations of their relationships, and in both cases the theories hang together pretty well.... Each waits to be torn down by a more satisfactory theory. This I fear Mr. Berlinski has not managed to do in his article, despite the eloquence of his prose.

Sheldon F. Gottlieb

Department of Biological Sciences
University of South Alabama, Mobile, Alabama

... DAVID Berlinski fails to understand the role of determinism in evolution, not in the sense of an Intelligent Designer but in the inherent properties of molecules themselves. Sidney Fox and his colleagues have demonstrated that amino acids, when heated (to speed up the rate of the reaction) at temperatures found on Earth give rise to "thermal proteins" (TP's). When placed in water, TP's organize themselves into protocells. Protocells demonstrate the major properties associated with living organisms, including metabolism, electrical excitation, reproduction, and so forth....

Fox's work shows that very early stages of evolution can occur rapidly, within minutes, and are highly reproducible. Sites where such reactions can and probably did occur include the hydrothermal vents found on the surface of the Earth and in the oceans. Around the ocean vents, highly complex ecosystems evolved....

Like other creationists, Mr. Berlinski seems to refuse to accept the fact that the introduction of the supernatural removes the discussion from the realm of science.... In the world of the supernatural, anything goes, and the only limitation is the extent of one's imagination. No evidence is required to substantiate any claims....

In science it is rarely appropriate to ask why. Scientists ask questions such as: what is the nature of _____?, or what is the mechanism whereby _____? Such questions take science out of the realm of the supernatural into the material world and provide the basis for constructing testable hypotheses and for obtaining factual evidence as to the physical, chemical, and biological makeup of the world.... But Mr. Berlinski asks: "Why did the firefly discover bioluminescence?" To the best of my knowledge there is no evidence to suggest that fireflies went into laboratories and discovered bioluminescence. Mr. Berlinski also asks: "Why is the Pitcher plant carnivorous but not the thorn bush?"... To which biologists respond: what are the evolutionary forces at work that resulted in the Pitcher plant adopting carnivory? In examining this question, biologists have found that carnivory is a means whereby Pitcher plants can grow in nutrient-poor soils: carnivory provides a means of obtaining phosphates and nitrates. The story of thorn bushes is different: for these species, found in dry, arid environments, the more pressing evolutionary forces apparently involved protection from desiccation and excessive herbivory. Thus, the more immediate survival mechanisms for each species evolved differently and conferred different survival advantages....

Finally, Mr. Berlinski asks: ". . . who on the basis of experience would be inclined to disagree" with his conclusion of deliberate design? I would!

Robert Shapiro

Department of Chemistry, New York University, New York City

In "The Deniable Darwin" David Berlinski has captured, in elegant prose, the complexity of living things. This complexity is displayed in both their structure and behavior, and is already overwhelming at the level of one-celled organisms. Mr. Berlinski quite justifiably scolds those biologists who exchange skepticism for dogma and proclaim that our current level of understanding is enough to provide a full account of the origin and development of life.

As his article develops, however, we recognize that the author... has a hidden agenda, the promotion of an idea that falls outside the bounds of scientific investigation: Intelligent Design. In advocating his cause, Mr. Berlinski juxtaposes Darwin and Design as if they were the only possible explanations for the intricacy of life, and then defines the Darwinian side to his own satisfaction. If we choose this answer, we must presume that natural selection is the only mechanism that drives evolution, and that only "sheer dumb luck" lies at its heart.

On the other side, no details are offered concerning the identity of the Designer or the means and mechanism by which He created. I suspect that the author does not wish to specify a power that operates within this universe and according to its laws, such as the intelligent silicon chip once proposed by Fred Hoyle. Mr. Berlinski prefers a supernatural explanation. He is not content to sustain his religious convictions by faith alone but, in the style of the creationists, harasses those who seek a natural explanation for origins and evolution. The creationists offer an even more polarized two-way choice, of course. They would have us select between Darwin and chance on the one hand, and a literal interpretation of the Bible with a 6,000-year-old universe on the other....

Fortunately, there is a third alternative.... In this view, an organizing principle exists in nature and governs the rise of complexity in life, and in many other natural phenomena. The roots of this idea lie in several philosophies, but it has taken on a new respectability with the development of mathematical theories of complexity.

An excellent popular account of this field and its application to evolution is given by the physicist Stuart Kauffman in his book, *At Home in the Universe*. Kauffman argues that both self-organization and natural selection have driven evolution; the development of increasing complexity in life is an expected result of these mechanisms.

Unlike Intelligent Design, self-organization can be subjected to the test of experiment. Few efforts have been made to explore this process, perhaps because complexity theory is new and scientists have been too devoted to the more traditional form of Darwinism. Yet I believe the concepts have considerable potential, particularly in the origin-of-life field, and I hope they will gain greater attention in the future. Questions of origins and evolution can then move out of the debating hall and into the laboratory, where they belong.

Paul H. Rubin

Emory University, Atlanta, Georgia

ONE REASON FOR READING *Commentary* is that it supplies a rational, intellectual analysis of important issues. Therefore, it was surprising and dismaying to read David Berlinski's anti-intellectual attack on Darwinism, one of the great intellectual accomplishments of the human mind....

"We never attribute the existence of a complex artifact to chance," says Mr. Berlinski. But evolution does not imply *only* chance. Mr. Berlinski makes it sound as if the alternatives were either chance or conscious direction; many systems, however, are complex and yet not planned. In my field of economics, for example, the entire structure of a market economy is neither designed nor planned.... Language itself, which Mr. Berlinski spends much time discussing, is not planned and yet is a highly complex artifact, and one which Stephen Pinker and others have shown to be a product of Darwinian processes.

Mr. Berlinski asks: "Could a system we do not completely understand be constructed by means of a process we cannot completely specify?" He answers, "we do not know." But it is not only evolutionary biology which presents this conundrum. All science is an attempt to explain systems we do not fully understand with processes we cannot completely specify. Unless we proceed as if the answer were yes, we cannot do science....

Mr. Berlinski calls the human mind "conscious, flexible, penetrating, inscrutable, and profound." He is of course correct. But today we are at the threshold of truly understanding the source of this mind, as the new science of evolutionary biology is beginning to explain it. Indeed, it is particularly perverse to publish a critique of Darwinism just as its practitioners are beginning to use it to explain human behavior in a remarkable and truly profound way....

John M. Levy
Blacksburg, Virginia

DAVID BERLINSKI DOES NOT like the philosophical and theological implications of Darwinism and so tells us that it plays the same role in the "economy of belief" as does the biblical account of creation. In that it supports a world view, perhaps that is so. But what is missing from his article, and what may not be apparent to the reader without some education in biology, is how massive is the evidence for an evolutionary scenario. The student in comparative anatomy cannot but be struck by how often similar forms appear in the structure of one species after another and how the same structure gradually changes form and sometimes function as one proceeds from lower to higher forms.

The student of embryology sees traces of evolution throughout the subject. At one stage human embryos have fishlike features. The early human embryo has an arterial pattern like the arteries in the gill arches of fish. As the embryo matures, that pattern changes to produce the mammalian pattern of the arteries in the thorax and neck. A comparable transition can be seen if one looks across species from fish to amphibian to reptile to mammal. Biology students are taught "ontogeny recapitulates phylogeny" (the history of the individual repeats the history of the phylum) because the evidence of it is so overwhelming....

Mr. Berlinski makes much of gaps in the fossil record, but consider the following: for many years biologists were convinced that aquatic mammals had evolved from terrestrial mammals but there were many gaps in the record. They could not provide the path. Then pieces of fossil evidence started turning up in Egypt, Pakistan, and elsewhere. Commenting on these, Stephen Jay Gould... states: "I am absolutely delighted to report that our usually recalcitrant fossil record has come through in exemplary fashion.... The embarrassment of past absence has been replaced by a bounty of new evidence—and by the sweetest series of transitional fossils an evolutionist could ever hope to find."

Terrestrial mammals returning to the sea to feed on fish and then evolving the aquatic features of whales and porpoises sounds like a "just-so" story. But the fossil evidence suggests that it really is "just so." To base one's rejection of Darwinism on gaps in the fossil record is to place it on a very erodable base....

Living creatures in their incredible complexity are mind-boggling. How a sequence of bases in a double helix can specify that you will have a dimple like your grandmother's is mind-boggling. But where the evidence for the hypothesis is so strong, and where various different types of evidence reinforce one another, I do not think that rejection-because-of-bogglement is the wisest course.

Martin Gardner
New York City

DAVID BERLINSKI'S SPIRITED ATTACK on evolution, like Phillip E. Johnson's earlier book, *Darwin on Trial*, contains one huge, glaring omission. Nowhere does he tell us what brand of creationism he supports. It is as if someone wrote an article blasting evidence for the earth's roundness but refused to say what shape he thought it was.

Like Johnson, Mr. Berlinski seems to think that the punctuated evolution of Stephen Jay Gould and his friends somehow damages the Darwinian view that all life evolved by gradual small changes. The jumps in Gould's theory are, of course, jumps only relative to the extremely long periods during which certain species remained stable—trilobites, for example. Gould's jumps are tens of thousands of years, arising from tiny mutations that for reasons still unclear occur more rapidly than usual. Darwin's bulldog, Thomas Huxley, was well aware of such jumps, and they have provided fuel for creationists from Darwin's day until now....

I assume that... Mr. Berlinski is not a young-earther who thinks all the fossils are records of life that perished in Noah's Flood. But what exactly *is* his creation scenario? There are three possibilities.

1. God created each individual species by fiat—that is, they had no ancestors. Because there are millions of insect species alone, this requires God to perform many millions of miracles. I cannot believe that... Mr. Berlinski favors such a view.

2. Creations occurred over millions of years but only for broad classifications of life, such as plants, fishes, reptiles, birds, mammals, and, above all, humans. Gradual evolution could then take place within those broad categories to give us different species of reptiles, birds, and so on.

3. Each mutation (Darwin, of course, did not know about mutations) is a creative act of God. Instead of relying on a combination of chance and natural laws, as almost all liberal Christians and Reform Jews believe, God's method of creation was to direct each mutation in a way it could not have occurred naturally. This is difficult to defend because, in the light of the beliefs of theists like myself that all natural laws were created and are upheld by a transcendent deity, there is no more need for God's special interventions than there is a need for Him to give periodic pushes to stars and planets, as Newton argued, to keep their orbits from deteriorating....

One final question. Does Mr. Berlinski think the first humans had parents who were beasts, or no parents at all?...

Herbert Gintis
University of Massachusetts, Amherst, Massachusetts

... WHAT is David Berlinski trying to say? He never denies evolution. He never denies that natural selection occurs, or that it is important. He never provides an alternative theory of genetic variation. Actually, his purpose is stated almost as an afterthought, in the final paragraph of his piece. Here he quotes Richard Dawkins's pronouncement, "Darwin made it possible to be an intellectually fulfilled atheist," and Mr. Berlinski himself remarks that "the theory of evolution" plays a "singular role in the life of our secular culture. . . ."

Fortunately, it is not necessary to deny Darwin in order to affirm the validity and importance of a spiritual (i.e., nonsecular) culture. Spiritual life has a validity of its own, neither supportive of nor in conflict with natural science. Faith, love, beauty, and other cherished dimensions of human life are experienced realities whose truth does not depend on this or that fact of nature.

Moreover, if one's belief in God depends on the inability of science to explain the world (I do not believe it should), science gives one ample support for one's faith. Even *with* Darwin, we are *far* from able to explain the emergence of consciousness from the history of the modification of the structure of DNA molecules....

I think Darwin made it *easier* to be a nonbeliever, undercutting one of the most powerful arguments for the existence of God—the so-called "argument from design." I remember personally, in first grade, being made to repeat every morning in our school prayer, "for it is God who has made us, and not we ourselves," and thinking how strong an argument that was. But by the time I became an adult, and came to appreciate the theory of natural selection for the brilliant monument to the human intellect that it is, I had already gone far beyond the argument from design in grounding my own spiritual life. Denying Darwin is not only intellectually impossible, it is spiritually unnecessary.

David P. Babcock
Boulder Creek, California

DAVID BERLINSKI IS REFRESHING in attacking evolution without assuming the correctness of its most frequently offered alternative, the literal truth of Genesis 1. Indeed, he is somewhat dismissive of creationism. So far, so good. But where Mr. Berlinski then gets into trouble is in failing to discuss what possibilities other than evolution remain after creationism is set aside. He seems to accept the general validity of science; fine. But given that, he might be expected to offer some explanation for the observation that species have arisen at different times. He does not.

Perhaps we are to infer, from his acceptance of the Argument from Design, that his designer from time to time flings new species onto the earth out of whole cloth. This is not an inherently ridiculous idea. It is, however, rather more embarrassed by lack of evidence than is the evolutionary hypothesis. Moreover, it is just the understanding of such evidence as does exist that caused evolutionary biologists to suspect that Piltdown Man was a fraud when their intellectual opponents had no scientific reason for doing so. And the evolutionary biologists were right. If their opponents have ever had such success with their inferences from the fossil record, history does not record it.

Eugenie C. Scott
National Center for Science Education, El Cerrito, California

. . . THE content of David Berlinski's article does not differ from more traditional creation-science material, though his tone is more genteel and his writing a lot more literate.... But true to the creation-science genre, his approach consists of constructing strawmen, then knocking them down with misinterpreted, faulty, or nonexistent data as well as carefully selected quotations from evolutionary scientists....

For example, the "intermediate/transitional-fossil" strawman is a staple of anti-evolutionist rhetoric.... Because their theology says God created all living things as separate "kinds," creationists assert that there can be no transitional fossils, and no descent with modification. As a result, they persist in their inability to recognize mosaic specimens. Show them a fossil like the part bird/part dinosaur *Archaeopteryx*, and they will claim that because it has feathers, it is "100-percent bird," regardless of the fact that specimens which lack clear feather imprints are indistinguishable from small, toothed, clawed, bipedal dinosaurs. Show them a hominid fossil with a small brain, teeth similar in some respects to those of an ape and in others to those of a human, and a pelvis requiring bipedal locomotion, and they will see "just an ape."...

Readers can write me at NCSE for a bibliography of articles and books that refute the Cambrian-explosion argument, refute the impossibility of amino acids coming together to form a protein, explain the difference between chance and evolution, discuss why William Paley's argument from design has been supplanted in science by arguments based on natural causes....

One can be a believer and accept evolution. Articles like Mr. Berlinski's confuse rather than elucidate issues. Mostly, however, they perpetrate bad science in a society that needs more science literacy, not ignorance.

Edward T. Oakes, S. J.

Regis University, Denver, Colorado

DAVID BERLINSKI'S SUPERBLY WRITTEN exposé of the loaded logic embedded in Darwin's "explanation" for the origin of species raises once more the issue of its replacement: if Darwinism is as inadequate as Mr. Berlinski claims, what *is* the engine of evolution?

I was raised in a Jesuit school system that pretty much took evolution for granted, and indeed seemed rather proud of the work of the Jesuit paleontologist Teilhard de Chardin in reconciling Christian doctrine with evolution....

I awoke from my own Darwinian dogmatic slumbers only late in life, when I first read Gertrude Himmelfarb's *tour de force* of a biography, *Darwin and the Darwinian Revolution*, now sadly out of print. Miss Himmelfarb makes many of the same points that Mr. Berlinski does and documents how these same objections were very much alive among Darwin's fellow scientists in the Royal Society.... Yet still, Darwin's theory won out, and that requires explanation, for which I can offer no other hypothesis than the even greater weakness of its perceived alternative: creationism.

According to the Himmelfarb thesis, nothing was more crucial to the eager hearing the Origin received than the appearance the year before of *Omphalos* by Philip Gosse, a professional geologist and (speaking *avant la lettre*) fundamentalist Christian minister. This peculiar book argued—for the first time and in explicit answer to Darwin, with whose theories Gosse was already professionally familiar—that there had been no incremental change in the earth's surface or gradual development of organic life. The *semblance* of change and development had been *imprinted* on the earth at the very moment of the act of creation. The reaction of the scientific and educated public to this bizarre, even grotesque, thesis was predictable. In Miss Himmelfarb's words: "The howl of execration that [might have] awaited Darwin was let loose on Gosse [instead]. . . ."

One of the most fair-minded historians of this controversy is Peter J. Bowler, who notes in his book *Evolution: The History of an Idea*, that Darwin's proposal *needed* creationism as a foil:

> Why, then, did paleontologists accept evolution? The answer seems to be that enough fossil evidence turned up to convince them of the futility of creationism, even if it did not prove evolution. The evolutionists could predict that if, for instance, birds have evolved from reptiles, then one might hope to find a fossil with characteristics intermediate

between the two classes. Creationists, by contrast, had a vested interest in maintaining the absolutely distinct characteristics of existing groups and were unwilling to postulate intermediates in the past....

Of course, as Mr. Berlinski points out so effectively, Darwinism lacks predictive power too, but I cannot help feeling that we are dealing with two different orders of unpredictability here. After all, meteorology and history lack true, rigorous predictive power, but no one accuses either discipline of alchemical chicanery just because the weatherman cannot predict a hurricane two weeks from now or a historian cannot predict where the next tin-horn dictator will overthrow a constitutional government. After all, evolutionary biology is, like history, a *retrospective* science, one whose search for causes must by definition terminate in the present pre-given status quo....

I can only say in closing that if I am to join the law firm of Himmelfarb, Berlinski & Bowler, I would first have to say: "Give me a replacement and I'm yours."

Kent Gordis
New York City

THE STRANGELY FLORID STYLE and lack of coherence of David Berlinski's article obscures a rather simple situation.

When Darwin's theory of evolution was formulated in the 1850s, it was a good example of a Popperian scientific hypothesis—in Karl Popper's words, "an imaginative or even mythological conjecture about the world." Today, this hypothesis has been convincingly refuted on both empirical and theoretical grounds. The empirical grounds are the total absence of fossil evidence of the transitional forms predicted by the theory, the existence of which is practically the definition of "evolution."

The theoretical grounds are the vanishingly low probability that random selection could explain the existence of life forms, the simplest of which is of extreme complexity. This point is vividly conveyed by Fred Hoyle's metaphor, quoted by Phillip E. Johnson: "that a living organism emerged by chance from a pre-biotic soup is about as likely as that a tornado sweeping through a junkyard might assemble a Boeing 747 from the materials therein."

Mr. Berlinski quotes Richard Dawkins, one of the leading contemporary Darwinists, as follows: "Darwin made it possible to be an intellectually fulfilled atheist." In fact, the refutation of Darwin's theory would threaten the faith of all those whose world view is based on "naturalism," the view that only natural or material phenomena are real, a class which includes not merely atheists like Dawkins but most agnostics, as well as many people professing "modern" religious beliefs. This, of course, is the reason Darwin's theory of evolution, although refuted as a scientific hypothesis, has become a naturalistic dogma whose transmission from generation to generation is seen as a sacred duty of the scientific clergy....

J. R. Dunn
New Brunswick, New Jersey

ONE POINT NOT MENTIONED in David Berlinski's article is the fact that natural selection is the sole scientific theory that cannot be superseded. Other theories, no matter how well-established, are commonly overthrown or subsumed in light of later discoveries, much as Newtonian mechanics became a special case within the framework of Einsteinian relativity. But natural selection is presented as the last word in its field, monumental and unassailable, with all ensuing work subordinated to the original thesis. This is a complete reversal of the accepted method of scientific inquiry, one that effectively turns Darwin into a prophet and his theory into a pseudo-religion—in a word, an ideology.

This explains why vicious infighting goes on among evolutionary biologists, so similar to doctrinal battles among religious sects; why natural selection is presented as a universal doctrine, dominant far beyond the boundaries of biology... and why contradictory findings are jammed into the basic theory with no attempt at consistency.... Lastly, it goes a long way toward explaining the incessant hostility that evolutionary biologists display toward religious faith—one does not see many astrophysicists or cosmologists crowing about their victories over the fundamentalists....

The vast majority of scientific theories prevalent in the mid-nineteenth century have been supplanted. How much longer will Darwin's immunity to fundamental scientific principles continue?

Arthur B. Cody
Santa Cruz, California

IT WAS COURAGEOUS OF *Commentary* to publish David Berlinski's much-needed questioning of evolutionary theory. It is important to have serious inquiry rather than religious dogmatism allowed into a debate over what seems to be a doctrine scientists are reluctant to examine. One unfortunate thing creationism has brought about is the hunkering down of biological scientists and others in a position which will allow no place for expressions of uncertainty. As a result, writers in all sorts of fields—biology, ethics, law—do not hesitate to justify their observations by teleological reasoning under the guise of Darwin's theory of natural selection, confident that none will propose an embarrassing query....

One hopes that David Berlinski's analysis of an important element in the neo-synthetic theory of evolution will be taken as needed ventilation, letting both light and air into this dark and darkening area of science. I fear instead what he will get are hysterical vituperation and angry shouts to close the window.

Tom Southwick
Milltown, New Jersey

IT IS SURELY BEYOND exasperation for science writers, being deeply invested in the conventional wisdom of blind evolution, to encounter doubts not only from a science-illiterate general public, but also challenges from their own ranks. They would prefer to identify all skeptics of evolution with young-earth biblical literalists. I am sure David Berlinski will be chastised for his brilliantly evocative tour through the evidential ruins of Darwinism, with its grand extrapolations from essentially trivial observable facts, and its simply unbelievable claim that chance— unguided natural processes—could organize the information requirements of even basic biological structures, let alone complete functioning organisms.

As Mr. Berlinski points out, chance ("pure dumb luck") is the "beating heart" of evolutionary theory, just as Jacques Monod clearly stated, and Richard Dawkins in turn has presumed to deny. The truth is that no hint of design is permitted; those who would propose a hybrid, as in "theistic evolution," are not really in the debate; and their accommodation to the regnant evolutionary view assumes that the "fact" (read, "proven fact") of evolution is inarguable.

But chance is not a satisfactory engine of the generation of life from nonlife, or the presumed grand procession of creatures from simpler antecedent forms. I appreciate the highly serviceable presentation Mr. Berlinski makes of the staggeringly great improbability involved: happening by blind chance upon the one useful configuration of chemicals, in a spot the size of a dime, on an earth-sized planet otherwise covered by uselessness.

I would add that this dime must be found for just one successful molecule to emerge; just one amino acid string to form one useful protein. To make a second protein, we must start over; and just happen to find that dime again, having no idea where we found it before, or even that we did find it; and having found it, our second spectacular success must also accidentally set aside our result such that it will combine usefully with our first success... and so on, passing quickly into the realm of plain impossibility.

Actually, it is even worse: even allowing for the eons of time required to give verisimilitude to the evolutionary model (typically four billion years), not one useful strand of DNA could have organized itself by accident. As the microbiologist Michael Denton puts it in *Evolution: A*

Theory in Crisis, to suppose that chance processes could create *life* (a far greater challenge) "is simply an affront to reason." In 1981, when Francis Crick (Mr. DNA) seriously proposed the it-came-from-outer-space explanation for life on earth, it was at least in part to expand (greatly) the time available to evolution....

It seems clear that intelligence designed life; and even the graybeards of the evolutionary cathedral must propose, as Mr. Berlinski puts it, a Head Monkey, overseeing vast numbers of ordinary monkeys banging uncomprehendingly at their keyboards, "striving" for that line from Shakespeare, and the Head Monkey discerning (accidental) "targets" being reached for, so as to preserve the results of successful aiming, thereby carrying out the selection process. Dumb luck *must* have also just blundered, an incalculably great number of times, into this highly sophisticated decision-making mechanism. Nothing recognizable as "evidence" is presented for this; evolution simply requires it, or something like it. This is philosophical musing, not science. But it is proposed, with no cognitive dissonance, as having the imprimatur of truth that science confers.

The debate over origins will persist, not because biblical creationism is convincing science, but because Darwinism (chance) is finally seen as an alluring impossibility. Even as it is passionately resisted, the God solution will stand as a reasonable interpretation of the emerging design paradigm....

John Wiester

Science Education Commission, Buellton, California

THANK YOU FOR DAVID Berlinski's insightful article. His examples of Darwinian "just-so" stories and fuzzy logic are both instructive and amusing. As a science educator and parent I am not amused, however, by the extent to which both science and reason are being abused by the Darwinist establishment in the education (?) of our children. The following excerpt from an official 1995 Statement on Teaching Evolution illustrates my point:

> Evolutionary theory, indeed all of science, is necessarily silent on religion and neither refutes nor supports the existence of a deity or deities. Accordingly, the National Association of Biology Teachers, an organization of science teachers, endorses the following tenets of science, evolution, and biology education: "The diversity of life [i.e., all life] on earth is the outcome of evolution: an unsupervised, impersonal, unpredictable, and natural process of temporal descent with genetic modification that is affected by natural selection, chance, historical contingencies, and changing environments."

Now that is just double-talk. Of course the claim that life is the product of an "unsupervised" process of evolution has important religious implications, and of course it goes way beyond the empirical evidence. The bogus claim of "silence" on religion serves only to protect these anti-religious statements from the critical analysis they deserve....

Nancy R. Pearcey
The Wilberforce Forum, Reston, Virginia

DAVID BERLINSKI'S PIECE ON Darwinism was colorful and witty, but his postscript topped it all, with its mock claim that all modern novels have descended from *Don Quixote* by a series of copying errors. Mr. Berlinski weakens his case unnecessarily, however, when he denies that DNA is "completely comparable" to a language. The application of information theory to biology reveals that the genetic code is structurally identical to human codes. That is, DNA is not merely like a language, it is a chemical language. To explain its functioning, biologists have been obliged to borrow the terminology of editors and linguists. They speak of DNA as a code or symbol system; they speak of molecules that copy and translate the message; they speak of proofreading functions and error correction.

The upshot is that the origin of life and its myriad forms must be recast as the origin of biological information. The DNA of even the simplest bacterium contains as much information as the entire *Encyclopedia*

Britannica. Mr. Berlinski's spoof puts the onus on Darwinists to explain how new information can emerge through the accumulation of errors in a text. And if that is a preposterous way to account for the origin of *Madame Bovary* or *Anna Karenina*, it is even more implausible when it comes to the origin of living things.

David J. Mandel
New York City

DAVID BERLINSKI'S ARTICLE, "THE Deniable Darwin," is brilliant, clear, witty, wonderful. His postscript on the evolution of *Ulysses* from *Don Quixote per Borges* is marvelous....

Larry Eubank
Jeffersonville, Indiana

DAVID BERLINSKI HAS MADE many good points about evolutionary theory. I am always taken aback by the unquestioning certainty displayed by evolutionists in the face of a theory so fundamentally, intuitively preposterous; by the solemn assurances of TV broadcasts on nature that their fairy-tale rationalizations are "science"; and by the intellectual thuggery of college professors who insist that evolution is scientific fact and that we *must* believe it....

Henry Sherman
New York City

DAVID BERLINSKI'S BRILLIANT "THE Deniable Darwin" riveted my attention. I have waited 40 years to read such an article.

Admittedly no better schooled than most other aspiring intellectuals in the fine points of Darwin's theory, I have never been able to accept its explanation of how the fantastic profusion of living and long-dead species evolved from that first lonely cell, aimlessly drifting in the primordial soup.

Something always stuck in the craw—it would just not go down. Why did the known fossil record fail to document a single example of the continuous evolutionary change proposed by the theory? If evolutionary mutation took place in response to changes in the external environment, how could one account for the bewildering abundance and variety of species surviving side by side for untold millennia in the *same* environment? Why, as Mr. Berlinski suggests in his closing sentence, does acceptance of Darwin's theory require an act of faith similar to that demanded of those who confidently place their trust in biblical (and other) accounts of creation?

For me, there was always too much that distinguished a trout from a tiger; an unfathomable distance between a mouse and a Mozart. Unhappily, there appeared to be nothing between the no-longer acceptable beliefs of the creationists and the Darwinism which seemed to have cleared the secular field of any other explanation of life on our planet. Bereft of the consoling beliefs of my fathers in divine intervention; dismissive of the pallid nostrums of the deists who see God in everything and thus, ultimately, in nothing; somehow sensing in Darwinism the illusionist's trick, which makes the impossible appear plausible (Mr. Berlinski's witty "*ta-da!*"), I found it rather lonely out there.

To be sure, Mr. Berlinski does not offer an alternative to the theory he so drolly declares deniable. Nor does he offer much comfort to those of us left out in the cold by our inability to accept any of the other answers to the perpetual questions: What are we doing here? How did we get here? Where are we going? However, by shaking the Darwinian tree with such vigor, he may have helped once again to make respectable to nonbelievers the proposition that man, after all, may have more soul than a snail. And, for that, we must thank him.

Tom Bethell
Hoover Institution, Stanford, California

DAVID BERLINSKI'S ARTICLE IS but a harbinger of the great brewing controversy over evolution. The truth is that, although we have long been told that evolution is a "fact," there is almost no evidence to support it. In 1981, Colin Patterson, senior paleontologist at the British Museum (Natural History), said that he did not know of any such evidence; subsequently he has amended this only slightly. The claim that evolution is a fact depends on a definition that is so weak as to be valueless: evolution becomes "change over time." The frequent reference to those *biston* moths in England—in a polluted environment, the dark or camouflaged variety became relatively more numerous than the light variety—shows how feeble the evidence really is. If the evolutionists had something more persuasive, we would have heard about it on a daily basis.

Darwin's great claim to fame was his discovery of the alleged mechanism of evolution—natural selection. But as Mr. Berlinski notes, this turns out to be a disguised tautology. The fittest survive, and survival is the criterion of fitness. Change was equated with improvement and the Darwinian discovery involved little more than the importation of progress into biology. Now we no longer believe in the idea of progress, and faith in biological evolution may be jeopardized as a result.

Rabbi Daniel Lapin
Toward Tradition, Mercer Island, Washington

PUBLISHING DAVID BERLINSKI'S "THE Deniable Darwin" was an act of great intellectual courage. You have fired a shot in what is becoming a great moral revolution, and it will be heard around the world. Discovering that Darwin is deniable might tell us a little of how primitive life began. It would tell us everything about how modern life should continue.

Today's greatest question is whether humans have been touched by the divine and thus possess moral judgment or whether we are just sophisticated animals. Upon the answer depend most important matters of public policy. Our view of criminals must depend upon whether we consider them to be acting naturally or in a depraved and evil manner. Seventh-grade sex-education classes would look very different if we knew of no reason to expect children to behave like animals. Come to think of it, all classes and indeed all schools would look different, too. The debate, thus, is not really over the mating habits of the male redback spider (*Latrodectus hasselti*), it is over the mating habits of American urban men (*Homo sapiens*).

On July 2, 1863, a thirty-four-year-old professor at Bowdoin College led his regiment to glory at Gettysburg where he ordered the brilliant charge that saved Little Round Top and avoided a Union catastrophe. After the war, Joshua L. Chamberlain was elected governor of Maine and later served as president of Bowdoin College. He wrote these words: "I do not fear these men of science, for after all they are following in God's ways.... I would say that [scientific] laws are God's ways seen by men." Remember that *The Origin of Species* had been published just four years earlier.

Today, Cornell University's William Provine, a leading historian of science, insists that the conflict between science and religion is inescapable. Specifically, he states: "Modern science directly implies that there are no inherent moral or ethical laws, no absolute guiding principles for human society... we must conclude that when we die, we die and that is the end of us."

Understandably, passions flare over this debate. It will ultimately resolve whether people must choose between science and religion or whether they can simultaneously rejoice in the Creator's morality and His methods.

Chamberlain and Provine cannot both be right. America will ultimately follow the vision of only one. Which one will be decided by the outcome of the intellectual and moral war you have courageously joined. Until now the war was being ignored. *Commentary*'s June 1996 issue forever eliminates that option.

Michael J. Behe

Department of Biological Sciences
Lehigh University, Bethlehem, Pennsylvania

DAVID BERLINSKI AND COMMENTARY both deserve much credit for shining a light on the woefully underexamined theory that supports much of modern thought. I found his arguments to be quite compatible with my own thinking concerning the systems I encounter daily as a professional biochemist and about which I have written in *Darwin's Black Box*. Indeed, just as the pleasing shape of a jetliner belies the complexity of its internal organization, so the complexity of life mushrooms as one gets closer to its foundation. The shape of the eye, which Darwin tried to explain, pales in comparison with the interactions of rhodopsin, transducin, arrestin, rhodopsin kinase, and other proteins in the visual cascade. Explaining how the swimming behavior of a whale might be produced gradually (if anybody ever succeeded in doing this) would be a walk in the park compared to explaining the bacterial swimming system—the *flagellum*, which requires more than 40 gene products to function.

Most people consider Darwinism, whatever its faults, to be science. Yet in an interesting way they are wrong—at least in my area of biochemistry, the study of life's foundation. Scientific results get reported in science journals, like *Nature*, the *Journal of Biological Chemistry*, and so forth. But if you search the science journals, as I have done, for detailed explanations of how particular, complex biochemical systems (such as, say, the blood-clotting system or intracellular-transport) were produced, you come up completely empty-handed. Astonishingly, science's own journals contain no explanations. Since science is found in science journals and Darwinian explanations are absent, Darwinism is not science.

Phillip E. Johnson
School of Law, University of California, Berkeley, California

THE MEDIA STEREOTYPE OF the evolution controversy is that it pits "science" against fundamentalist "religion." It would be more accurate to say that Darwinism is the most important of the materialist ideologies—Marxism, Freudianism, and behaviorism are others—which have had done so much damage to science and society in the twentieth century. The theory persists in spite of the evidence because Darwinism cannot survive without it. *Commentary* deserves great credit for helping to break the stereotype by publishing David Berlinski's fine essay.

Public understanding of the defects of Darwinism is limited because the educators think it their duty to indoctrinate, and prestigious propagandists like Richard Dawkins protect the theory effectively with ridicule and bullying. (This method of dealing with criticism is itself a hallmark of bad science.) Readers who are not intimidated should be sure to read Michael J. Behe's *Darwin's Black Box*, which shows why the mutation/selection mechanism has been all but abandoned as an explanation for the irreducible complexity found in the biochemistry of organisms. There is also a new journal called *Origins and Design*.

The fall of Darwinism will be the big story of the early twenty-first century—learn about it now, and be ahead of the curve!

Jeffrey Satinover
Westport, Connecticut

A THEORY BECOMES WIDELY accepted because it explains so much. This is exactly why Darwinian evolutionary theory has triumphed, especially its core proposition: that new living forms emerge over time exclusively via competitive, "natural" selection. The theory is so powerful that it has been able to account for variations at the level of DNA and protein morphology, even though it was formulated long before anyone knew anything about either. Indeed, "genetic algorithms," mathematical techniques modeled on competitive selection, have recently been developed and used to solve previously intractable problems in a growing number of nonbiological domains.

So massive is the accumulation of natural data adequately account-ed for by competitive selection, so impressive is its capacity to mimic goal-directed, problem-solving intelligence in spite of being a "dumb," mechanical method based upon random variation, that it should come as no surprise when critics of Darwin are perceived as willfully ignorant, or worse. This must certainly be the gist of what is being leveled at David Berlinski and at *Commentary*.

In fact, the challenges to the principle of competitive selection are of varying sophistication. The crucial challenge to Darwin is not the one which asks whether God created the world in 144 hours, and then clev-erly buried under its surface a bunch of rocks made to look like animal bones so as to ensure tenure to a group of professors 5,700 years later. The essential challenge is, rather, this: is there any evidence of anything at all at work in the world which is, however, not of the world? And if there is evidence of any such foundational anomalies, is there further ev-idence that this "something" has a central (not incidental or unnecessary) role in the processes by which the forms of life have come into existence? For if there is such evidence, then however machine-like much, or even most, of the universe is, it is not wholly a machine, and neither is man.

There are such anomalies, of both sorts. They may be put together as follows: the degree of unitary coherence in the stage-by-stage emergence of higher and higher levels of nervous-system organization (that is, brain structure) does not appear wholly explainable by purely physical influ-ences, however much can in fact occur in the way of self-assembly via nonlinear, stochastic (random) processes given (a) the amount of time since the emergence of life of earth and (b) the absence of any form of teleology or preprogramming (in which knowledge of a desired end in-fluences the choice of and application of means).

So much does competitive selection explain that it has long been presumed to be merely a matter of time before this anomaly will be reduced as well, and it has therefore been shunted aside as unimportant. After all, most of the interesting and fruitful projects in modern science have borne their fruit by presuming as a working hypothesis that the machine-model of existence is true, and by following up on the consequences, and there is simply no gainsaying the power of the successes thus achieved. But the fact remains that the most interesting discoveries of twentieth-century science have in fact provided evidence of exactly the opposite.

For 80 years now, ever since the great and utterly intractable anomalies of radiation and radioactive decay confounded the almost completed structure of determinism and forced the adoption of quantum mechanical principles (an oxymoron, since quantum, by definition, means not mechanical), the world's greatest scientific minds have been tying themselves in knots. They were quickly able to show just how outrageous the implications of quantum mechanics are; they waited patiently for a technology sophisticated enough to do the experiments they were certain would confirm the theory's falseness.... But in the last ten years, as the technology has reached the necessary level, many quantum hypotheses have been tested, and been proved correct. . .

Now, Darwinian theory and its variants are corollaries to the machine-model, and David Berlinski and the editors of *Commentary* deserve kudos for attacking head-on the notion that the fundamental principle of Darwinian theory is beyond rational dispute. The simple fact is that, however uncomfortable people may feel at having to take seriously points of view that remind them of religious fundamentalism, modern science itself has long recognized that the unquestioned hegemony of mechanical determinism is open to question, and this directly affects evolutionary theory....

The question may be put as follows: can one treat all of life as a gigantic random machine, a huge genetic algorithm "aimed" at solving the "problem" of survival and self-assembly, whose outputs (morphologic categories; species) can be sufficiently accounted for by purely random events operating according to the principles of "nonlinear" (chaotic) dynamics? Or are the results more akin to the solutions produced by the emerging, quite serious field of quantum computation: "impossible," noncomputable results that require "something more" than the physical universe to arrive at? Many serious scientists believe that the latter proposition is not on the level of the Scopes "monkey trial," but of the utmost significance.

Commentary is not a scientific journal, but let me give a brief taste, just to let it be known that the debate is real. In 1992, Koichiro Matsuno of the department of bioengineering at the Naguaka University of Technology in Japan published an article in *BioSystems* (a high-level journal of theoretical biophysics) entitled, "The Uncertainty Principle as an Evolutionary Engine." In it he demonstrates that there is a peculiar uncertainty relationship between local quantum fluctuations (these give rise to gene-level mutations) and final global morphologies. In brief, the "needs" of the final forms "influence" (in the peculiar meaning that this word takes on when dealing with quantum phenomena) the fluctuations necessary to generate the solutions.

It is precisely this kind of "impossible," nonmechanistic influence which quantum mechanics has repeatedly demonstrated at the heart of all physical systems. Serious scientists now entertain the possibility that the universe itself might be rather like a freely-acting agent (a pantheistic point of view, as it were); some consider that this "agency" might more precisely be spoken of as residing elsewhere, or being other, than any physical system, however large (a more theistic point of view, and less popular, though not for reasons of evidence).

The proposition that evolution is in some way guided by a kind of "intentionality" analogous to our own (not wholly determined: just regular nudges to keep it on this track instead of that one) is far from absurd.... Matsuno and others may be correct or they may be wrong, or partially both. But it is long overdue for readers to understand that they are not fools who insist that there is more at work in the universe than meets the modernist eye. Just what kind of central organizing principle will in fact arise to replace the emptiness of today's still-regnant Cartesian model remains to be seen, but construction is well under way.

Kudos again for the first public notice to appear in so prestigious a venue as *Commentary*; and thanks to David Berlinski for risking it.

Russell Roberts

Olin School of Business, Washington University, St. Louis, Missouri

I WANT TO THANK David Berlinski for his elegant exposé of the Darwinian emperor and his meager amount of clothing. Biologists and the rest of us often confuse what I would call micro-evolution (animals have coloring that makes them hard to spot in the wild) with macro-evolution (the tree of life that Mr. Berlinski mentions, where man makes his orderly climb out of the primordial soup). There is a great deal of evidence for the former and precious little for the latter. Where are those pesky intermediate forms?

People often speak about the theory of evolution as if it were a "fact" or "proven." Alas, it is only a theory, a useful way of organizing our thinking about the real world. When theories stretch too far to accommodate the facts, a paradigm shift is usually forthcoming. As Mr. Berlinski notes, biologists have trouble imagining an alternative paradigm to evolution. This may explain the vehemence with which they greet criticism. Perhaps they too are uneasy with the emperor's wardrobe.

M. P. Schützenberger

Academy of Sciences, Paris, France

MOLECULAR BIOLOGY HAS PROVIDED classical fields such as embryology with powerful new tools, but it has not brought any confirmation to the Darwinian theory of evolution. Therefore, many biologists have quietly turned away from this simplistic view of life.

Commentary is to be congratulated for inviting David Berlinski to present the most recent scientific argument against neo-Darwinism. In so doing it gives voice to a growing silent minority. This article will have a considerable impact.

David Berlinski Replies

SOME READERS SEEM TO have been persuaded that in criticizing the Darwinian theory of evolution, I intended to uphold a doctrine of creationism. This is a mistake, supported by nothing that I have written.

Confronted with a complex human artifact like a watch, William Paley inquired into the source of its complexity. Insofar as such a complex object is unlikely, Paley reasoned, its existence can be explained only in terms of a human act, one in which material objects (gears, springs, levers) are deliberately arranged in a particular configuration. The same pattern of observation and inference, Paley went on to argue, indicates that complex biological structures are likewise the products of a deliberate act of design, the designer in such cases being the Christian deity.

Darwinian theorists accept the first of Paley's inferences, but reject the second. Biological artifacts are complex, they say, but not designed. Their existence may be explained in terms of random variation and natural selection. I dispute this claim, without endorsing Paley's theological inference. It is not necessary to choose between doctrines. The rational alternative to Darwin's theory is intelligent uncertainty.

A number of letters raise similar points; I have distributed my comments over a number of responses.

IN MAINTAINING THAT EVOLUTION is a process that has not been observed, H. ALLEN ORR writes, I appear to have overlooked examples of evolution like the speckled moth, which undergoes mimetic changes in wing coloration as the result of environmental pollution, or the development of antibiotic resistance in bacteria. Mr. Orr is correct that there are such examples; I scruple only at the conclusions he draws from them. Changes in wing color and the development of drug resistance are intraspecies events. The speckled moth, after all, does not develop antlers or acquire webbed feet, and bacteria remain bacteria, even when drug-resistant. The most ardent creationists now accept micro-evolution as genuinely Darwinian events. They had better: such are the facts. But the grand evolutionary progressions, such as the transformation of a fish into a man, are examples of *macro*-evolution. They remain out of reach, accessible only at the end of an inferential trail.

In calling attention to "species [that] have even been recreated from their ancestors in the lab," Mr. Orr is, no doubt, referring to the recent work of L. H. Rieseberg and his co-workers ("Role of Gene Interaction in Hybrid Speciation: Evidence from Ancient and Experimental Hybrids," *Science* 272, 1996). The example is pertinent to my critique. Rieseberg and his co-authors reproduced under artificial conditions the genetic changes that have historically led from *H. annus* and *H. petiolaris* to *H. anomalus*. The plants in question are sunflowers. What is remarkable is the extent to which this experiment contravenes Darwinian doctrine. Given the crucial role played by random events in evolutionary theory, many biologists have drawn the conclusion that the tape of life, if rewound, would produce "a different array of evolutionary end products" (Stephen Jay Gould, *Wonderful Life*). The number of crossing schemes notwithstanding, the tape in this experiment ran to precisely the same genetic end product each time it was played.

Mr. Orr contends that in my discussion of the role played by randomness in formal systems, I appear to be upholding an analogy at the expense of the facts. Random events, he writes, *do* occur in molecular biological systems; so much the worse, then, for my analogy. But the interpretation of molecular biological facts in formal terms is hardly a matter of analogy. It is molecular biologists themselves who have found unavoidable the language of codes and codons, information, algorithms, organization, complexity, entropy, and the like. There is little by way of analogy in all this. DNA is not *like* a code; it *is* a code. It follows, then, that circumstances known to degrade meaning or information in formal systems should be the source of alarm in the context of theoretical biology.

No one denies that random events take place within molecular biological systems. The relevant question is how. In a formal context, the matter is not a mystery. Codes may be designed to remain robust in the face of background noise; what is required is redundancy, and the genetic code is, in point of fact, highly redundant. In communication systems, redundancy appears as a matter of design. It does not arise spontaneously. In the case of the genetic code, according to one commentator, "strong selection pressures" created the requisite redundancy. But this is to dispel one mystery by promoting another, as the familiar Darwinian circle again makes its appearance, the tail of one concept lodged firmly in the mouth of another. I will return to this point in a number of other responses, but let me repeat what I stressed in my essay, that I am not advancing an argument on this issue, only expressing intellectual unease.

In inserting a Head Monkey into Richard Dawkins's thought experiment, my aim was to show how the mechanism of design, purged on one level of Darwinian analysis, makes a stealthy reappearance at another. Mr. Orr is unpersuaded. "The monkey analogy," he believes, "shows that by saving favorable random changes, evolution can gradually build fancy structures." Such indeed is the perennial hope of Darwinian theorists, but Mr. Orr has, I believe, underestimated the force of my criticism. Favorable changes are one thing; changes that *will* be favorable, another. If the mechanism of Darwinian evolution is restricted to changes that are favorable at the time they are selected, I see no reason to suppose that it could produce any fancy structures whatsoever. If the mechanism is permitted to incorporate changes that are neutral at the time of selection, but that will be favorable some time in the future, I see no reason to consider the process Darwinian.

This is hardly a matter of semantics. A system conserving certain features in view of their future usefulness has access to information denied a Darwinian system; it functions by means of alien concepts. But this is precisely how Dawkins's experiment proceeds. My estimable Head Monkey conserves certain alphabetic changes because he *knows* where the experiment is going. This is forbidden knowledge; the Darwinian mechanism is blind, a point often stressed by Darwinian theorists themselves (see George C. Williams, *Natural Selection*, 1992). I develop this in more detail in responding to Randy M. Wadkins.

In his final argument, Mr. Orr repeats what is certainly the current orthodoxy, namely, that evolution has no targets and so like the rest of us is not going anywhere at all: "The only thing that 'guides' evolution is sheer, cold demographics." Although this is not a point I discuss in my essay, I remain unconvinced. There are certainly long-term trajectories visible in the progression of life. The development of neurological complexity is the obvious example.

RICHARD DAWKINS HAS SUCCUMBED to the endearing weakness of revising the history of an unpleasant encounter in one's own favor. I have done as much myself. But a public charge calls for a public response. In 1992, Mr. Dawkins, John Maynard Smith, and I did share a podium at Oxford University. His hands trembling with indignation, Mr. Dawkins proposed to attack organized religion; I proposed to attack Richard Dawkins; and John Maynard Smith, seeing that it was required, proposed to defend Mr. Dawkins from my attack. The intellectual drubbing that Mr. Dawkins imagines I received, I recall in distinctly different terms. But why argue over the past? I have a videotape of our encounter, which I would be happy to make publicly available. If he wishes to debate again, Mr. Dawkins need only set the time and the place.

In remarks that have by now become well known, Jacques Monod observed with some sorrow that under Darwin's theory, it is chance that plays the crucial role in the emergence and evolution of life. Mr. Dawkins proposes to deny this. His views and those of Monod are in conflict, a point clear to anyone able to read the English language. Mr. Dawkins's continuing insistence that two contradictory propositions are mutually consistent is evidence of an alarming logical deficiency.

In fact, Mr. Dawkins has simply misunderstood the fundamental character of the theory to which he has committed his passionate defense. Darwin's theory *is* both random *and* deterministic. True enough.

Mutations occur randomly, but once they have occurred, natural selection acts deterministically to cull the successes and discard the failures. By and large, true again. Nonetheless, Darwin's theory is *essentially* stochastic, a term which in statistics refers to a process involving a random sequence of observations.

Let me call a random mutation together with its deterministic consequences an evolutionary *episode*. The proto-tiger develops claws; he lives to mate successfully. Such is a single evolutionary episode. According to Darwin's theory, evolutionary episodes are independent. A snapshot of any given episode does not suffice to determine the character of future episodes. And for obvious reasons: future events are contingent on further random events. It follows that the episodes must themselves be represented by what probability theorists (and everyone else) call a random variable. And processes represented by a random variable are by definition stochastic. These facts are understood by anyone in possession of the requisite technical concepts. For all his flaws as a philosopher, Monod was quite clear about the character of Darwin's theory.

On one important matter, Mr. Dawkins's letter requires a word of reproof. At our debate, I was asked by a member of the audience whether I held to any creationist beliefs or doctrines. I replied unequivocally that I did not. My views of Darwinism, I said, were negative, but rational. On the videotape, as I utter these forthright words, Richard Dawkins may be seen sitting placidly on the podium, staring somberly into space.

DANIEL C. DENNETT is under the curious impression that the best rejoinder to criticism is a robust display of personal vulgarity. Nothing in his letter merits a response.

Still, one general point deserves attention. Both Daniel Dennett and Richard Dawkins have fashioned their reputations as defenders of a Darwinian orthodoxy. Their letters convey the impression of men who expect never to encounter criticism and are unprepared to deal with it. This strikes me as a deeply unhealthy state of affairs. Ordinary men and women are suspicious of Darwin's theory; Dennett and Dawkins hardly go far here in persuading them that their intellectual anxieties are in any way misplaced.

CONTRARY TO WHAT ARTHUR M. SHAPIRO asserts I do not doubt that evolutionary biologists are contentious; I said as much in my essay. What I deplored was their tendency to conceal their differences from the public. This is another matter entirely.

Evolutionary biologists have a habit of ignoring the most pertinent criticisms of their theory until they can decently call them out-of-date. My references to papers by M. P. Schützenberger and Murray Eden not surprisingly prompt Mr. Shapiro to the conclusion that my arguments are anachronistic. In fact, Schützenberger and Eden enter my essay un-obtrusively in largely a stage-setting role—Schützenberger to call atten-tion to a conceptual problem at the very heart of evolutionary theory and Eden to offer, for the first time, a quantitative assessment of the space within which evolutionary searches must be undertaken. Their papers are historically important, the points they make no longer controversial.

Does Mr. Shapiro doubt that randomness introduces an alien and discordant note into the dynamics by which very complex objects change? Or that combinatorial inflation blows up the space of possible proteins? These themes have been pursued in countless papers, monographs, and books. Thus, Hubert Yockey, arguing that the discovery by chance of a single molecule of iso-1-cytochrome c requires a miracle, continues the line initiated by Schützenberger and Eden (Hubert Yockey, *Informa-tion Theory and Molecular Biology*, 1992. But see also Gregory J. Chaitin, "Toward a Mathematical Definition of 'Life,'" in *The Maximum Entropy Formalism*, eds. R. D. Levine and M. Tribus, 1979; Francis Crick, *Life Itself, Its Origin and Nature*, 1981; Max Delbruck, *Mind from Matter?*, 1986; Robert Shapiro, *Origins: A Skeptic's Guide to the Creation of Life on Earth*, 1986). The language of choice has changed—a fragile consensus is emerging that Kolmogorov complexity is a natural measure of biologi-cal complexity or specificity—but the problems remain the same. The space of possible objects is entirely too large to be successfully searched by random means, a theme pursued yet again in Michael Denton's *Evolution: A Theory in Crisis* (1986).

I agree that much has happened in biology over the past 30 years—who could doubt it? Developments taking place within molecular genetics, molecular biology, and biochemistry *do* seem to me to be profoundly at odds with the Darwinian paradigm, and those within paleontology flamboyantly so. I mentioned some points of conflict in my essay; I refer to others here.

Consider, for example, the question of how an undifferentiated cell manages the task of specialization, becoming over the course of time a neuron or a muscle cell or any other particular and peculiar biological object. The requisite information is contained, of course, within the cell's genetic apparatus; the problem is one of specificity. What regulates the expression of some parts of that apparatus, while simultaneously suppressing the expression of other parts? One suggestion of long standing is that regulatory mechanisms are switched on and off by means of biochemical signals sent from neighboring cells.

It is this suggestion that experiments conducted by Chiou-Hwa Yuh and Eric Davidson seem to support (C.-H. Yuh and E. H. Davidson, "Modular *Cis*-regulatory Organization of Endo16, a Gut-specific Gene of the Sea Urchin Embryo," *Development* 122, 1996). A gene in the sea urchin *Strongylocentrotus purpuratus* contains over 30 binding sites for more than 13 regulatory factors. These regulatory factors turn parts of the genetic apparatus on and off. They are sensitive to signals from adjacent cells, and, what is more, they are clustered in discrete modules; different module combinations lead to different patterns of cellular development.

The system that results is one of extraordinary combinatorial intricacy and complexity. What of randomness in all this? We have no idea how the general mechanism for cell-specific transcription came into existence; to argue otherwise would be sheer dogmatism. But one might have thought that a system of such delicacy, once it came into existence, would be unusually sensitive to random perturbations. Not so. The regulatory apparatus seems designed to incorporate mutations. Certain sites within each module function as key switches, turning the module on or off; mutations have a specific, but highly discrete effect. Modules are largely independent, functioning as more or less complete sets of instructions, like blocks of code. As one commentator puts it, the transposition of cis-regulatory modules from one gene to the next "may be a convenient way for nature to develop novel patterns of development" (Wade Roush, *Science* 272, 1996).

As is so often the case in molecular genetics, the description of a specific biological system reveals a pattern at odds with the one demanded by Darwinian theory. Differentiation is a highly sophisticated, enormously complex, and stable process, one in which the system is protected from noise by its very design. Mutations play a role in developmental change, but not a driving role. Rather, the facts suggest a system in which there are a finite number of combinatorial possibilities, with mutations serving to initiate certain carefully stage-managed sequences. The possibilities for module combination are fixed from the first; random changes serve simply to throw the various switches.

Examples of this sort could be multiplied at length; as our knowledge increases, the crude Darwinian scheme seems progressively remote from the evidence. How, for example, to account for the astonishing fact that from the point of view of the informational macromolecules, human beings and chimpanzees are virtually identical, their sequences in alignment to 99 percent (M. C. King and A. C. Wilson, "Evolution at Two Levels in Humans and Chimpanzees," *Science* 188, 1975)? The remaining slight difference evidently has controlled the development of a bipedal gait, with the profound neurological changes that this requires; the fully formed hand; the formation of the organs of speech and articulation; and, of course, the elaboration of the human mind, an organ unlike anything else found in the animal kingdom. Nothing in this suggests even remotely a continuous accretion of small changes. The evolutionary promotion of our ancestors seems to have been under the control of powerful *regulatory* genes, instructional blocks capable of coordinating a wide variety of novel functions. We simply have no idea how any of this works.

Still, the real infirmities of Darwin's theory are conceptual and not empirical. By the standards of the serious sciences, Darwin's theory of evolution remains little more than a collection of anecdotal remarks. Criticism is often a matter of clarification. Thus, making a point unrelated to the evidence, I argued that Darwinian theory is logically troubled. When I maintain that both Messrs. Dennett and Dawkins introduce by means of the backdoor the element of design they have ostentatiously booted from the front, my criticisms are again intended to show that the theory is deficient if only because it fails to meet its *own* standards for success.

The redback spider gives Mr. Shapiro pause; my request that its dining habits be deduced from first principles—from any principles at all— strikes him as disingenuous. "The chain of causes," he remarks, is simply too long to be mastered by anyone other than an omniscient deity. I do not for a moment doubt that that is so, but precisely this circumstance prompts the request for scientific theory in the first place. The chain of causes is *everywhere* too long to be explored and then mastered; it is the purpose of a theory to abbreviate and compress the data. Imagine a biologist who, on the grounds that the chain of causes is too long, failed to explain the fact that while men develop arms, geese develop wings!

The fact that the redback spider commits sexual suicide is interesting; one wishes to know why it is so. In this, and in countless similar cases, evolutionary theory simply has no explanation. What good is it, then?

Finally, I am astonished that Mr. Shapiro should think to object to Romer's *Vertebrate Paleontology* as a reference. The subject at hand is not particle physics; the main lines of vertebrate development have been clear for more than a century. My aim in suggesting Romer was to point the reader toward his stratigraphic charts: these plot vertebrate groups against time, the very many dotted lines between groups indicating purely hypothetical phylogenetic relationships.

PAUL R. GROSS IS anxious lest in criticizing Darwinian theory I give comfort to creationists. It is a common concern among biologists, but one, I must confess, to which I am indifferent. I do not believe biologists should be in the business of protecting the rest of us from intellectual danger.

I did not say in my essay that the fossil record contains no intermediate forms; that is a silly claim. What I did say was that there are gaps in the fossil graveyard, places where there should be intermediate forms but where there is nothing whatsoever instead. No paleontologist writing in English (R. Carroll, *Vertebrate Paleontology and Evolution*, 1988), French (J. Chaline, "*Modalités, rythmes, mecanismes de l'evolution biologique: gradualisme phyletique ou equilibres ponctues?*," reprinted in *Editions du* CNRS, 1983), or German (V. Fahlbusch, "*Makroevolution, Punktualismus*," in *Palaontologie* 57, 1983), denies that this is so. It is simply a fact. Darwin's theory and the fossil record are in conflict. There may be excellent reasons for the conflict; it may in time be exposed as an artifact. But nothing is to be gained by suggesting that what is a fact in plain sight is nothing of the sort.

I do not doubt that Stephen Jay Gould and Niles Eldredge are convinced of the "essential truth of organic evolution"; or that they believe natural selection to be "at least one mechanism of it." The difficulty with this observation is that it is compromised by its qualifications. At any given moment, if the phases of the moon happened to be right, I might align myself with Mr. Gross's essential truth of organic evolution; as for natural selection, nothing at all remains at issue if it is demoted from its central position in Darwin's theory. The idea that evolution proceeds by means of many different forces is both unanswerable and uninteresting. To his credit, this is something Richard Dawkins recognizes.

It may well be true that my concerns for the logical niceties of Darwinian theory are out of date, as Mr. Gross suggests. So much the worse for evolutionary biology. To those of us on the outside, Darwin's theory will continue to seem seriously infected by conceptual circularity. (In

Concepts and Methods in Evolutionary Biology, the philosopher Robert Brandon begins by at least recognizing the problem.)

The pattern of self-deception that I mocked in my essay is on display in any number of publications. In the first chapter of *The Causes of Molecular Evolution* (1991), John H. Gillespie is concerned to determine why a certain kinetic parameter (the Michaelis constant, Km) reaches intermediate values in a certain class of fish, the ectotherms. "If natural selection is responsible for the evolution of Km," Gillespie writes, "we should be able to understand why it would be maladaptive to exhibit values that are much less or much greater than the intermediate value." In fact, the two biochemical explanations Gillespie considers are flatly in conflict. They cannot both be true, although both may be false. To Gillespie, though, it hardly matters. ". . . [A]t this point in our discussion," he writes, "it is important merely to accept that there are plausible reasons for Km to be evolutionarily adjusted to intermediate values. This forms the theoretical basis of our acceptance of the conservation of Km in ectotherms as evidence for the action of natural selection in response to different thermal environments" (emphasis added). If the discussion has proceeded beyond the observation that Km reaches intermediate values in ectotherms, the fact is not discernible by me.

On the matter of the eye, Mr. Gross has misunderstood me. I did not propose to champion Stephen Jay Gould's question, "What good is 5 percent of an eye?" Instead, I argued that the question is hopelessly premature. Without knowing how the visual system works, we cannot determine *whether* it is accessible to a Darwinian mechanism. This seems to me an incontrovertible point, if also one that in my experience evolutionary biologists indignantly deny.

Let me offer an analogy. Starting from one and adding by two's, I cannot hope to reach the number eighteen. So far, so good. Starting from one and adding by two's, can I hope to reach the number n, where n is some number or other? Who knows? The question is underdetermined. Starting from *some* principle of addition or other, could I expect to reach *some* number or other? The question is now doubly underdetermined, functioning as a single equation in two unknowns. So, too, the question of whether a Darwinian mechanism with unspecified properties could reach a mammalian visual system whose properties are not yet completely understood. It is only credulous philosophers who imagine that a Darwinian mechanism is universally competent.

I am in agreement with Mr. Gross when he refers to "new and astonishing evidence" about the origin of the eye. Herewith the facts. Halder, Callaerts, and Gehring's research group in Switzerland discovered that the *ey* gene in *Drosophila* is virtually identical to the genes controlling the development of the eye in mice and men. The doctrine of convergent evolution, long a Darwinian staple, may now be observed receding into the darkness. The same group's more recent paper, "Induction of Ectopic Eyes by Targeted Expression of the Eyeless Gene in *Drosophila*" (*Science* 167, 1988), is among the most remarkable in the history of biology, demonstrating as it does that the *ey* gene is related closely to the equivalent eye gene in Sea squirts (Ascidians), Cephalopods, and Nemerteans. This strongly suggests (the inference is almost irresistible) that *ey* function is universal (universal!) among multicellular organisms, the basic design of the eye having been their common property for over a half-billion years. The *ey* gene clearly is a master control mechanism, one capable of giving general instructions to very different organisms.

No one in possession of these facts can imagine that they *support* the Darwinian theory. How could the mechanism of random variation and natural selection have produced an instrument capable of *anticipating* the course of morphological development and controlling its expression in widely different organisms?

I deny that I have in any way misrepresented Jacques Monod's arguments; his words are clear and unequivocal, in English and in the original French. If the word "chance" has an idiosyncratic use among evolutionary biologists, the secret has been closely kept. As far as I can determine, evolutionary theorists make use of the same technical concepts as the rest of the mathematical community, a point evident in any current text—*Mathematical Evolutionary Theory*, ed. R. Feldman (1989), for example.

The notion of a *bauplan* (*der Bauplan, die Bauplane*), or body plan, has had some currency in the English-speaking world; Stephen Jay Gould and Richard Lewontin use the term and so does J. W. Valentine. But it is in general dismissed as a concept by Darwinians—George C. Williams, for example. The idea of a body plan gained currency in the work of the great turn-of-the-century embryologists (Dreisch, Child, Boveri, Spemann), and before that in the writings of pre-Darwinian French biologists (Saint-Hilaire, Cuvier, Serres). I cannot imagine why Mr. Gross thinks it a term I have misused. (For reasons that are obscure to me, both he and Daniel Dennett carelessly assume that they are in a position to instruct me on a point of usage in German, my first language.)

I do not for a moment suppose, nor have I ever written, that biologists are in *conspiracy* "to hide from outsiders the bankruptcy of the central principle of biology." For one thing, the theory of evolution is hardly the central principle of contemporary biology. That description must surely be reserved for the thesis that all of life is to be understood in terms of the "coordinative interaction of large and small molecules" (James Watson, *The Molecular Biology of the Gene*). Rather, the theory of evolution functions as biology's reigning ideology. And no conspiracy is required to explain the attachment of biologists to a doctrine they find sustaining; all that is required is Freud's reminder that those in the grip of an illusion never recognize their affliction.

WHAT RANDY M. WADKINS affirms about the pre-Cambrian era is true enough (but see E. H. Davidson, K. Peterson, and R. Cameron, "Origin of Bilaterian Body Plans: Evolution of Developmental Regulatory Mechanism" (*Science* 270, 1995), for a real sense of the inadequacy of our grasp of fundamental facts concerning the Cambrian explosion). I only wonder why he imagines that his observations are in conflict with anything I have written.

In the same spirit, Mr. Wadkins calls me to task for failing to cite "the specific cases where transitional fossil forms are found in abundance." The fossil record does contain many intermediate forms; a recent publication on the Internet (Kathleen Hunt, *Transitional Vertebrate Fossils*, FAQ [Frequently Asked Questions], jespah@u.washington.edu) lists more than 250. But Mr. Wadkins has misunderstood the nature of the argument. My concern was to state the obvious: the fossil record contains gaps, places where the continuity assumptions of Darwinian theory break down. That there are places where the gaps are filled is interesting, but irrelevant. It is the gaps that are crucial.

Classical physics suggests that the spectral distribution of intensity in black-body radiation should be a continuous function of temperature. Experiments conducted at the end of the nineteenth century indicated otherwise. Continuity is an essential aspect of classical mechanics, impossible to discard without discarding the entire theory. Since the anomaly of black-body radiation could not be understood within classical mechanics, physicists sensitive to the evidence were persuaded to attach their allegiance to the new quantum theory.

The classical Darwinian theory of random variation and natural selection requires a continuous distribution of animal forms, one that must be reflected in the fossil record. The assumption of continuity is a crucial aspect of Darwinian theory; it cannot be carelessly discarded. This again is something that Richard Dawkins has rightly emphasized. The fossil record does not appear to support the assumption of evolutionary continuity, or anything much like it. Why is it that evolutionary biology is immune to evidence of the sort that falsified classical physics?

It is upon the horse that Mr. Wadkins pins his best hopes: its evolution, he suggests, comprises an unassailable sequence, one bright bursting beast after the other. To a certain extent, however, the neat evolutionary progression from *Eohippus* to *Equus* is an artifact of selection. The original groupings of species are far thicker (or bushier, to use the term of choice) than first thought, so that the sequence depends on a judicious selection of horse-like organisms at each stage of development.

It is rather as if one were to explain the emergence of the word cat by selecting the "c" from cattle, the "a" from abattoir and the "t" from tattle. Although almost 50 years old, G. G. Simpson's discussion in *The Major Features of Evolution* still repays study; for the modern point of view there is B. J. MacFadden, "Horses, the Fossil Record, and Evolution: A Current Perspective" (*Evolutionary Biology* 22, 1988).

But setting these reservations aside, what follows if the *Equus* sequence is accepted as a Darwinian progression? Very little. There are no more than three or four evolutionary sequences that, under the best of circumstances, suggest a complete progression of forms. By contrast, there are thousands upon thousands of species whose significant morphological features are not explained by complete or even highly suggestive sequences. Are the existing evolutionary sequences representative, or anomalous? In view of the striking discontinuities in the fossil record, I urge that they are anomalous; it would be interesting to know why Mr. Wadkins demurs.

When I observed that Richard Dawkins was unable to write a computer program that simulated his linguistic thought experiment, I did not mean that the task at hand was difficult. It is impossible. Mr. Wadkins commends the discussion in Keen and Spain's *Computer Simulation in Biology* as a counterexample; it is no such thing. What Keen and Spain have done is transcribe Dawkins's blunder into the computer language Basic. Here are the steps they undertake. A target sentence is selected—basic biological modeling is fun. The computer is given a randomly derived set of letters. The letters are scrambled. At each iteration, the computer (or the programmer) compares the randomly derived sequence with the target phrase. If the arrays—sequences on the one hand, target phrase on the other—do not match, the experiment continues; if they do, it stops.

There is nothing in this that is not also in Dawkins, the fog spreading from one book to the next. The experiment that Keen and Spain conduct is successful inasmuch as the computer reaches its target; but unsuccessful as a defense of Darwinian evolution. In looking to its target and comparing distances, the computer is appealing to information a biological system could not possess.

This point seems to be less straightforward than I imagined, so let me spell out the mistake. Starting from a random string, suppose the computer generates the sequence "bndit disne sot sodiswn toswxmspw sso." Comparing the sequence with its target, it proposes to conserve the initial "B." But why? The string is gibberish. Plainly, the conservation of vagrant successes has been undertaken with the computer's eye fixed firmly on its future target, intermediates selected not for what they are (gibberish, after all), but for what they *will* be (an English sentence). This is a violation of the rule against deferred success. Without the rule, there is nothing remotely like Darwinian evolution. What the computer has in fact done is to match randomly selected items to a template, thus inevitably reintroducing the element of deliberate design that was banished from the Darwinian world.

KARL F. WESSEL BELIEVES that Marshall Horwitz and Lawrence Loeb have provided an experimental refutation of my main argument. He has misunderstood both their argument and mine. In "Promoters Selected from Random DNA Sequences" (*Proceedings of the National Academy of Science*, October 1986), Horwitz and Loeb reported that by substituting randomly derived DNA sequences for original promoters in certain bacterial plasmids, they were able to discover additional promoters that maintained or enhanced drug resistance. Promoter sequences, I should say, are not highly active; their function is to be bound by the proteins of the transcription complex. What is more, base pair sequences in promoters are very often low in specificity: only a few base pairs are crucial to the promoter's function.

What Horwitz and Loeb were after may be understood by a linguistic analogy. Consider an English sentence: "The cat sleeps on the mat." Suppose one wished to discover which words might be substituted for the word "cat" while still preserving either the sense of the original sentence or a sense close to the original. One method might be to combine

English letters at random to form three-letter words (rat, tar, aim, tic, etc.), the words inserted in turn into the original sentence. Sentences in which meaning is preserved would be counted as successes. But such an experiment—similar in fact to experiments used in linguistics—would be a device for generating substitution classes, *not* a demonstration that meaning is preserved under arbitrary substitutions. Ditto the experiment reported by Horwitz and Loeb.

In addition to missing the point of the experiment, Mr. Wessel conveys an erroneous impression of its quantitative structure. "Of the approximately 10^{11} possible sequences of this type," he writes, referring to the total number of possible sequence substitutions, "it turns out that many promoted the function of the deleted natural sequence. . . ." In fact, Horwitz and Loeb report that ". . . very roughly, 2 percent of the 3×10^{11} possible recognition sequences present in this construction may duplicate promoter recognition site activity." This, of course, is a statistical estimate, one based on a small sample. In the case of plasmids deficient in adenine, the relevant number of successes drops to 0.2 percent. The vast majority of random sequences thus *failed* to duplicate promoter recognition site activity.

Citing with satisfaction the work of John Koza ("Genetic Programming: A Paradigm for Genetically Breeding Populations of Computer Programs to Solve Problems," Technical Report STAN-CS-90-1314, Department of Computer Science, Stanford University), Mr. Wessel would argue that genetic algorithms embody the abstract properties of robustness in the face of randomness that I claim could not be a feature of living systems. In fact, I made no such claim. As their name suggests, genetic algorithms are structures designed to incorporate (or mimic) certain biological operations. Typically, strings or sets of strings are introduced as fundamental data structures and manipulated by operators that reproduce the effects of random variation, genetic crossing, and natural selection. Work in this area was initiated in the 1970s by John Holland in *Adaptation in Natural and Artificial Systems*.

Mr. Wessel's claim that "[m]any of these evolved programs perform their optimizing tasks better than the best intentionally designed ones" is surely not incontrovertible; some computer groups argue that genetic algorithms do not outperform hill-climbing algorithms (*Proceedings of the Second International Conference on Genetic Algorithms and Their Applications*, ed. J. J. Grefenstette, 1987). But let that pass. I do not for a moment deny the possibility that a controlled random search might be an effective way in which to explore a large space. In referring to the Face Print algorithm in my essay, I said as much. What is at issue is the nature of the controls. The Face Print algorithm may be a fine method for prompting a crime victim's memory; it may, in fact, be far superior to traditional methods in which the victim offers a police artist a description of the malefactor ("Uh, let me see, big nose, yes, it was a big nose, I think, but no, not that big . . ."); but the algorithm fails to capture an essential feature of a Darwinian mechanism, for fitness is evaluated in terms of an ever-progressing match between what the algorithm produces and what the crime victim remembers, with the crime victim's memory functioning as—once again—a forbidden design or template.

The larger question posed by genetic algorithms is whether they can reach any interesting lifelike structures. Although genetic algorithms are new, they make use of an old mathematical concept, a Markov chain. It is worth noting that mathematical models based on Markov chains cannot in principle generate the sentences of a natural language. This is something known since the 1950s; I offer it as an observation.

In speaking of redundancy, Mr. Wessel has appeared to misunderstand the basic technical facts. There are two theories loitering in the background. One is Shannon's theory of information, which was developed in the 1940s (see A. I. Khinchin, *Mathematical Foundations of Information Theory*); the other is the Kolmogorov-Chaitin theory of algorithmic complexity, which was developed in the 1960s and 1970s (see G. Chaitin, "Information-theoretic Computational Complexity," in *IEEE Transactions on Information Theory*, IT-20, 1974).

Information theory makes use of concepts drawn from the classical theory of probability; the theory of algorithmic complexity does not. When Mr. Wessel speaks of "maximally compressed programs, in the sense of algorithmic information theory," he is incoherently attributing to one theory the concepts of another. Code compression is an information-theoretic concept, not a concept of algorithmic information theory at all. What information theory reveals is that in order to determine the best code compression for a given text, all that need be considered is the entropy of the text—the information it contains—and the number of its symbols. The maximally compressed codes are, indeed, rare, but only because they are maximally compressed. (The fact that they are rare follows from the fundamental Shannon-Macmillan theorem, but it follows trivially.) Contrary to what Mr. Wessel imagines, the genetic code is "optimal," a fact demonstrated by Cullmann and Labouygues ("*Le Code genetique, instantane et absolument optimal*," *Compte rendue* 301, 1985).

Redundancy, too, is an information-theoretic concept. Natural languages are redundant by virtue of their structure and their error-correcting mechanisms; the informational macromolecules are redundant by virtue of their structure and error-correcting mechanisms. This accounts for their stability. The problem at hand, however, is not to explain why the informational macromolecules have stayed the same, but how they might be generated by random means and how they might change by random mutations. Mutations used to be thick on the ground; now the beneficial ones are considered unlikely events. I am all for biological stability: I simply wonder, given all those stable macromolecules, that anything ever happens.

In addition to being redundant, the informational macromolecules are complex or highly specified. (*These* are terms drawn from the theory of algorithmic complexity and not from information theory.) Complexity is a form of *in*compressibility. A program, for example, is a linear string: so many bits of 1's and 0's. Complex linear strings are those that cannot be generated by strings shorter than themselves; the simple strings have some give. Thus, a string of randomly strung-out 0's and 1's cannot be generated by a shorter string; they are what they are, and in conveying their nature, I must convey the strings themselves. A string of 100 1's, on the other hand, may be generated by a string that simply specifies that 1 be written 100 times. Such strings are simple.

Far from being rare, as Mr. Wessel appears to claim, it is the complex strings that are in the majority. For example, of 1,000 sequences of a given length, typically only *one* may be compressed into a sequence shorter than itself. The informational macromolecules are thus buried in an enormous set of complex, utterly random sequences. It is very difficult to see how they might have been discovered by chance; and difficult again to see how chance might be the instrument of their change. In passing from the information macromolecules to complex biochemical *systems*—the Cori cycle, the mammalian immunodefense systems—the problems become infinitely more difficult.

I am unimpressed by computer models demonstrating punctuated equilibrium, whether elegantly or not. A typical problem in applied mathematics is to discover an equation that will describe a set of data points. A good deal depends on what the mathematician permits himself. As Enrico Fermi noted long ago, with five free parameters, an equation may be made to represent data points resembling an elephant. So, too, with computer models.

I AM TAKEN WITH the analogy proposed by PHILIP H. SMITH, JR. between living creatures and human languages, if only for prophylactic reasons. Just as the striking properties of human language cannot, as far as we know, be derived from considerations of engineering or communications optimality, so it seems to me that many of the most striking properties of living systems will fail to reflect properties of adaptation or optimality.

But contrary to what Mr. Smith believes, I did not ask for predictive laws for biology, or laws comparable to those found in physics. I simply observed that evolutionary theories quite typically fail to answer any interesting questions, whether historically or not.

Mr. Smith's remonstrations on thermodynamics point to another misunderstanding. In writing that living creatures appear to offer at least a temporary rebuke to the second law of thermodynamics, my operative word was "appear." I am familiar with the scenario, standard from Boltzmann to Monod, according to which life represents a statistical fluctuation in the scheme of things. It was this scenario that I meant to evoke by writing that both the second law of thermodynamics and the theory of evolution explain things by an appeal to a turn of the same cosmic wheel—chance.

Still, I would not wish to overstate my agreement with the standard line. Living systems do constitute an open system, the sun affording them energy which they then degrade. But the sun shines alike on the living and the dead. And so the question inevitably returns to its old familiar haunts: how did living creatures acquire the mechanisms needed to exploit all that free energy? I have no idea; but then, neither does anyone else.

In general, the relationship between the principles of biology, if there are any, and the laws of physics seems to me wide open. There are, after all, three possibilities. Those principles may prove consistent with the laws of physics; inconsistent; or independent. We have no idea at present which version of events is true. I see no reason to assume that the problem will prove any simpler than the problem of establishing that the continuum hypothesis and the axiom of choice are independent of the axioms of set theory, an enterprise requiring for its resolution the genius of Kurt Gödel and Paul Cohen.

IF SHELDON F. GOTTLIEB really believes that creationist doctrines follow as natural inferences from any remark of mine, he needs to show how.

There is no widely accepted, remotely plausible scenario for the emergence of life on earth. The proteinoid hypothesis of Sidney Fox and his colleagues (S. W. Fox, "Molecular Evolution to the First Cells," *Pure and Applied Chemistry* 34, 1973) has not persuaded the biological community of its strength. Criticisms of it are overwhelming (W. Day, *Genesis on Planet Earth*; K. Dose, "Ordering Processes and the Evolution of the First Enzymes," in *Protein Structure and Evolution*, eds. J. L. Fox, Z. Deyl, A. Blazy, 1976; C. E. Folsome, "Synthetic Organic Microstructures and the Origin of Cellular Life," *Die Naturwissenschaften* 7, 1976; C. Ponnamperuma, "Cosmochemistry and the Origin of Life," in *Cosmochemistry and the Origin of Life*, ed. C. Ponnamperuma, 1983; and so forth).

In commenting on my discussion of the thorn bush and the Pitcher plant, Mr. Gottlieb topples straight into a trap I never dared imagine would lure a biologist. I know as well as anyone that once the facts are available, an evolutionary story may be concocted by which those facts may be explained. But if the geographical facts were reversed, does Mr. Gottlieb doubt that an imaginative biologist could provide an account of the Pitcher plant showing why it thrives in nutrient-*rich* soil? I am asking for the stories to follow from general principles *before* the facts are known.

When I asked who on the basis of experience would be inclined to disagree with the account of creation given in the Book of Genesis, my aim was rhetorical; but I stand by the question. Our experience is overwhelmingly in favor of the thesis that complex objects arise as the result of a deliberate act of design. It may well be that the thesis is false; but it is what experience *suggests*.

Mr. Gottlieb has come to the conclusion that in science it is rarely pertinent to ask *why?* Why are the equations of physics expressed as quadratic forms? Why is the orbital spectrum of the hydrogen atom discrete? Why does darkness fall so quickly in the tropics? Why does the earth not spiral into the sun? Why do women outlive men by seven years? Why do cat's eyes contain a nictitating membrane? And why did Sheldon Gottlieb not think more carefully before conveying himself into print?

Robert Shapiro has modestly withheld from readers the fact that he is the author of a penetrating work on pre-biotic evolution, which I have already cited: *Origins: A Skeptic's Guide to the Creation of Life on Earth.* As I might have hoped, Mr. Shapiro is with me for nine-tenths of my argument. He jumps ship at the thought that our options may have narrowed to a choice between Darwin and what he calls Intelligent Design. But, contrary to what Mr. Shapiro supposes, I entertain no supernatural explanations for the complexity of living systems. The thing is a mystery, and if there is never to be a naturalistic explanation, I shall forever be content to keep on calling it a mystery. The two of us might have gone on together to the end.

A skeptic about so much, Mr. Shapiro now feels compelled to commend theories of self-organization and complexity as solutions to the problems that vex us. For me, the papers, books, and monographs coming from the Santa Fe Institute, with which Stuart Kauffman is prominently identified, convey something familiar. I once spent a good deal of time demolishing the set of fashionable mathematical theories collected under the generic term of systems analysis (*On Systems Analysis: An Essay on the Limitations of Some Mathematical Methods in the Social, Political, and Biological Sciences,* MIT Press, 1976). Reading Kauffman's *The Origins of Order* (1993), I was flooded with memories. Had I time, I would go after Santa Fe with gusto; but soon the night comes, Dr. Johnson reminds us, wherein no man can work. I find nothing of value in various theories of self-organization; the very idea is to my mind incoherent; but I leave it to others to make the case.

In my essay, I endorsed Paley's inference *from* the complexity of a human artifact to the design, and hence the designer, that brought it into being. As a counterexample, Paul H. Rubin offers the neoclassical market, complex but not designed.

Let me draw a distinction between the institutional structure of a market and the behavior of its economic agents. Neoclassical economic theory suggests that even though there are very many agents in a given market, its overall properties may be explained on the basis of relatively simple economic assumptions. They come together, those agents, each to maximize his utility; there follows a process of *tâtonnement*, of feeling one's way, followed in turn by equilibrium. Such is the vision given expression, for example, by Léon Walras.

The intellectual structure of micro-economic theory is very similar to that of theories controlling the behavior of perfect gases in physics. But to the extent that simple laws prevail, there is no reason to describe this aspect of a market as complex.

The institutions of a market comprise the law of contracts, accounting practices, various regulatory bodies, long standing traditions, and the like. And here Paley's original inference does come into play. Complex human institutions as well as complex human artifacts arise as the result of deliberate design. On the most general level, Paley's question— whence the origin of complexity?—draws a connection between complexity and intelligence, a connection preserved in the case of markets.

In *The Language Instinct* (1994), Steven Pinker has written an engaging book about Noam Chomsky's revolution in linguistics. Chomsky has from time to time expressed his impatience with Darwinian doctrine, and it is this Chomsky whom Pinker proposes to instruct. His argument is nothing more than a weak solution of Richard Dawkins. In fact, little in our understanding of language even hints at a Darwinian development. "Any progress toward" the goal of understanding the language system, Chomsky writes, "will deepen a problem for the biological sciences that is far from trivial: how can a system such as human language arise in the mind/brain, or for that matter in the organic world, in which one seems not to find anything like the basic properties of human language" (*The Minimalist Program*, 1995)?

As it has done to so many others, epistemology brings Mr. Rubin low. "All science," he writes, "is an attempt to explain systems we do not fully understand with processes we cannot completely specify." Who could doubt it? We pause, grope, stare in perplexity, pause, grope again. But Mr. Rubin has confused the facts of life with the conditions for knowledge. The rational answer to the question of whether a system we do not completely understand might be constructed by means of a process we cannot completely specify must be that we do not know. Let us have the details and we shall see. The bouncy assumption that every biological structure *must* be accessible to a Darwinian mechanism serves only further to empty Darwinian theory of its empirical content.

The new science of evolutionary biology, Mr. Rubin believes, has carried us to the threshold of "truly understanding" the source of the human mind. I see no evidence that this is so. What has that zestful new science told us about the origins of human language; the human ability to do mathematics and construct far-reaching scientific theories; human culture, with its attendant mysteries; human art, music, and poetry; the human sense of time, or those spiritual urges which baffle and torment us? Not much, I am afraid.

I was concerned in my essay to question the thesis that random variation and natural selection are the mechanisms by which the appearance, development, and organization of life on earth are to be explained. John M. Levy is concerned to defend what he calls an "evolutionary scenario"—descent with modification, as it happens. To a certain extent, his letter and my essay have passed each other like ships in the night.

There is a great deal of evidence in favor of descent with modification; the pattern is illustrated in a thousand different textbooks. And it is possible to accept descent with modification as an overall description of the spatial and temporal organization of living creatures without in any way agreeing that a Darwinian mechanism explains the pattern. Such seems to have been the position of the great French zoologist, Pierre Grasse (*L'Evolution du Vivant*).

Still, the evidence in favor of descent with modification is not without its troubles. *Hyman's Comparative Vertebrate Anatomy* (ed. Marvalee H. Wake), for example, offers a useful corrective to the idea that the facts of comparative anatomy are either straightforward or unequivocal. This magnificent discipline does make it overwhelmingly clear that chordate anatomical structures are in many cases similar; but the crucial issue is not whether such structures are similar but whether they are *homological*, in the sense that, of two given structures, one is ancestral to the other.

It does little good to say, as *Hyman* does, that "homology *means* intrinsic similarity that indicates a common evolutionary origin." If this is what homology means, there is little point in asking empirically what it signifies. Without a sharp criterion distinguishing analogical from homological structures, biologists have no way of determining that, say, the fusiform shape of the seal and the tuna is of no evolutionary significance.

The line separating analogical from homological structures has traditionally been drawn with respect to three features: (a) anatomical structure; (b) topographic relationships of anatomical structures to animal bodies; and (c) embryological development. As the criterion for homological relationships is clarified, many examples adduced in favor of the hypothesis of descent with modification become dubious. The vertebrate occipital arch and the vertebrate kidney provide well-known examples. There are many others.

In general, it remains true that while the anatomical facts are rarely in dispute, their interpretation remains both difficult and tentative. G. De Beer's *Homology: An Unsolved Problem* (1971) still merits study, while Mark Ridley's *The Problem of Evolution* (1985) offers instructive evidence of the ease with which the matter of homology may be subordinated to an adroit verbal shuffle.

In objecting to my claim that the fossil record contains gaps, Mr. Levy would have me appreciate more robustly "the pieces of the fossil record that started turning up," especially in connection with the ungulate-to-whale transition. He is referring to *Abulocetus natans*, and he is correct in observing that this makes the hypothesis that aquatic mammals evolved from terrestrial mammals more plausible than it was. I am concerned only to ask whether the sequence is representative or anomalous—the same question I posed in my response to Randy Wadkins. In *Evolution: A Theory in Crisis*, Michael Denton questions the ungulate-to-whale transition from a different perspective.

UNLIKE MARTIN GARDNER, I do believe that punctuated equilibrium damages the Darwinian viewpoint; so does everyone else. By compressing the time available for speciation, Stephen Jay Gould has eliminated an accretion of small changes as its mechanism. The result is a theory which very nicely fits the facts, but a theory that all the same leaves the mechanism of change entirely in the dark.

As for Mr. Gardner's last question: for many years I have been puzzling over whether the first humans had parents; sad to say, I still have no answer.

FINDING NO CONVENIENT POINT of affirmation in my essay, HERBERT GINTIS wishes to know what I am trying to say. I never deny the facts of evolution, he remarks, apparently with some vexation. And it is true, I never do, if only because evolution has come to cover *any* process of biological change explicable in naturalistic terms. As for natural selection, I do not deny that it occurs, either, insofar as it is occupied in establishing that what survives, survives. (There is no denying the inescapable.) But denials being called for, I do deny that *theories* of random mutation and natural selection explain much about the emergence or development of life on earth. Mr. Gintis seems to feel that such denials are intellectually impossible. Evidently not.

DAVID P. BABCOCK IS correct at least in this: when it comes to the major problems of biology, I have nothing better to offer than the theories I dismiss. The pugilistic wisdom that you can't beat something with nothing seems to have become a staple of the philosophy of science; it now functions as a reactionary force, making it difficult to bang away at deficient or defective theories. This is surely an unhappy state of affairs. I am not a biologist and so have no theory to offer in place of the one I criticize. But neither am I a chef. On being presented an oversalted dish, should I refrain from upbraiding the cook because I cannot prepare anything better?

I AM HAPPY TO salute *Archaeopteryx*, recognizing the little monster as half-bird and half-reptile (or anything EUGENIE C. SCOTT wishes).

Strawmen? As long as I am at it, let me knock down a few more. There is no "Cambrian-explosion argument"; there is the fact that an explosion of biological forms took place in the Cambrian era. Far from being impossible, amino acids come together to form proteins all the time; no one has provided a plausible account of the *origin* of the informational macromolecules. And there is no need to explain the difference between evolution and chance to any native speaker of English: the words mean different things.

I ENDORSE EDWARD T. OAKES's endorsement of Gertrude Himmelfarb, but find myself unable to appreciate his counsel when the discussion passes from history to the philosophy of science. Like David Babcock, Father Oakes wishes that I would exchange a bad theory for a better one. The wish betrays a confusion of genres; my business is criticism and not biological theory-making. But may I voice my unhappiness with the formulation of the request itself? In asking for the engine of evolution, Father Oakes has already determined the form an answer must take. There is evolution, on the one hand, and a dynamic theory, on the other. This is entirely too narrow a vision.

Like a number of other writers, Father Oakes believes that the gravamen of my essay lies with the failure of Darwin's theory successfully to predict the future. Not so. I ask only that evolutionary explanations follow from general principles.

RESPONDING TO KENT GORDIS, I find myself in the lunatic position of offering a few words in defense of Darwin's doctrine. To speak of the "total absence" of intermediate fossil forms is to speak too strongly. There are plenty of intermediate forms, as I point out in my response to Randy Wadkins. The trouble is more subtle but in the end more perplexing: the absence of crucial intermediate forms, and the fact that the overall fossil record seems so very strongly biased in favor of a snooze-alarm pattern in which stasis is followed by speciation.

The metaphor Mr. Gordis attributes to Fred Hoyle is a simile, and was invoked not by Hoyle but by his collaborator, the astrophysicist N. C. Wickramasinghe.

I AGREE WITH J. R. DUNN that Darwinian theory is essentially nineteenth-century in its cast, and that it has not undergone the systematic development of other nineteenth-century scientific concepts. In two respects Darwin was attempting to solve a problem he could not state. Knowing nothing of biochemistry, he was unable to suggest how random variations and natural selection might account for biochemical complexity. In his provocative new book, *Darwin's Black Box*, Michael J. Behe has investigated a number of biochemical systems or machines and concluded that more than a century after Darwin became common dogma, we still have no idea whether a Darwinian mechanism was responsible for the bacterial *flagellum*, the cellular protein transport system, the immunodefense system, or the blood-clotting mechanism.

By the same token, Darwin was unable to determine whether his theory could, even in principle, account for high-level biological systems—language, for example, or the mammalian visual system. In the case of biochemistry, we can at least describe the systems; in the case of higher-order systems, we cannot even do that, so that asking whether they are accessible to a Darwinian mechanism is an exercise in irrelevance.

In observing that natural selection is often presented as a universal doctrine, Mr. Dunn has put his finger on a tender nerve in contemporary thought. The great ideological structures of the twentieth century lie in ruins. Outside the academy, no one would think to identify himself as a Marxist. But the ideological *impulse* is as strong as ever, and there yet remains in Darwin's thought an unsullied temple.

ARTHUR B. CODY IS right twice: first in calling attention to the way writers in any number of fields have adopted a careless and largely incoherent form of Darwinism, and second in predicting the way many Darwinian theorists would respond to criticism.

As Tom Southwick notes, there are daunting improbabilities associated with the spontaneous creation of life, or specifically with the spontaneous creation of the informational macromolecules. These improbabilities are what drove Francis Crick to his inspired suggestion that life arose elsewhere in the universe and was simply sent here. But the situation, as Mr. Southwick indicates, is even worse. Life must not only have found the dime's worth of area in which things work, it must also have stayed there once the initial happy discoveries were made. Some biologists estimate that there are fewer than 1,000 working protein *families* in nature (see *Science* 273, 1996, for a fascinating new discussion by Hao Li, Robert Helling, Chao Tang, and Ned Wingreen).

When we pass from simple macromolecules to complex biological structures, the situation is always the same. There is Mount Improbable, to use Richard Dawkins's fine metaphor, and there is Something Eager, puttering around at the base. In time, Something Eager makes it to the top of Mount Improbable. But how? The usual approach taken by Darwinian theorists to avoid the invocation of a miracle is to suggest that Something Eager need not climb Mount Improbable directly: a slow ascent by an ever-turning spiral will do as well. Something Eager need only acquire a toehold, which the odds do not prohibit, and thereafter natural selection acts to conserve stray successes.

As I have said repeatedly, I think this is a mistake. When natural selection is given its proper Darwinian interpretation as a force or property denied access to the future, it loses its power to seal off random success. It is this point that my Head Monkey was intended to establish. Once he has been dismissed from the scene, the Darwinian mechanism again acts randomly. In one way or another, all Darwinian scenarios involve an attempt to bring that monkey back.

Like so many others, Tom Southwick finds theology an awkward business. Me, too.

The 1995 Statement on Teaching Evolution cited by John Wiester strikes me as a superb example of the gibberish to which biology teachers are so often prone. I can just imagine the furious wrangling in the committee room as the document underwent its eleventh and final revision. Still, I would not take the result too seriously. Gibberish is its own punishment.

As NANCY R. PEARCEY suggests, there is no way to understand the organization of, say, the bacterial cell without such concepts as code and codon, transcription, translation, information, and the like—the concepts, that is, that are left over when biochemistry is subtracted from molecular biology. Still, the fact that DNA is a code does not yet mean that it is a language or even much like a human language. A natural language is organized to express human thoughts, its concepts constrained by human needs and purposes. This is not true of DNA.

The central role attributed to information by so many theoretical biologists seems to me problematic, if only because it is endorsed so warmly by Richard Dawkins. Information is a property of strings of symbols. It is a mistake to regard the information resident in the genome as somehow a measure of a quantity resident in the whole of a complex organism, the more so since we have only the haziest idea how the genome translates its fantastically complex message from one dimension into three. "[T]he central and still unsolved problem is, how do genes direct the making of an organism?" (Rudolf A. Raff and Thomas C. Kaufman, *Embryos, Genes, and Evolution*, 1983). Until we know that, I, for one, would hold off on claims that "the origin of life and its myriad of forms must be recast as the origin of biological information."

I VERY MUCH APPRECIATE DAVID J. MANDEL's kind words.

THE CONTRAST WHICH LARRY EUBANK points out, between the very tentative nature of the evidence supporting evolutionary theory and the offensive self-confidence of many evolutionary biologists, is one of those mysterious coils that history in its cunning simply leaves lying about. What makes the situation so strange is that evolutionary biologists seem utterly determined to maintain positions from which physicists have long fled. It is the physicists, after all, who have been struck by evidence of design in the cosmos; books pour from the press explaining how this or that feature of the physical world—the fundamental constants, for example—simply could not have been an accident. The biologists will have none of it. Theirs is a world unyieldingly material, with Daniel Dennett, for one, prepared to extend natural selection to the very firmaments themselves, whole universes mutating joyfully or winking out of existence after having failed to make the cosmic cut.

But this is to explain the smugness of evolutionary biologists by an appeal to intellectual lag. No doubt, other forces are at work. The defense of Darwin's theory by writers like Dennett or Dawkins is at least in part a calculated *political* act, a statement about who is to control the ideologies of a democratic state.

I HAVE SOME SYMPATHY for HENRY SHERMAN's sense of his own plight, and sympathy as well for the sense of betrayal his letter conveys, perhaps at an unconscious level. It is a measure of the utterly immoderate claims made by evolutionary biologists that we should all feel the intellectual universe emptied when those claims are challenged or found wanting.

As TOM BETHELL NOTES, there is a feeling both among biologists and the public that the intellectual foundations of biology may be about to crack. Many biologists are unhappy with what is sometimes called ultra-Darwinism; they imagine that they can camouflage their distress by attaching their allegiance to a version of evolution that has been emptied of controversy by being evacuated of content. It is by means of this tactic, Mr. Bethell observes, that evolution has become a sonorous synonym for change.

Evolution is but one issue. Contemporary biology has also been one of the damp breeding spots from which the mosquito of materialism arises. If we are in general disposed to identify the human mind with the human brain, or to look into the eyes of an ape and find reflected there no absolute difference between species, we are simply giving expression to the reach and influence of biological thought. So, too, if we assume, against all evidence to the contrary, that living creatures emerged spontaneously from the world of matter, the first bug clambering from some warm little pond somewhere, eager to get on with the business of infecting millions.

Like Robert Shapiro, Tom Bethell has not mentioned his own contribution to the debate. His articles (in *Harper's* and *The American Spectator*) were unique in having offered readers a lucid objection to Darwinian theory at a time when the gesture itself was certain to invite ridicule.

The contrast that Daniel Lapin draws between Joshua Chamberlain and William Provine is both painful and poignant. Chamberlain was convinced that the laws of nature are God's ways seen by man. In this, he echoed Gerard Manley Hopkins: "The world is charged with the grandeur of God. . . ." There is no question that civil society would be much improved if these sentiments were widely shared. But an innocent conviction of grace, once lost, cannot easily be regained.

William Provine's remarks, by way of contrast, seem enviably hardheaded, almost brutal. It is perhaps this brutality that persuades so many readers that what he says is correct. It is surely not the claims themselves: not one of them is true.

To Rabbi Lapin, a choice between Chamberlain and Provine is inescapable. I demur. While Chamberlain and Provine cannot both be right, they may well both be wrong.

Certain biochemical systems, Michael J. Behe argues, are not only complex but *irreducibly* complex. Irreducibly complex systems have two crucial properties. First, only one of the possible arrangements of their parts is apt to allow the system to function; it is this feature that distinguishes complex objects, such as watches or the human kidney, from such structures as a modern city, which are not complex in this sense even though they contain many parts or actors. Second, the parts of an irreducibly complex biological object are mutually dependent.

Darwin's theory requires complex structures to be built by accretion—one small change after another. But if a system is irreducibly complex, there is no advantage to be gained from intermediates; usefulness arises only when many interacting parts play their roles simultaneously. Looking at a number of biochemical systems, Mr. Behe argues in *Darwin's Black Box* that they could not arise by a Darwinian mechanism.

Surely, one would think, this claim has already been investigated by the biological community, the matter settled. Not at all. What Mr. Behe has done, apparently for the first time, is to show how much sheer bluff has gone into contemporary versions of Darwin's theory.

SINCE IT SEEMS MY pleasant duty to toot other men's horns, may I point out that PHILLIP E. JOHNSON is the author of *Darwin on Trial*, a remarkable book indicating, if nothing else, that evolutionary biologists have held to a shockingly low standard of argument and evidence.

I DISAGREE WITH JEFFREY SATINOVER's assessment of Darwin's theory. If there is nothing it cannot explain, it follows that it cannot explain anything. In any case it is not Darwin's theory that accounts for variations "at the level of DNA." Genetic variability is an open problem *within* Darwinian theory. Whatever the capabilities of genetical algorithms, they are only remotely connected to the biological world.

But all this seems beside the point in the context of Mr. Satinover's long and fascinating letter. There are, he is convinced, certain anomalies in the scheme of things. The emergence of ever more complex neurological systems is an example—the pattern of increasing complexity, he believes, is not entirely explicable in physical terms. True; but *nothing* in the physical world is entirely explicable in physical terms. Physics is a profoundly nonphysical discipline, making use of such abstract objects as forces, fields, functions, and the like.

Quantum mechanics is without doubt the strangest, most counterintuitive theory ever devised by the mind of man, and Mr. Satinover is right to be astonished. He and I part company when he attempts to draw, from the history of twentieth-century physics, lessons for evolutionary biology.

Can one, he asks, treat all of life as a gigantic random machine, operating according to the principles of nonlinear (chaotic) dynamics? Or is life best understood in terms of quantum computations involving "impossible" noncomputable results? Whatever Mr. Satinover may be after here, the alternatives he offers are unclear. Nonlinear or chaotic dynamics? Surely not: these are mathematical descriptions and must be used with care. Quantum computations involving noncomputable results? Huh?

I have not read the work of Koichiro Matsuno, but I would like to know how quantum fluctuations bear in any way on genetic mutations. Quantum fluctuations are quantum events; mutations are not. I am intrigued by the idea that the global morphology of a complex biochemical structure has an influence on the pattern of its development—this is René Thom's leading idea (*Structural Stability and Morphogenesis*, 1975), and indeed it is as old as Aristotle—but I simply see no reason to involve quantum mechanics in the whole business.

Still, there are more things in heaven and earth than are dreamt of in any philosophy. This is Mr. Satinover's real point. I hope that it is true.

A GREAT DEAL OF the evidence for evolution, RUSSELL ROBERTS suggests, arises from a grand and unsupported extrapolation. The speckled moth changes its wing coloring; bacteria develop drug resistance. Why should this count in favor of the thesis that whales are derived from ungulates, or men from fish? Plodding steadily upward on any given mountain, the Darwinian climber (a.k.a. Something Eager) is bound to find that there are certain places forever out of reach—the surface of the moon, for example. The Darwinian argument of evolution by accretion is itself missing a crucial step, one that would demonstrate either from first principles or from close observation that complex biological structures are accessible to a Darwinian mechanism, and so function as a mountain peak rather than as the surface of the moon.

THE TRIUMPH OF DARWINIAN thought, M. P. SCHÜTZENBERGER often reminded me, occurred later in France than in the United States;

Darwin's thought did not fully establish itself in official intellectual circles until the development in the late 1940s and 1950s of a powerful French school of molecular biology. Although profoundly impressed by the achievements of that field, M. P. Schützenberger also deplored the coarsening of thought that it brought about. A physician as well as a mathematician, he saw himself as a defender of the great French tradition of speculative biology (or natural philosophy), a tradition stretching from Cuvier to Lucien Cuenot. His criticisms of Darwinian theory were in many ways an attempt to express the insights of that tradition by means of the modern mathematical tools that he himself introduced into French intellectual life. Recently, he had come to believe that the fundamental distinction between the physical and the biological world was that biological creatures live in time, a medium, or dimension, that does not exist in the physical world. Einstein and Gödel both thought time an illusion; Schützenberger took their insight and reinterpreted it so that the illusion became a characteristic of biological systems.

Marco Schützenberger and I spent a year working together day in and day out on a book devoted to evolution; we accumulated a great many notes, but in the end, circumstances compromised our plans and our book was never finished. M. P. Schützenberger died on July 29 of this year.

KEEPING AN EYE ON EVOLUTION

THE THEORY OF EVOLUTION IS THE GREAT WHITE ELEPHANT OF contemporary thought. It is large, almost entirely useless, and the object of superstitious awe. Richard Dawkins is widely known as the theory's uncompromising champion. Having made his case in *The Blind Watchmaker* and *River out of Eden*, Dawkins proposes to make it yet again in *Climbing Mount Improbable* (W. H. Norton & Company, 1996). He is not a man given to tiring himself by repetition.

Darwin's theory has a double aspect. The first is the doctrine of descent with modification; the second, the doctrine of random variation and natural selection. Descent with modification provides the pattern; random variation and natural selection, the mechanism. Dawkins's concern is with the mechanism; the pattern he takes for granted.

Biological structures such as the mammalian eye are complex in the sense that they contain many parts arranged in specific ways. It is unlikely that such structures could have been discovered by chance. No one, the astrophysicist N. C. Wickramasinghe once observed with some asperity, expects a tornado touching on a junkyard to produce a Boeing 747. This may suggest—it has suggested to some physicists—a disturbing gap between what life has accomplished and what the theory of evolution can explain. The suggestion provokes Dawkins to indignation. "It is grindingly, creakingly, crashingly obvious," he writes, mixing three metaphors joyously, that the discovery by chance of a complex object is improbable; but the Darwinian mechanism, he adds, "acts by breaking the improbability up into small manageable parts, smearing out the luck needed, going round the back of Mount Improbable and crawling up the gentle slopes...."

This is a fine image, one introduced originally by the American bio-mathematician Sewall Wright. Random variation offers the mountain-eer an allowance of small changes. Chance is at work. Natural selection freezes the successful changes in place. And this process owes nothing to chance. In time, the successful changes form a connected path, a stair-case to complexity.

The example that Dawkins pursues in greatest detail is the eye. Darwin himself wondered at its complexity, remarking in a letter to an American colleague that "the eye... gives me a cold shudder." That shud-der notwithstanding, Darwin resolved his doubts in his own favor; the eye, he concluded, was created by a single-step series of improvements, what he called "fine gradations." Where Darwin went, Dawkins follows.

It is one thing, however, to appeal to a path up Mount Improbable, quite another to demonstrate its existence. Dawkins has persuaded himself that because such a path might exist, further argument is un-necessary. Impediments are simply directed to disappear: "There is no difficulty"; "there is a definite tendency in the right direction"; "It is easy to see that..."; "it is not at all difficult to imagine...."

IN FACT, THE DIFFICULTIES are very considerable. A single retinal cell of the human eye consists of a nucleus, a mitochondrial rod, and a rectan-gular array containing discrete layers of photon-trapping pigment. The evolutionary development of the eye evidently required an increase in such layers. An inferential staircase being required, the thing virtually constructs itself, Dawkins believes, one layer at a time. "The point," he writes, "is that ninety-one membranes are more effective... than ninety, ninety are more effective that eighty-nine, and so on back to one mem-brane, which is more effective than zero."

This is a plausible scheme only because Dawkins has considered a single feature of the eye in isolation. The parts of a complex artifact or object typically gain their usefulness as an ensemble. A Dixie Cup consists of a tube joined to a disk. Without the disk, the cup does not hold less water than it might; it cannot hold water at all. And ditto for the tube, the two items, disk and tube, forming an irreducibly complex system.

What holds for the Dixie Cup holds for the eye as well. Light strikes the eye in the form of photons, but the optic nerve conveys electrical impulses to the brain. Acting as a sophisticated transducer, the eye must mediate between two different physical signals. The retinal cells that figure in Dawkins's account are connected to horizontal cells; these shuttle information laterally between photoreceptors in order to smooth the visual signal. Amacrine cells act to filter the signal. Bipolar cells convey visual information further to ganglion cells, which in turn conduct information to the optic nerve. The system gives every indication of being tightly integrated, its parts mutually dependent.

The very problem that Darwin's theory was designed to evade now reappears. Like vibrations passing through a spider's web, changes to any part of the eye, if they are to improve vision, must bring about changes throughout the optical system. Without a correlative increase in the size and complexity of the optic nerve, an increase in the number of photoreceptive membranes can have no effect. A change in the optic nerve must in turn induce corresponding neurological changes in the brain. If these changes come about simultaneously, it makes no sense to talk of a gradual ascent of Mount Improbable. If they do not come about simultaneously, it is not clear why they should come about at all.

The same problem reappears at the level of biochemistry. Dawkins has framed his discussion in terms of gross anatomy. Each anatomical change that he describes requires a number of coordinate biochemical steps. "[T]he anatomical steps and structures that Darwin thought were so simple," the biochemist Michael Behe remarks in his provocative book *Darwin's Black Box*, "actually involve staggeringly complicated biochemical processes." A number of separate biochemical events are required simply to begin the process of curving a layer of proteins to form a lens. What initiates the sequence? How is it coordinated? And how controlled? On these absolutely fundamental matters, Dawkins has nothing whatsoever to say.

IN ADDITION TO THE eye, Dawkins discusses spiders and their webs, the origin of flight, and the nature of seashells. The natural history is charming.

Dawkins is a capable if somewhat dry prose stylist, although such expressions as "designoid" and "wince-makingly" are themselves wince-making. The science throughout is primitive. Difficulties are resolved by sleight-of-hand. "In real life," Dawkins remarks in a representative passage, "there may be formidable complications of detail." Yes? What of them, those formidable complications? "These emerge simply and without fuss."

Is the elephant's large nose truly the result of an evolutionary progression? Then some demonstration is required showing that intermediate-sized noses are valuable as well. None is forthcoming. "If a medium sized trunk were always less efficient," Dawkins writes, "than either a small nose or a big trunk, the big trunk would never have evolved." Indeed. The emergence of powered flight is treated as an engaging fable, one in which either arboreal animals glided downward from the tree tops or a primitive dinosaur hopped upward toward the sky. "The beauty of this theory," Dawkins affirms, commending the hopping scenario, "is that the same nervous circuits that were used to control the center of gravity in the jumping ancestor would, rather effortlessly, have lent themselves to controlling the flight surfaces later in the evolutionary story." It is the phrase "rather effortlessly" that gives to this preposterous assertion its antic charm.

A final note. In a book whose examples are chosen from natural history, it is important to get the details right. Hawks may soar or sail, but they cannot hover like helicopters. Not all organisms share precisely the same genetic code. And Gary Kasparov was defeated by IBM's Big Blue, and not a program entitled Genius 2.

THE END OF
MATERIALIST SCIENCE

Simply the thing you are shall make you live.

—OLD SPANISH PROVERB

FOR THE MOMENT, WE ARE ALL WAITING FOR THE GATE OF TIME TO open. The heroic era of scientific exploration appears at an end, the large aching questions settled. An official ideology is everywhere in evidence and everywhere resisted. From the place where space and time are curved to the arena of the elementary particles, there is nothing but matter in one of its modes. Physicists are now pursuing the grand unified theory that will in one limpid gesture amalgamate the world's four far-flung forces.

For all of its ambitiousness, it is hardly an inspiring view. And few have been inspired by it. "The more the universe seems comprehensible," Steven Weinberg has written sourly, "the more it seems pointless." Yet even as the system is said to be finished, with only the details to be put in place, a delicate system of subversion is at work, the very technology made possible by the sciences themselves undermining the foundations of the edifice, compromising its principles, altering its shape and the way it feels, conveying the immemorial message that the land is more fragrant than it seemed at sea.

Entombed in one century certain questions sometimes arise at the threshold of another, their vitality strangely intact, rather like one of the Haitian undead, hair floating and silvered eyes flashing. *Complexity*, the Reverend William Paley observed in the eighteenth century, is a property of things, one as notable as their shape or mass. But complexity, he went on to observe, is also a property that requires an explanation.

It was a shrewd, a pregnant, a strangely modern observation. The simple structures formed by the action of the waves along a beach—the shapely dunes, sea caves, the sparkling pattern of the perishable foam itself—may be explained by a lighthearted invocation of air, water, and wind. But the things that interest us and compel our fascination are different. The laws of matter and the laws of chance, *these*, Paley seemed to suggest, control the behavior of ordinary material objects, but nothing in nature suggests the appearance of a complicated artifact. Unlike things that are simple, complex objects are logically isolated, and so they are logically unexpected.

Paley offered examples that he had on hand, a pocket watch chiefly, but that watch, its golden bezel still glowing after all these years, Paley pulled across his ample paunch as an act of calculated misdirection. The target of his cunning argument lay elsewhere, with the world of biological artifacts: the orchid's secret chamber, the biochemical cascade that stops the blood from flooding the body at a cut. These, too, are complex, infinitely more so than a watch. Today, with these extraordinary objects now open for dissection by the biological sciences, precisely the same inferential pattern that sweeps back from a complex human artifact to the circumstance of its design sweeps back from a complex biological artifact to the circumstance of its design.

What, then, is the *origin* of their complexity? This is Paley's question.

It is ours as well. At century's end, the great clock ticking resolutely, complexity seems more complex than ever. Along with chaos, it is the convenient explanation for every conceivable catastrophe; given the prevalence of catastrophes, it is much in vogue. Civil and computer codes are complex; so is the air transport system, diseases of the liver, and the law of torts. Ditto for modern automobile engines, even the mechanic at my shop complaining richly that "this here gizmo's too complex for me, Mister B'linski." Semiconductors are complex and so are the properties of silicon, Bose-Einstein condensates, and Japanese *kanji*. The life of the butterfly is marvelously complex, the thing born a caterpillar and fated thus to crawl upon the ground, immolated later in its own liquid, and then majestically reemerging as a radiant, multicolored insect, bright master of the air.

Everything in nature, the French mathematician René Thom observed with a certain irony, is complex in one way or another. And not the least of the problems with this forthright if overstated observation (some things, after all, must be simple if anything is to be complex) is the fact that in most cases we have no definition of complexity to which we can appeal, no sense, beyond the superficial, of what complexity *means*.

IN DEVELOPING HIS ARGUMENT, Paley drew—he intended to draw—the curtain of a connection between complexity and design, and so between complexity and *intelligence*. Whether inscribed on paper or recorded in computer code, a design is, after all, the physical overflow of intelligence itself, its trace in matter. A large, a general biological property, intelligence is exhibited in varying degrees by everything that lives. It is intelligence that immerses living creatures in time, allowing the cat and the cockroach alike to peep into the future and remember the past. The lowly paramecium is intelligent, learning gradually to respond to electrical shocks, this quite without a brain, let alone a nervous system. But like so many other psychological properties, intelligence remains elusive without an objective correlative, some public set of circumstances to which one can point with the intention of saying *there*, that is what intelligence is, or what intelligence is *like*.

The stony soil between mental and mathematical concepts is not usually thought efflorescent, but in the idea of an *algorithm* modern mathematics does offer an obliging witness to the very idea of intelligence. They arise, these objects, from an old, a wrinkled class of human artifacts, things so familiar in collective memory as to pass unnoticed. A simple recipe for *bœuf a la mode* is an algorithm; it serves to control the flow of time, breaking human action into small and manageable steps (braising, browning, skimming). Those maddeningly imprecise computer manuals, written half in Korean, it would seem, and half in English, they, too, are algorithms, although ones that are often badly organized.

The promotion of an algorithm from its humdrum historical antecedents to its modern incarnation was undertaken in the 1930s by a quartet of gifted mathematicians: Emil Post, a Polish American logician, doomed like Ovid to spend his years in exile, his most productive work undertaken at City College in New York; the great Kurt Gödel; Alonzo Church, the American Gothic; and, of course, the odd, utterly original Alan Turing, his spirit yet restless and uneasy more than forty years after his unhappy death (an enforced course of hormone therapy to reverse his homosexuality, an apple laced with cyanide).

The problem they faced was to give precise meaning to the concept of an *effective procedure* within mathematics. The essential idea of an algorithm blazes forth from any digital computer, the unfolding of genius having passed inexorably from Gödel's incompleteness theorem to Space Invaders rattling on an arcade Atari, a progression suggesting something both melancholy and exuberant about our culture.

The computer is a machine, and so belongs to the class of things in nature that *do* something, but the computer is also a device dividing itself into aspects: symbols set into software to the left, and the hardware needed to read, store and manipulate the software to the right. This division of labor is unique among man-made artifacts: It suggests the mind immersed within the brain, the soul within the body, the presence *anywhere* of spirit in matter. An algorithm is thus an ambidextrous artifact, residing at the heart of both artificial *and* human intelligence. Computer science and the computational theory of mind appeal to precisely the same garden of branching forks to explain what computers do or what men can do or what in the tide of time they have done.

It is this circumstance that in the 1940s prompted Turing to ask whether computers could *really* think, a question that we have for more than forty years evidently found oddly compelling. We ask it time and again. Turing was prepared to settle his own question in favor of the computer if the machines could satisfactorily conceal their identity, in-

souciantly passing themselves off, when suitably disguised, as human beings. Every year the issue is somewhere joined, a collection of computer
scientists facing a number of curtained wooden booths, trying to determine from the cryptic messages they are receiving—*My name is Bertha
and I am hungry for love*—whether Bertha is a cleverly programmed machine or whether, warm and wet, Bertha is herself resident behind the
curtain, stamping her large feet and hoping for a message in turn.

It is a deliciously comic spectacle. So, too, the contests between human and computer chess champions, as when, with his simian brow
furrowed nobly, Gary Kasparov defended the honor of the human race
against Deep Blue, a dedicated chess-playing computer with an engaging willingness to topple into tactical traps. Artificial *intelligence?* The
idea provokes anxiety, of course, especially among those who shuffle
symbols for a living (*me*, come to think of it), but there it is: if an ability
to execute ordinary arithmetic operations is a form of thought, then intelligence has *already* been localized in a device that fits within the palm
of one's hand.

Everything else is a matter of detail. Or denial.

AN ALGORITHM IS A scheme for the manipulation of symbols—a garden
of branching forks, but to say this is only to say what an algorithm does.
Symbols do more than suffer themselves to be hustled around; they are
there to offer their reflections of the world. They are instruments that
convey *information*.

The most general of fungible commodities, information has become
something shipped, organized, displayed, routed, stored, held, manipulated, disbursed, bought, sold, and exchanged. It is, I think, the first entirely abstract object to have become an item of trade, rather as if one of
the Platonic forms were to become the subject of a public offering. The
superbly reptilian Richard Dawkins has written of life as a river of information, one proceeding out of Eden, almost as if a digital flood had evacuated itself at its source. Somewhere in the wheat-stabbed fields of the
American Midwest, a physicist has argued for a vision of reincarnation

in which human beings may look forward to a resumption of their activities after death on the basis of simulation within a gigantic computer. Sexual pleasures are said to be unusually keen in the digital hereafter.

The American mathematician Claude Shannon gave the concept of information its modern form, endowing an old, a familiar idea with an unyielding but perspicacious mathematical structure. The date is 1948. His definition is one of the cornerstones in the arch of modern thought. Information, Shannon realized, is a property resident in symbols, his theory thus confined from the start by the artifice of words. What Shannon required was a way of incorporating information into the noble category of continuous properties that, like mass or distance, can be represented by real numbers. Philosophers have from time immemorial talked of symbols and signs, their theories turning in an endless circle. Shannon regarded their domain with a mathematician's cold and cunning eye. He saw that symbols function in a human universe where things in themselves are glimpsed through a hot haze of confusion and doubt. Whatever else a message may do, its most important function is to relieve uncertainty, *Rejoice, we conquer* making clear that one side has lost, the other won, the contest finally a matter of fact. It is by means of this superb insight that Shannon was able to coordinate in a limpid circle what symbols do, the accordion of human emotions wheezing in at doubt and expanding outward at certainty, and the great, the classical concepts of the theory of probability. A simple binary symbol, its existence suspended between two equally likely states (*on* or *off*, zero or one, yes or no), holds latent the promise of resolving an initial state of uncertainty by half. The symbol's information, measured in bits, is thus one half.

Simple, no? And so very elegant.

The promotion of information from an informal concept to the mathematical Big Time has provided another benefit, and that is to bring *complexity* itself into the larger community of properties that are fundamental because they are measurable. The essential idea is due, ecumenically, to the great Russian mathematician Andrey Kolmogorov and to Gregory Chaitin, an American student at City College in New York at the time of his discovery (the spirit of Emil Post no doubt acting ectoplasmically). The cynosure of their concerns lay with strings of symbols. Lines of computer code, and so inevitably algorithms, are the obvious examples, but descending gravely down a wide flight of stairs, their black hair swept up and clasped together by diamond brooches, *Anna Karenina* and *Madame Bovary* in the end reduce themselves to strings of symbols, the ravishing women vanishing into the words, and hence the symbols, that describe them. *Adieu, Mesdames.*

In such settings, Kolmogorov and Chaitin argued, complexity is allied to compressibility and then to length, a simple counting concept. Certain strings of symbols may be expressed, and expressed completely, by strings that are shorter than they are; they have some give. A string of ten H's (H,H,H,H,H,H,H,H,H,H) is an example. It may be replaced by the command, given to a computer, say, to print the letter H ten times. Strings that have little or no give are what they are. No scheme for their compression is available. The obvious example is a random string—H,T,T,H,H,T,H,H,H,T, say, which I have generated by flipping a coin ten times. Kolmogorov and Chaitin identified the complexity of a string with the length of the shortest computer command capable of generating that string. This returns the discussion to the idea of information. *It* functions in this discussion (and everywhere else) as a massive gravitational object, exerting enormous influence on every other object in its conceptual field.

The definition of algorithmic complexity gives the appearance of turning the discussion in a circle, the complexity of things explained by an appeal to intelligence and intelligence explained, or at least exhibited, by an appeal to a community of concepts—*algorithms, symbols, information*—upon which a definition of complexity has been impressed. In fact, I am not so much moving aimlessly in a circle as descending slowly in a spiral, explaining the complexity of things by an appeal to the complexity of strings. Explanation is, perhaps, the wrong, the dramatically *inflated* word. Nothing has been explained by what has just been concluded. The alliance between complexity and intelligence that Paley saw through a dark glass remains in place, but the descending spiral has done only what descending spirals can do, and that is to convey a question downward.

MOLECULAR BIOLOGY HAS REVEALED that whatever else a living creature may be—God's creation, the locus in the universe of pity and terror, so much blubbery protoplasm—a living creature is *also* a combinatorial system, its organization controlled by a strange, a hidden and obscure text, one written in a biochemical code. It is an algorithm that lies at the humming heart of life, ferrying information from one set of symbols (the nucleic acids) to another (the proteins). An algorithm? How else to describe the intricacy of transcription, translation, and replication than by an appeal to an algorithm? For that matter, what else to call the quantity stored in the macromolecules than information? And if the macromolecules store information, they function in some sense as symbols.

We are traveling in all the old familiar circles. Nonetheless, molecular biology provides the first clear, the first *resonant*, answer to Paley's question. The complexity of human artifacts, the things that human beings make, finds its explanation in human intelligence. The intelligence responsible for the construction of complex artifacts—watches, computers, military campaigns, federal budgets, this very essay—finds *its* explanation in biology. This may seem suspiciously as if one were explaining the content of one conversation by appealing to the content of another, and so, perhaps, it is, but at the very least, molecular biology represents a place lower on the spiral than the place from which we first started, the downward descent offering the impression of progress if only because it offers the impression of movement.

However invigorating it is to see the threefold pattern of algorithm, information, and symbol appear and reappear, especially on the molecular biological level, it is important to remember that we have little idea how the pattern is amplified. The explanation of complexity that biology affords is yet largely ceremonial. A living creature is, after all, a concrete, complex, and autonomous three-dimensional object, something or someone able to carry on its own affairs; it belongs to a world of its own—our world, as it happens, the world in which animals hunt and hustle, scratch themselves in the shedding sun, yawn, move about at will. The triple artifacts of algorithm, information, and symbol are abstract and *one*-dimensional, entirely static; they belong to the very different universe of symbolic forms. At the very heart of molecular biology, a great mystery is in evidence, as those symbolic forms bring an organism into existence, control its morphology and development, and slip a copy of themselves into the future.

These transactions hide a process never seen among purely physical objects, although one that is characteristic of the world in which orders are given *and* obeyed, questions asked *and* answered, promises made *and* kept. In *that* world, where computers hum and human beings attend to one another, intelligence is always relative to intelligence itself, systems of symbols gaining their point from having their point gained.

This is not a paradox. It is simply the way things are. Some two hundred years ago, the Swiss naturalist Charles Bonnet—a contemporary of Paley's—asked for an account of the "mechanics which will preside over the formation of a brain, a heart, a lung, and so many other organs." No account in terms of *mechanics* is yet available. Information passes from the genome to the organism. Something is given and something read; something is ordered and something done. But just who is doing the reading and who is executing the orders, this remains unclear.

The triple concepts of algorithm, information, and symbol lie at the humming heart of life. How they promote themselves into an organism is something of a mystery, a part of the general mystery by which intelligence achieves its effects. But just how in the scheme of things did these superb symbolic instruments come into existence? Why should there be very complex informational macromolecules at all? We are looking further downward now, toward the laws of physics.

Darwin's theory of evolution is widely thought to provide purely a materialistic explanation for the emergence and development of life. But even if this extravagant and silly claim is accepted at face value, no one suggests that theories of evolution are in any sense a fundamental answer to Paley's question. One can very easily imagine a universe rather like the surface of Jupiter, a mass of flaming gases too hot or too insubstantial for the emergence or the flourishing of life. There is in the universe we inhabit a very cozy relationship between the fundamental structure of things and our own boisterous emergence on the scene, something re-marked by every physicist. The theory of evolution is yet another station, a place to which complexity has been transferred and from which it must be transferred anew.

The fundamental laws of physics describe the clean, cool place from which the world's complexity ultimately arises. What else *besides* those laws remains? They hold the promise of radical simplicity in a double sense.

Within the past quarter century, physicists have come to realize that theories of change may always be expressed in terms of the conservation of certain quantities. Where there is conservation, there is symmetry as well. The behavior of an ordinary triangle in space preserves three rota-tional symmetries, the vertices of the triangle simply changing positions until the topmost vertex is back where it started. And it preserves three reflectional symmetries as well, as when the triangle is flipped over its altitude. A theory of how the triangle changes its position in space is at the same time—it is one and the same thing—a theory of the quantities conserved by the triangle as it is being rotated or reflected. The appropri-ate object for the description of the triangle is a structure that exhibits the requisite symmetry. (Such are the groups.) The fundamental laws of physics—the province of *gauge* theories—achieve their effect by ap-pealing to symmetries. The physicist looks upon a domain made simple because it's made symmetrical. This sense of simplicity is a sense of sim-plicity in things; whether captured fully by the laws of nature or not, symmetry is an objective property of the real world, *out there*, in some sense of "out" and some sense of "there."

And yet, the world in which we moan and mate is corrupt, fragmented, hopelessly asymmetrical, with even the bilateral symmetry of the human body compromised by two kidneys but one comically misplaced heart, two lungs, one oblate liver. This, too, the fundamental laws of physics explain, those laws miraculously accounting for both purity *and* corruption in one superb intellectual gesture. The basic expressions of mathematical physics are mathematical equations. Like the familiar equations of high school algebra, they point to an unknown by means of an adroit compilation of connected clues—specification in the dark.

But while the equations respect nature's perfect symmetries, their various *solutions* do not. How can this be? Simple. The equation that some number when squared is four has two matched and thus symmetrical solutions (two and minus two). The equation preserves a certain symmetry, but once a particular solution is chosen, some unavoidable bias is introduced, whether positive or negative. So too with the fundamental equations of physics. Their bias is revealed in spontaneous symmetry-breaking—our own breathtakingly asymmetrical world the result of a kind of random tilt, one that obscures the world's symmetries even as it breaks them.

Such is the standard line, its principles governing the standard model, the half-completed collection of principles that, like a trusted scout, physicists believe point toward the grand unified theory to come. The laws of nature are radically simple in a second sense. They are, those laws, simple in their *structure*, exhibiting a shapeliness of mathematical form and a compactness of expression that itself cannot be improved in favor of anything shapelier or more compact. They represent the hard knot into which the world of matter has been compressed. This is to return the discussion to symbols and information, the cross of words throwing a queer but illuminating lurid red light onto the laws of physics.

The fundamental laws of physics capture the world's patterns by capturing the play of its symmetries. Where there is pattern and symmetry, there is room for compression, and where room for compression, fundamental laws by which the room is compressed. At the conceptual basement, no further explanation is possible and so no further compression. The fundamental laws of physics are complex in that they are *incompressible*, but they are simple in that they are *short*—astonishingly so in that they may be programmed in only a few pages of computer code. They function as tense, tight spasms of information.

It is with the fundamental laws of physics that finally Paley's question comes to an end; it is the place where Paley's question *must* come to an end if we are not to pass with infinite weariness from one set of complicated facts to another. It is thus crucial that they be simple, those laws, so that in surveying what they say, the physicist is tempted no longer to ask for an account of *their* complexity. Like the mind of God, they must explain themselves. At the same time, they must be complete, explaining *everything* that is complex. Otherwise what is their use? And, finally, the fundamental laws must be material, offering an account of spirit and substance, form and function, *all* of the insubstantial aspects of reality, in terms (metaphorically) of atoms and the void. Otherwise they would not be fundamental laws of *physics*.

To THOSE COMMITTED TO its success, contemporary physics has seemed a convergent sequence, a fantastic synthesis of a scientific theory and a moral drama. The grand unified theory is not only a fixed point but a place of blessed if somewhat delayed release.

And yet many of the most notable movements in twentieth-century thought suggest something rather different, the conclusion of almost every grand intellectual progression blocked by incompleteness, or daunting complexity, or a wilderness of mirrors.

Any mathematical system rich enough to express ordinary arithmetic, Gödel demonstrated in 1931, is incomplete, some simple arithmetic proposition lying beyond the system's reach. In a result only slightly less fantastic, the Polish logician Alfred Tarski demonstrated that the concept of truth could not be defined within any language in which it is expressed. For that, a retreat to a richer language is required. That language, in turn, requires a still *further* language for the requisite definition, and so upward along an endless progression, the complete definition of truth beckoning forever and forever out of reach.

Until recently, the physical sciences have seemed immune to the retrograde currents so conspicuous elsewhere. But much against their will, physicists have come to suspect that there is a three-way tug, a kind of tension between materialism, on the one hand, and the idea of an unutterably complete but superbly simple theory, on the other.

It is at the far margins of speculation that, like oil, anxiety is leaking from the great speculative structures. The universe, so current theory runs, erupted into existence in an explosion. One minute there is nothing, the next, *poof...* something arises out of nothing, as space and time come into existence and the universe goes on with the business of getting on.

An explosion? An explosion in what?

Nothing.

Nothing?

That's right. *Nada.* Just as there is no point north of the geomagnetic North Pole, Stephen Hawking has observed with some asperity, so there is no time *prior* to the big bang, no space *within* which the big bang took place.

Thoughtful men and women have from time immemorial scrupled at this scenario; theirs is an ancient argument. Something cannot arise from nothing. At the place where the event occurs, the intellect simply goes blank. Explanations rarely help. "In the latest version of string theories," Steven Weinberg writes, "space and time arise as derived quantities, which do not appear in the fundamental equations of the theory."

This is odd, but what follows is odder yet. "In these theories, space and time have only an approximate significance; it makes no sense to talk about any time closer to the big bang than about a million trillion trillion trillionth of a second." This lighthearted remark suggests that within a certain interval of *time* it makes no sense to talk of time at all. The effect is rather as if a genial host were loudly to assure a guest that his money is no good while simultaneously endeavoring to take it. Still other physicists repose their confidence in the *laws* of physics, attributing a strange mystic power to the symbols themselves. Existing in the same undiscoverable realm as Plato's forms, they nonetheless function, those laws, to bring the world into existence. This may seem an exercise in metaphysics of just the sort that, with a fierce wet snort, some physicists routinely deride; it is what other physicists commend. The origin of the big bang lies with the laws of physics, Paul Davies insists, and "the laws of physics don't exist in space and time at all. They describe the world, they are not 'in' it."

The image of the fundamental laws of physics zestfully wrestling with the void to bring the universe into being is one that suggests very little improvement over the accounts given by the ancient Norse in which the world is revealed to be balanced on the back of a gigantic ox.

If this is how explanations come to an end, what of materialism? The laws of physics are sets of symbols, after all—these now vested with the monstrous power to bring things and urges into creation, and symbols belong to the intelligence-infused aspects of the universe. They convey information, they carry meaning, they are part of the human exchange. And yet, it was to have been intelligence *itself* (and thus complexity) that the laws of physics were to explain, symbols and signs, meanings and messages—all accounted for by the behavior of matter in one of its modes.

TRIAGE IS A TERM of battlefield medicine. That shell having exploded in the latrine, the tough but caring physician divides the victims into those who will not make it, those who will, and those who might. The fundamental laws of physics were to provide a scheme of things at once materialistic, complete, and simple. By now we know, or at least suspect, that materialism will not make it. And not simply because symbols have been given a say-so in the generation of the universe. *Entre nous soit dit*, physics is simply riddled with nonmaterial entities: functions, forces, fields, abstract objects of every stripe and kind, waves of probability, the quantum vacuum, entropy, and energies.

There remains completeness and simplicity. Completeness is, of course, crucial, for without completeness, there is no compelling answer to Paley's question at all. What profit would there be if the laws of physics explained the complexity of plate tectonics, but *not* the formation of the ribosomes? Absent completeness, may not the universe break apart into separate kingdoms ruled by separate gods, just as, rubbing their oiled and braided beards, the ancient priests foretold? It is a vision that is at issue, in part metaphysical in its expression, and in part religious in its impulse—the fundamental laws of physics functioning in the popular imagination as demiurges, potent and full of creative power. And if they *are* potent and full of creative power, they had better get on with the full business of creation, leaving piecework to part-time workers.

What remains to be completed for this, the most dramatic of visions to shine forth irrefrangibly, is the construction of the inferential staircase leading *from* the laws of physics *to* the world that lies about us, corrupt, partial, fragmented, messy, asymmetrical, but our very own, beloved and irreplaceable.

No one expects the laws of physics by themselves to be controlling. "The most extreme hope for science," Steven Weinberg admits, "is that we will be able to trace the explanations of all natural phenomena to final laws *and* historical accidents." Why not give historical accidents their proper name—chance? The world and everything in it, Weinberg might have written, has come into being by means of the laws of physics and by means of chance.

A premonitory chill may now be felt sweeping the room. "We cannot see," Richard Feynman wrote in his remarkable lectures on physics, "whether Schrödinger's equation [the fundamental law of quantum mechanics] contains frogs, musical composers, or morality."

Cannot see? These are ominous words. Without the seeing, there is no secular or sacred vision, and no inferential staircase. Only a large, damp, unmortgaged claim.

And a claim, moreover, that others have regarded as dubious. "The formation within geological time of a human body," Kurt Gödel remarked to the logician Hao Wang, "by the laws of physics (or any other laws of similar nature), starting from a random distribution of elementary particles and the field, is as unlikely as the separation by chance of the atmosphere into its components."

This is a somewhat enigmatic statement. Let me explain. When Gödel spoke of the "field" he meant, no doubt, the quantum field; Schrödinger's equation is in charge. And by invoking a "random distribution of elementary particles," Gödel meant to confine the discussion to *typical* or *generic* patterns—what might reasonably be expected. Chance, again.

Under the double action of the fundamental laws and chance, Gödel was persuaded, no form of complexity could reasonably be expected to arise. This is not an argument, of course; it functions merely as a claim, although one made with the authority of Gödel's genius. But it is a claim with a queer prophetic power, anticipating, as it does, a very specific contemporary argument.

"The complexity of living bodies," Gödel observed, "has to be present either in the material [from which they are derived] or in the laws [governing their formation]." In this, Gödel was simply echoing Paley. Complexity must come from somewhere; it requires an explanation.

And here is the artful, the hidden and subversive point. The laws of physics are simple in that they are short; they function, they can *only* function, to abbreviate or compress things that display pattern or that are rich in symmetry. They gain no purchase in compressing strings that are at once long and complex—strings of random numbers, for example, the very record in the universe of chance itself. But the nucleic acids and the proteins are precisely such strings. We have no idea how they arose; filled with disturbing manic energy, they seem devoid of pattern. They are what they are. Their appearance by means of chance is impossible; their generation from the simple laws of physics ruled out of court simply because the simple laws of physics are indeed simple.

Gödel wrote well before the structure of the genetic code was completely understood, but time has been his faithful friend (in this as in so much else). The utter complexity of the informational macromolecules is now a given. Using very simple counting arguments, Hubert Yockey has concluded that an ancient protein such as "cytochrome c" could be expected to arise by chance only *once* in 10^{44} trials. The image of an indefatigable but hopelessly muddled universe trying throughout all eternity to create a single biological molecule is very sobering. It is this image that, no doubt, accounted for Francis Crick's suggestion that life did not originate on earth at all, but was sent here from outer space, a wonderful example of an intellectual operation known generally as fog displacement.

It would seem that in order to preserve the inferential staircase some compromises in the simplicity of the laws of nature might be in prospect. Luck cannot do quite what luck was supposed to do. Roger Penrose has argued on thermodynamic grounds that the universe began in a highly unusual state, one in which entropy was *low* and organization *high*. Things have been running down ever since, a proposition for which each of us has overwhelming evidence. This again is to explain the appearance of complexity on the world's great stage by means of an intellectual shuffle, the argument essentially coming down to the claim that no explanation is required if only because the complexity in things was there all along.

It is better to say frankly—it is intellectually more honest—that either simplicity or the inferential staircase must go. If simplicity is to join materialism in Valhalla, how, then, have the laws of physics provided an ultimate answer for Paley's question?

ALTHOUGH PHYSICISTS ARE QUITE sure that quantum mechanics provides a complete explanation for the nature of the chemical bond (and so for all of chemistry), they are sure of a great many things that may not be so, and they have provided the requisite calculations only for the hydrogen and helium atoms. Quantum mechanics was, of course, created before the advent of the computer. It is a superbly linear theory; its equations admit of exact solution. But beyond the simplest of systems, a computational wilderness arises—mangrove trees, steaming swamps, some horrid thing shambling through the undergrowth, eager to engage the physicists (and the rest of us) in conversation.

Hey, fellas, wait!

The fundamental laws of physics are in control of the fundamental physical objects, giving instructions to the quarks and bossing around the gluons. The mathematician yet rules the elementary quantum world. But beyond their natural domain, the influence of the fundamental laws is transmitted only by interpretation.

As complex systems of particles are studied, equations must be introduced for each particle, the equations, together with their fearful interactions, settled and then solved. There is no hope of doing this analytically; the difficulties tend to accumulate exponentially. Very powerful computers are required. Between the music of Mozart and the fundamental laws of physics, the triple concepts of algorithm, information, and symbol hold sway. The content of the fundamental laws of physics is *relative*—it *must* be relative—to the computational systems needed to interpret them.

There is a world of difference between the character of the fundamental laws, on the one hand, and the nature of the computations required to breathe life into them, on the other. The statement that a group of rabbits is roughly doubling its size suggests to a mathematician an underlying law of growth. The rabbits are reproducing themselves exponentially. The law provides a general description of how the rabbits are increasing, one good for all time. It is an *analytical* statement; it strikes to the heart of things by accounting for what the rabbits are doing by means of a precise mathematical function.

Observing those same rabbits over a certain period of time, the computer tracks their growth in a finite number of steps: There are two rabbits to begin with, and then four and then eight, and so upward to sixty-four, when this imaginary exercise ceases on my say-so. The finite collection of numbers is a *simulation*, one in which the underlying reality is suggested, but never fully described.

Law and simulation have very different natures, very different properties. The law is infinite, describing the increase in rabbits to the very end of time, and the law, moreover, is continuous, and so a part of the great tradition of mathematical description that is tied ultimately to the calculus.

The simulation is neither infinite nor continuous, but finite and discrete. It provides no deep mathematical function to explain anything in nature. It specifies only a series of numbers, the conversion of those numbers into a pattern a matter for the mathematician to decide.

If the law describes the hidden heart of things, the simulation provides only a series of stylized snapshots, akin really to the succession of old-fashioned stills that New York tabloids used to feature: the woman, her skirt askew, falling from the sixth-floor window, then passing the third floor, a look of alarm on her disorganized features, finally landing on the hood of an automobile, woman and hood both crumpled. The lack of continuity in either scheme makes any interpolation between shots or simulation a calculated conjecture. The snapshots, it is worthwhile to recall, provide no evidence that the woman was falling *down*, but we see the pictures and we neglect the contribution *we* make to their interpretation.

Computational schemes figure in any description of the inferential staircase; the triple concepts of algorithm, information, and symbol have made yet another appearance. They are like the sun, which comes up anew each day. And they serve to sound a relentlessly human voice. The expectation among physicists, of course, is that these concepts play merely an ancillary role, the inferential staircase under construction *essentially* by means of the fundamental laws of physics. But the question of who is to be the master and who the mastered in this business is anything but clear.

To the extent that the fundamental laws of physics function as premises to a grand metaphysical argument, one in which the universe is to appear as its empyrean conclusion, the triple concepts—algorithm, information, symbol—function as additional premises. Without those premises, the laws of physics would sit simply in the shedding sun, mute, inglorious, and unrevealing.

A third revision of Steven Weinberg's memorable affirmation is now in prospect: The best that can be expected, the most extreme hope for science, is an explanation of all natural phenomena on the basis of the fundamental laws of physics, and chance, and a congeries of computational schemes, algorithms, specialized programming languages, techniques for numerical integration, huge canned programs (such as Mathematica or Maple), computer graphics, interpolation methods, computer-theoretic shortcuts, and the best efforts by mathematicians and physicists to convert the data of simulation into coherent patterns, artfully revealing symmetries and continuous narratives.

A certain sense of radiant simplicity may now be observed dimming itself. The greater the contribution of the triple concepts, the less compelling the vision of an inferential staircase. Matters stand on a destructive dilemma. Without the triple concepts, the fundamental laws of physics are incomplete, but with the triple concepts in place, they are no longer simple. The inferential staircase was to have led from the laws of physics to that sense of intelligence by which complexity is explained; if intelligence is required to construct the very stairs themselves, why bother climbing them? The profoundly simple statements that were to have redeemed the world must do their redeeming by means of the very concepts that they were intended to redeem.

THERE SHOULD BE IN this no cause for lamentation; the destabilizing relationship between the technology of information and the fundamental laws of physics represents a form of progress, an improvement in our understanding. We have always suspected, and now we know, that while there are things that are simple and things that are complex in nature, the rich variety of things is not derived from anything simple, or if derived from something simple, not derived from them completely.

Complexity may be transferred; it may be shifted from theories to facts and back again to theories. It may be localized in computer derivations, the fundamental laws kept simple, or it may be expressed in a variety of principles and laws at the expense of the idea of completeness. But ultimately, things are as they are for no better reason than they are what they are. We cannot know more without making further assumptions, those assumptions like a dog's maddening tail—in front of our nose, but forever out of reach.

What is curious, because it seems to have been something that we expected all along, is that it does not seem to matter. It does not matter at all...

...*They have gathered by the gate of time, the quick and the dead and those eager to be born; some come to remember, others to forget. A cool gray fog is blowing. The spidery hands of the great clock measuring millennia instead of minutes are crawling toward midnight. Techno-wizards with pale green eyes have thronged the lounge of paradise. Elvis is there, singing in a sulky baritone. Everything has been forgiven, but nothing has been forgotten. The Ancients tell stories of why time began and how space was curved. There is the sound of chimes, the smell of incense. A woman with flaming cheeks remarks in a calm, clear voice that she has been kidnapped by space aliens. They have probed her every orifice with lights that burn like fire.*

FULL HOUSE FOLLIES

Stephen Jay Gould's little book *Full House: The Spread of Excellence from Plato to Darwin* (Harmony Books, 1996) is intended to correct the popular impression that "progress and increasing complexity" are characteristics of life's course on Earth. Progress has, in the twentieth century, already been punched silly; but paleontologists seem genuinely more complex than paramecia, a point that Gould concedes, if only for reasons of professional pride. His doubts arise whenever the twig of a trend is taken as typical or representative. The underlying error evidently runs deep. "[W]e are still suffering," he writes, "from a legacy as old as Plato, a tendency to abstract a single idea or average as the 'essence' of a system, and to devalue or ignore variation among the individuals that constitute the full population."

If essentialism is the affliction, evolution is the remedy. "Darwin's revolution," Gould believes, "should be epitomized as the substitution of variation for essence as the central category of natural reality." The revolution complete, there will be individuals and their variances out there where the central categories lie. Nothing else. This seems for all the world like nominalism, a doctrine current in twelfth-century Paris, where Abelard may be seen stealing kisses from Heloïse, drab Darwin dutifully waiting his turn at bat. Gould is off by centuries; his essential quarrel, of course, is with Aristotle and the doctrine of real essences.

No matter, I would ordinarily say, drawing down the curtain of charity. Like paleontology, philosophy is an acquired taste. But having been tempted by philosophy, Gould is tempted again by statistics, prompting me to yank that curtain up again. Witness thus his discussion of normal distributions, the familiar bell-shaped curve. "We regard the normal distribution as canonical," Gould writes, "because we tend to view systems as having idealized 'correct' values, with random variations on either side—another consequence of lingering Platonism."

I am afraid that Plato, poor pooch, has nothing to do with it. Whatever the Form of the Good may be, it is not constructed by counting noses. We regard the normal distribution as canonical because it is canonical, a fact readily demonstrated in elementary texts, the distribution of distributions itself following a bell-shaped curve. Perhaps that curtain had better stay up. Although drawn incorrigibly to the Big Picture, the focal plane of Gould's attention is best adapted to short distances. Ted Williams was the last of the big-time hitters, his 1941 record now receding into memory and myth. Why should this be, Gould asks? It is not clear that in a world bursting with sin and suffering this question cries out for an answer, but the question illustrates a trend, and trends are what tend to interest Gould.

The disappearance of the four-hundred hitter might suggest that in baseball, as in so much else, things have gotten worse. In fact, Gould argues, play has in general improved, those rumpy players now scampering across the field better-conditioned and surely better-paid than their stubble-chinned predecessors. It is this circumstance, curiously enough, that provides Gould with an answer to his question.

What follows as the standard of play improves? Nothing on average, better hitters encountering better fielders, their simultaneous improvements canceling one another. But athletic performance is bounded by a wall marking the limitations of the human body. Batters may hit so far, and no further; there are some balls that no ballboy ever shags. The wall stands to the right of the bell curve measuring athletic performance like the north face of Anapurna, grim, forbidding and uncrossable. As play improves, the bell curve shifts itself closer to the wall. Variations shrink, the right tail of the curve steepening. Without such steepening, the curve would cross the wall into the forbidden zone. The disappearance of the four-hundred hitter now follows inexorably, if only because his room for maneuver has been vacated, the batter squeezed between better competition and the wall.

THIS IS AN ELEGANT analysis, but one that tempts Gould anew into the badlands of metaphysics. "Hitting 400," Gould concludes, "is not a thing, but the right tail of the full house for variations of batting averages." With this, who would quarrel, but then who has ever been persuaded that a batting average is a thing? The assimilation of a very particular batting average to the full house of variations is a mistake as well. An average is a number; the full house a curve embodying a multitude of points and so a continuum of numbers. Trends arise when specific points on the curve are tracked.

Gould has confused the shadow of one distinction for the substance of another. If he could have continued to play unaged, Ted Williams would have seen his batting average drop year by year, even though he stood forever in the sunshine of his youth. Batting averages are relative to the circumstances of play. In this, a man may be diminished without changing, his relative decline nonetheless marking a real trend. Not so in the case of height or weight. No man becomes fat because other men have grown thin. It is this distinction between relative and absolute properties that is the cynosure of Gould's concerns. The contrast between variations and trends that Gould means to collapse remains healthy as a horse. "Darwin's revolution will be completed," Gould writes, "when we smash the pedestal of human arrogance and own the plain implications of evolution for life's nonpredictable non-directionality." A great many biologists seem keen to smash that pedestal, seeing forever an ape's haunted eyes peering from every human face; but the more that Darwinian thought shows that life is going nowhere, the less it explains about where life has gone, the net effect akin to division by zero. Writing about the evolution of the horse, Gould juxtaposes evolutionary fable to the fossil record. High school textbooks propose that, desiring an increase in stature, the rabbit-sized *Eohippus* (but not curiously enough, the rabbit-sized rabbit), commenced his move up through the evolutionary ranks, one incremental step after the other. This might suggest a trend, the familiar museum displays showing the beasts getting bigger and better, with even their coats becoming glossier over time.

The fossil record shows something different: rather a thick bush, with horse-like species entering the record at one time, leaving it at another. The high school progression is an artifact; a great many intermediates are absent from the record, their trace only a doleful dotted line. The facts are discrete. There is no hint of gradual change, no hint either of selective advantages accumulating. "Throughout the history of horses," Gould recounts (quoting Prothero and Shubin's well-known monograph), "the species are well-marked and static over millions of years."

All this confirms Gould in his conviction that variations are crucial and trends an epiphenomena; but having in this case destroyed the notion of an evolutionary progression, Gould runs the risk of destroying the notion of an evolutionary explanation as well. The stratigraphic charts make the heresy irresistible. There is variance, to be sure, but no change in variance, and so no evident trend. Why imagine that there is anything to the record beyond what one sees—one damn horse after the other? If the horses are going nowhere, not so the *Foraminfera*—single-celled creatures, protozoa, in fact, that popped into existence at the beginning of the Cretaceous some 136 million years ago. Starting out small, some species increased in size, at least until an obliging mass extinction served to reduce the population, with essentially the same story replayed during the earlier and later Cenozoic eras. The statistics reflect a bell-shaped curve, but one that is skewed in the directing of increasing size.

An effect demands a cause. Why the skewing? The classical Darwinian explanation involves a trip of three steps: random variations, followed by natural selection, followed in turn by biological change. Gould proposes to cut out the middleman entirely, moving from random variations to biological change wholesale.

WHAT IS GIVEN, THEN, are *Foraminfera* with an inherent variability, a repertoire of potential changes. As the Cretaceous commences, they are, those protozoa, collected by a wall marking the lower limits of their size. Thereafter, they simply drift, taking themselves to whatever place their variations allow. The wall functions to enforce a principle of elastic reflection; as protozoa hit the wall, they bounce forward. This serves to impart a preferred direction to their drifting, the simple structure of a stochastic set-up producing a simulacrum of a trend. Some species get bigger over time simply because they have the chance. "Size increase," Gould writes, "is really random evolution away from small size, not directed evolution toward large size."

Such is Gould's theory of expanding stochastic variation. It is a theory that suggests again to Gould the enduring reality of variations, the trends in this case arising from a failure to properly distinguish passive from driven systems. The driven systems move forward with a vengeance, natural selection on the classical view positively hustling species over an adaptive landscape; the passive systems just sit there and drift. This distinction seen, trends tend to dwindle and disappear, leaving behind only the *Foraminfera*, smiling enigmatically, like so many cats. Does the combination of drift and bounce suffice in the case of *Foraminfera* to explain their increase in size? We cannot be sure. Size has been measured by a single numerical parameter; this lends a false plausibility to a picture of drift—false because lacking analytical precision. *Foraminfera* never become large as whales. Their variations are bounded. How so? And by what? The mechanism of variation that allows *Foraminfera* to increase in size, too, is left largely in the dark, with no coordination drawn between what the *Foraminfera* do and how they do it. The statistical pattern that Gould sees among the *Foraminfera* he sees again in life as a whole. There is no progress in the evolutionary record; indeed, no overall trends. Life is dominated by bacteria, the little bugs slithering through every interstice and filling every conceivable crevice, the bacterial mode remaining unchanged over time, stable as a rock and about as interesting. There is, Gould acknowledges, the fact that "the most complex creature has tended to increase in elaboration through time," but increasing complexity reflects nothing more than the same pattern seen locally among the

Foraminfera as they increase in size. The mean measuring complexity increases, the mode does not; whatever trends are tending reflect "changes in variation rather than things moving somewhere."

By NOW THE POINT has become familiar while remaining unpersuasive, like an advertisement for breakfast cereal. The provocative and plausible part of Gould's thesis lies elsewhere with the idea of expanding stochastic variation. But where the doctrine is plausible, as in the case of *Foraminfera*, it is not provocative, and where provocative, not plausible.

Looking backwards some 3.5 billion years, Gould observes bacteria crouched by an initial wall, one marking what he calls "minimal complexity." Movement away from the wall is movement toward increasing complexity. "As life diversified," he writes, "only one direction stood open for expansion." Some species have grown progressively more complex; but "the vaunted progress of life," Gould writes with satisfaction, "is really random motion away from simple beginnings, not directed impetus toward inherently advantageous complexity."

In coming to this conclusion, Gould has been misled by his own analogy between size and complexity. Both increasing size (among *Foraminfera*) and increasing complexity (in life) yield right-skewed distributions: bell-shaped curves with long right tails. Such is the statistical picture, but it is a picture that describes as well the distribution of height in human habitation since the fifteenth century. Most buildings have remained small, some have gotten bigger; and no one imagines that taller buildings arise because they have simply "wandered into a previously unoccupied domain." This is a point to which Gould is sensitive, but the statistical tests that he invokes to distinguish passive from driven systems seem to me far too coarse to be of interest.

Expanding stochastic variation may well carry the *Foraminfera* to an increase in size; the problems associated with complexity are of a different order entirely. Biological objects are made of many parts and only a particular arrangement of those parts serves to realize a biological function. A probability threshold, a wall, to continue Gould's image, separates complex objects from any mechanism of random search. No arbitrary rearrangement of the eye is apt to see; no reconfiguration of the heart to pump blood. This is to speak only at the most superficial level of analysis. An organism comprises any number of very complex systems coordinated to achieve a variety of familiar but poorly-described biological functions. Even the simplest of biological systems—the bacterial cell, in fact—suggests an exponentially increasing order of complexity, as complex structures interact with one another in ways that are themselves complex.

The great evolutionary trajectories span many dimensions and relatively short periods of time. Causes evoke effects, but effects in turn influence their causes; things are strongly non-linear and analytically intractable. The structures that result occupy points in a space that no mathematician has ever glimpsed, let alone described. The examples of complexity that Gould invokes, by way of contrast, are one dimensional and almost hopelessly trivial, a point that with a good-natured shrug, Gould himself comes close to acknowledging.

Gould is right to scruple at progress, but he has added his voice late to the chorus. Everyone else has already left off shouting. He is right again to be skeptical of traditional explanations in biology. Richard Dawkins's *The Blind Watchmaker* and Daniel Dennett's *Darwin's Dangerous Idea* are books in which ignorance is made an epistemological principle. But he is wrong to suppose that the only alternative to systems driven by natural selection are systems that are not driven at all. Random mechanisms cannot in principle explain the nature of biological complexity, the doctrine of expanding stochastic variation returning evolutionary thought to the dread place from which it has always tried to flee. By endeavoring to deconstruct a distinction between trends and variances, Gould has inadvertently come to express a more profound and ultimately more explosive truth: that the issue of why things in biology change as they do remains what it has always been, which is to say an abysmal mystery.

1998

GÖDEL'S QUESTION

The affirmation of being is the purpose and the cause of the world.

—KURT GÖDEL

DESPITE A RICH AND MARVELOUSLY UNINTERESTING BODY OF WORK devoted to mathematical population genetics, the claims of Darwin's theory (and of Darwinian theories generally) have often seemed fantastic, especially to mathematicians. "[T]he formation within geological time of a human body," Kurt Gödel remarked to the logician Hao Wang, "by the laws of physics (or any other laws of similar nature), starting from a random distribution of elementary particles and the field, is as unlikely as the separation by chance of the atmosphere into its components."[12]

"The complexity of living bodies," Gödel went on to say, Hao Wang listening without comment, "has to be present either in the material [from which they are derived] or in the laws [governing their formation]." In this, Gödel seemed to be claiming without argument that complexity is subject to a principle of conservation, much like energy or angular momentum. Now every human body is derived from yet another human body; complexity is in reproduction transferred from one similar structure to another. The *immediate* complexity of the human body is thus present in the matter from which it is derived; but on Darwin's theory, human beings as a species are, by a process of random variation and natural selection, themselves derived from less complicated structures, the process tapering downward until the rich panorama of organic life

12. Hao Wang, "On 'Computabilism' and Physicalism: Some Problems," in *Nature's Imagination*, ed. John Cornwell (Oxford University Press, 1995), p. 173.

is driven into an *in*organic and thus a relatively simple point. The origin of complexity thus lies—it *must* lie—with laws of matter and thus with laws of physics.

In these casual inferences, Gödel was in some measure retracing a version of the argument from design. The manifest complexity of living creatures suggested to William Paley an inexorably providential designer. Cancel the theology from the argument and a connection emerges between complexity and some form of intelligence.[13] And with another cancellation, a connection between complexity and the laws of physics.

These algebraic operations shift the argument's shape without entirely altering its conclusions. They suggest that on Gödel's view, complexity has no ultimate metaphysical source. It may be transferred, as when an artist fabricates a painting or a statue, endowing durable pigment or stone with the complexity already latent in his mind; it may be shuffled by various combinatorial processes, but it cannot be created. The inferential trail that begins in wonder ends not with an infinitely accomplished designer, but with a principle instead, one that assimilates complexity to mass, energy or angular momentum in mechanics.[14] This is not, needless to say, a position to which the Reverend Paley would have given his ungrudging assent.

IN EXPATIATING GRANDLY ON biology, Gödel chose to express his doubts on a cosmic scale, his skepticism valuable as much for its sharp suggestiveness as anything else; but the large question that Gödel asked about the discovery of life in a world of matter has an inner voice within biology itself, interesting evidence that questions of design and complex-

---

13. I have pursued this theme in "The End of the Matter," *Forbes ASAP*, December 1996. It has seldom been noted, but criticism of Darwin's theory forms a chapter in the history of idealist thought.

14. Physicists often say that information is neither created nor destroyed; it is a thesis that would appear to have a counter-example in the case of a physical signal passing beyond the event horizon of a black hole. If complexity is defined in terms of information—a generous assumption, to be sure—then the conservation of complexity is no less plausible than the conservation of information. *No less*, note, but unfortunately, *no more* either.

ity are independent of scale. Living systems have achieved a remarkable degree of complexity within a very short time, such structures as the blood clotting cascade or the immunodefense system or human language suggesting nothing so much as a process of careful coordination and intelligent design.

Darwin's theory of evolution is widely thought to have exploded this suggestion, complexity in nature arising by means of random variation and natural selection. It is instructive to regard these claims from an oblique, a mathematical perspective, if only for the sake of the harsh revealing light that mathematics may sometimes provide.

Evolution is a stochastic process, one that moves forward by means of inconclusive humps. Those humps having for a time humped themselves, the mathematician observes the outline of a Markov process or a Markov chain, the standard mathematical structure for the depiction of conditional change.[15] At any given time, a Markov process occupies one of a finite or a countably infinite number of states $\Omega = \varepsilon_1, \varepsilon_2, \dots, \varepsilon_n$. These function as bursts of information, a form of memory. In the context of theoretical biology, each state is the buoyant expression of some organism sliced sideways in the stream of time. Starting at $t = 0$, the system changes randomly at $t = 1, 2, \dots$ and times thereafter. The evolution of the system is described by the crabwise movement of a random variable $\xi(t)$:

$$\xi(0) \to \xi(1) \to \xi(2) \to \dots$$

Further details: an initial probability distribution defined over Ω recounts the likelihood $P\{\xi(0) = \varepsilon_j\}$ that at $t = 0$, the system is in state ε_j. A scheme of transition probabilities $P\{\xi(n + 1) = \varepsilon_j \mid \xi(n) = \varepsilon_k\}$, ($j, k = 1, 2, \dots$), describes the chances that a system in state ε_k will in a single enlightened step change to the state ε_j. A state ε_j is *accessible* from some other state ε_k if there is some probability $P(n) > 0$ of going from ε_k to ε_j in n steps.

---

15. Genetic algorithms, currently a subject of interest among theoretical biologists, are themselves finite state Markov chains. See David E. Goldberg, *Genetic Algorithms* (Addison-Wesley, 1989), for details.

There is no very deep theoretical content to the concept of a Markov process. Its purpose is to stiffen an intellectual structure with a tendency otherwise to sag. The metaphor of a skeleton is valuable nonetheless; skeletons are load bearing structures and skeptical doubts about Darwinian theories almost always resolve themselves to the question whether such structures can bear their intended weight. In *Darwin's Black Box*, Michael Behe has argued that with respect to the *irreducibly complex* systems the answer is unequivocally no. Such structures are comprised of parts designed to work with one another or not to work at all.[16] A common padlock is an example, the key matched to a specific locking cylinder and vice versa. Lacking functional intermediates, the padlock remains inaccessible to any process of random variation and natural selection. So, too, Behe argues, many biochemical systems.

EXAMPLE[17] Eukaryotic replication takes place at the rate of perhaps 50 nucleotides per second. The process thus must take place at many thousands of points simultaneously—to lay down 3 billion nucleotides at 50 nucleotides per second would take 16,667 hours or about 9 months. That means that specific enzymes must make thousands of cuts at predetermined forks; other enzymes must straightened twists in the resulting strands, still other enzymes acting to disentangle them; it is then that an RNA primer roughly five nucleotides in length has to be attached to the replicating strand. This, too, requires the action of additional enzymes. Energy for the process must be supplied by the convenient release of a pyrophosphate group from each incoming nucleotide. That group is either added to the old nucleotides (by one process) or made with new ones (by another process). Because the two separated strands are now moving in opposite directions, two different enzymes

16. This is not Behe's definition. A system is irreducibly complex, Behe argues, if it is critically contingent upon its parts: withdraw any of them from the system and the system fails. See *Darwin's Black Box* (Free Press, 1996), for details. On Behe's definition, an ordinary three legged foot stool would appear to be irreducibly complex. The definition at hand entails Behe's definition.

17. The discussion that follows is almost entirely Arthur Cody's work; I have adjusted his remarks just slightly so that they agree in style with my own.

are pressed into service to lay down new nucleotides on each strand, but in opposite directions, one starting from the crotch of the forking strand, the other from an initial tip. Thousands of these nucleotides must be joined. RNA primer must be removed; this requires the action of yet another enzyme. The ends have to be collected and put together in the right order; still another enzyme is required. Mistakes occur despite the pairing rule. In prokaryotes, the polymerase that catalyzes synthesis of a new DNA strand also checks errors and fixes them.[18] In eukaryotes it is not known how checking and repairing are done.[19]

IRREDUCIBLY COMPLEX PROCESSES SEEM quite beyond the reach of any mechanism of change requiring functioning intermediates and for this reason beyond the reach of any mechanism governed by Darwinian principles. These are strong and daring claims: but they are not self-evident. The mathematician having provided the big picture, the biologist must provide the details.

Which follow: the set of states Ω—this remains, but each ε in Ω now admits of a dissection $\varepsilon = \{\rho_1, \rho_2, \dots, \rho_n\}$ into parts, the parts themselves arranged in various complicated ways.

Now Darwin's theory envisages complexity emerging by means of the double action of chance and selection. Things happen. Good accidents survive; the bad are washed away. Fitness plays a prominent role in this scheme; it is the curiously featureless Factor X that accounts for biological success and explains biological failure. For purposes at hand, the content to the concept may be expressed by a function $f: \Omega \rightarrow R$, one taking elements in Ω to their numerical reward.

For every $\varepsilon = \{\rho_1, \rho_2, \dots, \rho_n\}$ in Ω, that f is

$$\text{Additive } f(\varepsilon) = f(\rho_1) + f(\rho_2) + \dots + f(\rho_n).\text{[20]}$$

18. Where, one might ask, does its information come from?

19. In correspondence, a distinguished mathematician once remarked to me that "organic chemicals have an affinity for one another that chemistry cannot explain."

20. It is obviously possible to specify a more complex relationship. Exploring the so-called NK model of epistatic interaction, Stuart Kauffman defines the fitness of a genotype as the *average* of the fitness of its parts. This has the somewhat bizarre consequence that a

This is the first of three assumptions required to bring about the transmogrification of the skeleton of a Markov process into the torso of a Darwinian mechanism.[21]

Within the ambit of that torso, biological structures and their parts are selected, and so retained, because they make a contribution to the whole. A Darwinian system is in this respect unlike a social organism: there are no free riders. For every $\varepsilon = \{\rho_1, \rho_2, \dots, \rho_n\}$ in Ω, fitness is thus

Partwise Positive $f(\varepsilon) > f(\{\rho_1, \dots, \rho_{n-1}\}) > \dots > f(\{\rho_1\})$.

This, too, is an assumption, but one inexpungible from Darwinian theories.[22]

A third and mercifully final assumption deals not with fitness but with chance and distance. Whatever the fossil record may say, within a Darwinian universe, organisms change one step at a time if they are likely to change at all. Alterations in any given state arise in a neighborhood of that state. Given ε_j and ε_k, let μ represent their degree of difference. At $\mu = 0$, ε_j and ε_k, are entirely the same; at $\mu = 1$, entirely different. The details, which can prove vexing, are left as an exercise, but however

---

structure of N parts interacting K ways could be less fit than some N − k of its parts. Stuart Kaufmann, *The Origins of Order* (Oxford University Press, 1993), p. 42.

21. Change within Darwinian systems is *always* in the direction of increasing fitness: $P\{\xi(n + 1) = \varepsilon_j \mid \xi(n) = \varepsilon_k\}$ $(j, k = 1, 2, \dots) = 0$ if $f(\varepsilon_j) < f(\varepsilon_k)$; this is an assumption collapsing fitness and selection. It is, by the way, almost certainly false as stated and probably false when stated any other way. It is the central assumption of any mountain-climbing metaphor. It is not needed in what follows.

22. It may seem at first cut that I am confusing two quite separate processes: the discovery by evolution of certain biological structures, and the creation by morphogenesis of an organism from an initial cell. Morphogenesis is, in fact, a concession to the twin requirements of miniaturization and replication. It is hardly needed conceptually. The problem of evolutionary discovery is logically prior. If this seems paradoxical, the reader might imagine evolution as a process $k_1 \rightarrow k_2 \rightarrow \dots \rightarrow k_n$, extending over n fully formed generations such that for every i, j, either $k_i = k_j$ or k_j is derived from k_i by means of augmentation of parts *or* by means of a change in the nature or structure of one of its parts. These are in fact the only logical possibilities for any evolutionary process driven by point mutations. In this sense, the question how a complex structure is *assembled* and how a complex structure *evolves* turn out to be one and the same.

those details are settled, the scope of change must be conservative, various probabilities

Serially Ordered $\mathbf{P}\{\xi(n + 1) = \varepsilon_j \mid \xi(n) = \varepsilon_k\}$ $(j, k = 1, 2, \dots)\to 0$ as $\mu\to 1$.

A Darwinian mechanism $\mathcal{DM}$ thus emerges on the scene, all torso but no head, just as its creators have always claimed.[23]

What of irreducible complexity now that a Markov process has been provided with independent counsel? Intuition reports itself unclouded. If ε_j is irreducibly complex it is unlikely to be accessible from any other state. A Darwinian mechanism, mathematicians say, is *closed* to irreducible complexity.

A formal argument requires a number of details, but the argument's force, its large-limbed suggestiveness, may be conveyed by a cooked-up case, one in which a state $\varepsilon = \{\rho_1, \rho_2\}$ consists of two and only two parts.

A definition first: that ε is irreducibly complex if and only if $f(\rho_1) = 0$ for every γ in Ω such that $\rho_2 \not\in \gamma$. Ditto for $f(\rho_2)$.

A mini-theorem next.[24] If ε is irreducibly complex

THEOREM *There is no state γ in Ω such that ε is accessible from γ.*

The argument proceeds by contradiction. Suppose that ε were accessible from γ. It follows that $\varepsilon = \{\rho_1, \rho_2\}$ and $\gamma = \{\zeta_1, \zeta_2\}$ differ by only a single part—ρ_1, say—so that $\rho_2 = \zeta_2$. Otherwise $\mu = 1$ and the probability that ε is accessible from γ drops to 0. Now $f(\gamma) = f(\zeta_1) + f(\zeta_2)$. By the definition of irreducible complexity, $f(\zeta_2) = 0$. But $f(\gamma) > f(\zeta_1)$ and thus $f(\zeta_1) + f(\zeta_2) > f(\zeta_1)$, whence $f(\zeta_1) > f(\zeta_1)$.

Disorderly impulses are plainly at work, the tension evident even in Richard Dawkins's *Climbing Mount Improbable*, a book otherwise free of intellectual tension altogether. In speculating on the evolutionary de-

---

23. The three assumptions just given do not logically specify a theory or anything like a theory. They must be taken for what they are—three assumptions merely.

24. A *theorem*, note. The argument is entirely conceptual. I am not for a moment arguing that there are no intermediates *in fact* for irreducibly complex systems. I are arguing that such inter mediates *could not logically* exist. Suggesting nothing so much as a man proposing gamely to trisect an angle, H. Allen Orr has argued the contrary in the *Boston Review* (February 1996).

velopment of the mammalian eye, Dawkins imagines improvements occurring one at a time, the eye's photoreceptive array acquiring new cells incrementally. This satisfies a principle of serial order. Changes are concentrated in the neighborhood of a given structure. But the mammalian eye is irreducibly complex, or so it would appear, the eye's parts working together in a tightly coordinated hierarchical system, the structure as a whole ranging from the biochemical to the neurological.[25] The blue shadow of a contradiction may now be observed falling over the discussion.

The definition of irreducible complexity makes strong empirical claims: it is foolish to deny this, and foolish as well to suggest that these claims have been met. The argument having been forged in analogy, it remains possible that the analogy may collapse at just the crucial joint. The mammalian eye *seems* irreducibly complex; so, too, eukaryotic replication, and countlessly many biochemical systems, but who knows? Still, the concept of irreducible complexity is surely not empty. A modern airplane, a wristwatch, a nuclear reactor, or even a mousetrap all contain a core of critical parts designed specifically to work with other parts designed specifically to work with them. This *we* know because *we* have done the designing. Gödel's words may now be heard again both as an overtone and as a warning: "The complexity of living bodies has to be present either in the material [from which they are derived] or in the laws [governing their formation]." What holds for human bodies holds for human artifacts, irreducible complexity transferring itself backward, *from* the inorganic world of things and artifacts *to* the biological world in which plans are made and things and artifacts designed.

A MARKOV PROCESS IS in the context of Darwinian thought useful because it expresses the stochastic character of Darwinian theory. Whatever else it may be, evolution is, on Darwin's theory, an essentially serial process—one damn thing after the other. A finite state automaton is the simplest of devices that expresses the serial order of behavior and useful

---

25. See p. 18 *et. seq.*

thus in precisely this regard; but a finite state automaton is also a device whose memory is fixed and so is independent of its inputs, indifferent to its fate. And this, too, is a characteristic of Darwinian theories.[26] Another mathematical model is now in prospect.[27]

An automaton is defined first in terms of a finite alphabet Σ.[28] This reflects the three-way interpretation of the subject in logic, linguistics and computer science, but an alphabet in this context is no more intrinsically linguistic than symbols are intrinsically symbolic in the context of symbolic dynamics; the origins of finite state devices in neural modeling should suggest the relevance of a biological interpretation of those very mutable symbols.[29]

Given a finite alphabet Σ, it is the business of an automaton either to recognize or to generate a finite sequence of symbols drawn on the alphabet.[30] Such are the words,

26. Daniel Dennett has argued that evolution is an algorithmic process; but beyond saying that it exists, he has not specified the algorithm; see *Darwin's Dangerous Idea* (Simon & Schuster, 1994). If by an algorithmic process, he means one that is effectively computable, the claim is uninteresting. As a philosopher, Dennett seems unaware of the very significant issues raised when a physical process is explained by an appeal to an algorithm. If we were to notice that the rivers of the world express a correlation between the position of their names in the alphabet and their size, so that rivers whose names begin with A are invariably larger or longer than rivers whose names begin B, would we conclude that we had explained the phenomenon in question by appealing to a sorting algorithm?

27. Markov process and finite state automata are, of course, closely related objects.

28. The discussion is taken from S. Eilenberg, *Automata, Languages & Machines*, Volume B (Academic Press, 1974). The subject has in recent years been infused with algebraic techniques and concepts. Readers familiar with early discussions from formal linguistics will require a short period of orientation before the new notations become compelling.

29. An alphabet consists of a finite list of discrete items; and syntax is pursued in terms of shapes. There is no reason whatsoever that those symbols may not receive an interpretation in terms of any other finite and discrete set of structures—tissues or organs, for example, or more generally, "elements of an abstract set of events in the case of an analysis of the behavior of a system," as Dominique Perrin puts it in his essay, "Finite Automata," in J. van Leeuwen, ed., *Handbook of Theoretical Computer Science*, Volume B (MIT Press, 1994), p. 4. It is for this reason that I use *languages*, *events* and *structures* with a certain lack of specificity.

30. In what follows, I draw no distinction between recognition and generation, swallowing both concepts in the all purpose notion of a word being *accessible* to an automaton.

$$x = a_1 a_2 \dots a_n;$$

these are structures entirely determined by their shape or internal configuration. The set of all words on Σ is recorded as Σ^*. It has the structure of a free monoid.[31]

An automaton $\mathcal{A}$ over Σ comprises four parts

States	A finite set Q of elements
Initial States	A subset I of Q
Terminal States	A subset T of Q
Edges	A subset E of $Q \times \Sigma \times Q$

A particular edge (p, σ, q) begins at p and ends at q. The symbol σ serves as the label of the edge. An edge may also serve as the expression of a function $\sigma : p \rightarrow q$, or as the expression of a function from symbols and states to states themselves: $\sigma : Q \times \Sigma \rightarrow Q$, or as the expression of a rule, one in which the automaton is instructed to move from a state and a symbol to a new state.

Edges first, paths next. A path c in $\mathcal{A}$ is a finite sequence of consecutive edges

$$c = (q_0, \sigma_1, q_1)(q_1, \sigma_1, q_2) \dots (q_{k-1}, \sigma_k, q_k).$$

If a path $c: i \rightarrow t, i \in \mathbf{I}, t \in \mathbf{T}$, conducts itself from an initial to a terminal state, it is said to be successful; the successful paths form a subset $|A|$ of Σ^*. They embody, they *represent*, the behavior of $\mathcal{A}$.

The behavior of a finite state automaton may be illustrated by a simple schema. Small circles indicate states of the system. Initial states are marked by an inward (or facing) arrow; terminal states by an outward arrow.

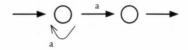

---

31. A free monoid is a group-like object, but one without an inverse. DNA is a word, of course; so, too, any of the proteins. In dealing with such structures, their associated free monoid is of an astonishing cardinality, an interesting and important fact, one pertinent to any assessment of the plausibility of Darwinian claims.

THIS AUTOMATON MAY DO one of two things: it may move directly from an initial to a terminal state, or it may loop through its initial state. Its extraordinary simplicity notwithstanding , this machine is capable of recognizing all structures of the form: *a, aa, aaa, aaaa,* this up to a^m.

There is, finally, the crucial concept of recognizability. Herewith the obvious definition. A subset A of Σ^* is recognizable if there exists an automaton A over Σ such that A = $|\mathcal{A}|$. With the larger project of interpreting finite state automata in the context of evolutionary theories, recognizability is one step from accessibility. Thus for some biological organism or structure $\varepsilon = \{\rho_1, \rho_2, \dots, \rho_n\}$ in Ω, ε is accessible from some Darwinian mechanism $\mathcal{DM}$, if and only if (i) $\mathcal{DM}$ and $\mathcal{A}$ are isomorphic (or otherwise the same) for some automaton $\mathcal{A}$, (ii) $\{\rho_1, \rho_2, \dots, \rho_n\}$ is a subset of Σ^*, and (iii) $\{\rho_1, \rho_2, \dots, \rho_n\}$ is recognizable to $\mathcal{A}$. Nothing more.

Finite state automata are interesting for the promise they hold of permitting the mathematician to draw a limpid connection between what a finite state automaton is and what it does.[32] In this, the theory of finite state automata is decisively characterized by a single absolutely fundamental theorem, one known for more than half a century.

Let Σ be an alphabet. The rational expressions[33] on Σ are defined with respect to its subsets. There are just three pertinent constructions:

$$\text{Union } X \cup Y$$
$$\text{Product } XY = \{xy \mid x \in X, y \in Y\}$$
$$\text{Star } X^* = \{x_1, x_2 \dots x_n \mid n \geq 0, x_1 \in X\}.$$

The rational subsets of Σ are just those that can be derived from the finite subsets of Σ using the rational operations of union, product, and star formation.

32. Before any evolutionary theory may rationally be addressed, two logically prior questions need to be asked. The first: what are the properties of living systems that must be explained? The second: what are the capabilities of the devices invoked to explain them? Neither question is ever pursued in a serious way; but it is only when both questions have received an answer that anyone is in a position to determine whether in principle a particular theory could account for a particular fact or evolutionary development. The point is widely understood in other disciplines, but stoutly resisted in biology.

33. These are also known as regular expressions or events.

Now let Σ_S designate the subsets of Σ. That absolutely fundamental theorem:

THEOREM *(Kleene) A set* $X \subset \Sigma_S$ *is recognizable if and only if* X *is rational.*

Rationality is the mark of those structures accessible to a finite state automaton; and a finite state automaton is that device capable of recognizing them. It is sobering to think that nowhere else in evolutionary theory is there anything like the precise matching of a device to what it can do that is available in this context and by means of this result.

HAVING COME THIS FAR through the badlands of symbolism, the reader may well wonder what the connection—the *precise* connection—between Darwinian mechanisms and finite state automata. The question, although fascinating, is, of course, ill posed. The finite state automata have received a precise definition. Not so the Darwinian mechanisms. They are half Markov chain and half some shambling Other. There is still room for conjecture, guesses playing over the undiscovered country in which everything is ultimately made clear. My conjecture, recorded in the definition of accessibility, is that any finite state automaton may be approximated by some Darwinian mechanism.

The case is circumstantial, but nonetheless compelling, or if not compelling then at least suggestive. Finite state automata are inherently serial. They achieve their effects by looping and by iteration, but lacking any memory beyond their states, they are incapable of intricate acts of deferral and subordination. Hoping to electrify the community of linguists, Patrick Suppes demonstrated in the late 1960s that any finite state automaton could be asymptotically approximated by some stimulus sampling theory. His theorem rested on a connection between Markov processes and finite state automata that was antecedently obvious; but it had the merit of fixing stimulus sampling theories in psychology precisely in a region of space.

THEOREM *(Suppes) Given any connected finite automaton $\mathcal{A}$, there is a stimulus response model $\mathcal{SR}$ such that in the limit $\mathcal{SR}$ and $\mathcal{A}$ are isomorphic.*[34]

Now stimulus sampling theories in psychology stand to Darwinian theories in biology as sister to long lost sister. They are both theories embedded in Markov chains; they share the same vocabulary; and they appeal to the same mechanism, one in which the trigger of chance is followed by the filter of selection or reinforcement.[35] This suggests a revision of Suppes's theorem, one in which psychology gives way to evolutionary biology. The perspective afforded is, of course, oblique, but an oblique perspective is all that the mathematician can ever offer, so while little may be gained, absolutely nothing has been lost.

CONJECTURE *Given any connected finite automaton $\mathcal{A}$, there is a Darwinian model $\mathcal{DM}$ such that in the limit $\mathcal{DM}$ and $\mathcal{A}$ are isomorphic.*

Although shallow mathematically, theorem and conjecture do send a shaft of sunshine into a discussion that might otherwise remain dark. Two facts stand illuminated. Kleene's theorem specifies, and specifies precisely, what a finite state automaton can do. Suppes's theorem and the conjecture that follows establish that whatever a finite state automaton may do, this may *also* be done by a device randomly shooting changes into the wilderness and trusting in the wilderness to select those changes that do some good.

Theorems and conjecture settle only half the vexing issue of theory. They describe what a particular mechanism or device can do. The other half remains. What is it that such devices *need* do? Suppes evidently believed that by demonstrating his theorem for stimulus-response models in psychology he was coming to their defense. Another more medical interpretation is possible. The identification achieved between structures serves only to permit a certain contagion to spread, the limitations of

34. See Patrick Suppes, *Studies in the Methodology and Foundations of Science* (Humanities Press, 1969), pp. 411-453.

35. The connection between behavioral theories in psychology and evolutionary theories in biology is one that typically goes unremarked by biologists, no doubt because of the stain of the association.

finite state automata infecting the stimulus response theories and so, of course, the Darwinian mechanisms as well.

IF RATIONALITY OFFERS AN absolute standard of accessibility, it suggests—and it *must* provide—a correlative concept of inaccessibility. And so it does. The inaccessible structures are those that *cannot* be recognized by a finite state automaton. This makes for a voluptuously general scheme of classification, simplicity identified with the rational events, and complexity with everything else. Behe's irreducible complexity now emerges as a special case of the more wide ranging concept of *general irreducible complexity*.

Surprisingly enough, examples often turn on properties of symmetry.[36] A partial list follows, given in terms of strings drawn on a finite alphabet or directly in terms of languages:

$L = \{a^n b^n\}$, *where $n > 0$*
The set of strings over a finite alphabet in which parentheses are mated
Parenthesis-free propositional calculus in prefix notation
The set of palindromes over a finite alphabet
$L = \{aXa\}$, *where X is a point of symmetry.*

Though short, this list reveals sharp and unsuspected limitations on what finite state devices can do. A finite state machine cannot produce *arbitrary* structures of the form $\{a^n b^n\}$. This may suggest that it is the promiscuity of an arbitrary n that is at fault. Not so. The same finite state machine can generate structures of the form $\{a^n b^p\}$, where p is a fixed integer. Difficulties lie elsewhere.

A finite state machine is inherently serial—a point in its favor. Now imagine n for the moment fixed, time frozen at t. There is no way that by t, a serial device could have realized $\{a^n b^n\}$; at any particular point in its deliberations, it can have reached only $\{a^n b^p\}$, where $p < n$. Before winding things up with a grunt of satisfaction, the automaton must loop

---

36. This should provoke an intellectual twitch. From the point of view of Kolmogorov complexity, symmetry is the mark of structures that are simple; symmetry in physics marks the appearance of a local group. Just how these concepts are to be integrated is yet unknown.

through b: $b \rightarrow b^{2k} \rightarrow b^{ik}$, the effect of the routing indicated by placing a symbol in context: $b \rightarrow Rb^{2k} \rightarrow Rb^{ik}$. The index i counts the number of loops. But the automaton must bring some Rb^k to an end at, say, b^l, if b is ever to count as a terminal symbol.

It is thus that at some point the automaton contemplates a structure of the form $a^m b^p R$. It has reached m of the a's, and p of the b's, and completed who knows how many R's, and is set to loop for the last time. But this means that the automaton is poised to recognize infinitely many strings of the form $\{a^m, b^{p+ik+l}\}$—a gloomy monstrosity, not at all the intended result.

This tight little argument has a large, a percussive charge, one that explodes a convenient superstition. General irreducible complexity has nothing whatsoever to do with size or indeed with any of the standard measures of complexity, such as that "degree of irregularity" to which Stephen Jay Gould appeals in *Full House*. There are very simple symmetrical structures whose organization lies beyond the reach of *any* serial process.

It is the push down storage automata that are capable of recognizing or generating structures such as $\{a^n b^n\}$—capable, in fact, of recognizing the entire class of context-free events or languages or structures. This little fact yields the pleasurable sense that something is under construction, a hierarchy of sorts, the finite state machines neatly matched to the regular languages, machines that are slightly more powerful matched to languages that are slightly more complicated. From the perspective of automata theory or algebra, it is, no doubt, the hierarchy that is of interest; but from the perspective of theoretical biology, there is something else at work here, some faint but ineffably important conceptual line that is crossed when finite state machines give way to push down storage automata.[37]

A push down storage automaton is a finite state device, one augmented by a stack or memory. Three objects are required:

---

37. I have been calling attention to this point for years on end. See my essay, "Philosophical Aspects of Molecular Biological Systems," *The Journal of Philosophy*, Vol. LXIX (June 1972).

$$X = \Gamma^* \times \Sigma^*$$
$$\alpha\colon \Sigma^* \to \Gamma^* \times \Sigma^*$$
$$X_\omega = \Gamma^* \times 1.$$

Γ^* denotes the stack or memory. It is derived algebraically from Γ, which is an output alphabet. Whereas a finite state automaton had only one alphabet in Σ, now it has two, Σ of old functioning now as an initial or input alphabet. Both Γ^* and Σ^* denote sets of words and both have the structure of a free monoid. The set X denotes the Cartesian product of the two sets of words. The function $\alpha\colon \Sigma^* \to \Gamma^* \times \Sigma^*$ maps words from the input alphabet onto X. It serves to instruct the automaton whether symbols or structures should go to memory or pass onto the output. The stack is unbounded: hence X_ω.

The (superficial) complexity of notation notwithstanding, the action of a push down storage automaton is simple. It is given a finite string of symbols, these having a particular shape—*aaabbb*, say. It is often convenient to imagine those symbols being conveyed to the automaton by a tape. The machine acts via the function α to place symbols in memory or to assign them to its output. This, too, can be thought of as a tape, so that the action of the machine is in the end a curious concourse between two tapes, the machine acting to evacuate one and fill the other. Detours occur when some symbols are sent to the stack; but when the stack is empty, the automaton has done its work.

A context-free language or structure is any language or structure accessible to a push down storage automaton.[38] This easy association serves the intellectually valuable purpose of illuminating the powers of a device by reference to the objects that it can recognize. But unlike the regular events, the class of context free structures has remained undefined, their appearance so far a matter only of unobtrusive examples. Whatever the definition to come, general irreducible complexity is nonetheless an *ab-*

---

38. Context free language may also be defined directly in terms of grammatical rules or systems of equations.

solute property of certain systems.[39] The burden of these remarks is suggested by two classical results, both of them given in the old-fashioned language of formal linguistics

THEOREM *(Chomsky, Schützenberger) If φ is any set of finite state productions, and γ, any set of context free productions, then φ is a proper subset of γ.*

This theorem establishes the first step on the full, the famous ascending Chomsky hierarchy, which stretches from the finite state automata to the Turing machines on the left, and the regular languages to the unrestricted rewriting systems on the right; but what gives the theorem its intellectual force is, of course, the unequivocal nature of the distinction that it draws.

The second theorem goes further by establishing that context free structures cannot in general even be *approximated* by finite state structures.

THEOREM *(Parikh) There exists a push down storage automaton PDS, whose productions are PDS(Pr), with the following property: given a finite state automaton FSA_1, whose productions are $FSA_1(Pr) \subset PDS(Pr)$, it is possible to construct a finite state automaton FSA_2, with productions $FSA_2(Pr)$, such that (i) $FSA_1(Pr) \subset FSA_2(Pr)$; and moreover (ii) $FSA_2(Pr)$ contains infinitely many structures not in $FSA_1(Pr)$.*

It follows from the asymptotic identification of Darwinian mechanisms with finite state automata that

CONJECTURE *There exists a push down storage automaton PDS that cannot be approximated by any Darwinian mechanism $\mathcal{DM}$.*

Despite the number of assumptions that have been built into this conjecture, it is nonetheless a disturbing, a controversial claim, suggesting as it does that an empirical theory might be fatally compromised by a chain of logical circumstances. The association between stimulus sampling and evolutionary theories is a source of irony as well as insight,

---

39. The use of the term "absolute" is meant to suggest Gödel's own well known remarks about recursiveness.

behavioral theories in psychology having themselves been brought low by arguments very similar to those that I am now advancing.

BEHE'S IRREDUCIBLY COMPLEX STRUCTURES lie beyond the reach of any Darwinian mechanism, and for this a waterproof argument suffices. Doubts crept in on the argument's far side, when biological structures were sifted for signs of irreducible complexity. The pattern repeats itself in the case of general irreducible complexity, but with little sense of real scruples being scrupled. Biology is *essentially* the study of non-serial behavior, the point overwhelmingly clear, of course, in the case of natural languages, where the facts by now are unassailable. Elsewhere, more of the same.[40]

In addition to the creation of proteins, DNA creates proteins in a significant order. In some cases, it is the order in which the proteins are coded that is essential for the proper development of some feature of the organism. But this sort of development is very limited in the overall construction of the organism. So far as we know, it is limited only to items such as the formation of insect exoskeletons. In these instances the actual development is, like the DNA plan itself, serial; it goes from front to back or top to bottom. Most of the organism for one or another very different sorts of reasons *cannot* be serially organized, and certainly not the whole of an organism, as though it were laid down in a series of planes like the slices seen in an MRI scan. This is no doubt obvious, but if we look at some of the reasons against such a suggestion, we see the nature of the problem:

1. Embryological development is not serial. Where it is visible, it seems to be systematic; that is, the physical systems, whether skeletal, circulatory, or nervous, have their own schedules and procedures. What little truth there is to the notion of ontology recapitulating phylogeny shows not serial planar construction but rather some process of system building and integration.

---

40. The examples that follow are again due to Arthur Cody.

2.   There are not only many systems within every living creature that
     have to be integrated into an organic whole—they must be put to-
     gether in some sort of balance. The legs have to be the right size
     and strength to support the body and to carry out the appropriate
     activities of the animal; the same can be said of a stem or a leaf or
     the cilia of a paramecium. And this is to say that the whole has a
     superordinate influence on its parts. One is likely to imagine that
     the parts just naturally fall into the right scale and proportion, but
     in the purely chance driven mechanism of the DNA molecule, there
     is nothing to justify such a conclusion.

3.   An animal crawls, walks, swims or flies by a characteristic coordina-
     tion of appendages. Hence there are central nervous specifications
     to this end. The central nervous system is a physical organ, to be sure,
     and as such might be serially accessible to some Darwinian device.
     However, it is the function of the nervous system that characterizes
     it as an organ, and, unlike the leg, its function is not apparent in its
     movement. It makes no sense to speak of the executive properties of
     the nervous system in serial terms, as though the various programs
     were set up in and because of an order corresponding to the physi-
     cal relationship of the neurons (starting from the top of the head or
     some other point of origin). It makes no sense because while there
     is only one physically realizable serial order to parts of the nervous
     system, there are—there *must* be—an extraordinary number of dis-
     tinct executive programs; for example, if there is a walking program,
     it must be modified each time the organism walks so as to adjust to
     the ground and the destination.

4.   All animals have sense organs of some sort even if primitive ones. A
     sense organ is of no use by itself. Its purpose comes from its informa-
     tional contribution. Typically doing x is preceded by perceiving x or
     signs of x as warnings or indications. Hence, those organs of sense
     have to be tied to other circuits in the central nervous system that
     are first evaluative and then executive. Putting it this way makes

it sound as though only higher animals are included, but a worm which turns abruptly from contact with a dry rock has evaluated the problem and withdrawn from it as part of its program, much as the human somatic nervous response to touching a hot stove does not involve consciousness, not even figuratively. If the central nervous system, the peripheral nervous system and the muscles all operated together because they were constructed in the same serial order, then the events in the world would have to cooperate by placing demands on them in the same serial order.

5. Within the higher organisms, the geometry and mechanical arrangement of the muscles and appendages of movement do not prescribe their use. The antelope does not have to run faster, the eagle does not have to fly. Their tendency to run or to fly is located in a program quite separate from the one that expresses their powers of running or flight. The eagle must develop the will to fly, along with its muscles and its wings. In concentrating on the higher organisms, let us not forget the remarkable coordination DNA codons impose on the design of protein groups in the manufacture of insects, where will is not involved. The wasp equipped with an ovipositor and some special faculties detects the presence of larva under the ground and then pierces the ground and the body of the larva to the appropriate depth, whereupon she inserts her eggs. There is no way to understand this talent as developed piecemeal and no way to justify thinking that the various coordinated components are close to one another in the DNA metric.

Such are the various facts. What of their significance? The mammalian eye is an organ that has exerted a magnetic influence on Darwinists and doubters alike. It is plain that the eye is organized as a hierarchical structure, one extending over at least three levels: [41] that of the visual

---

41. There is, of course, a fourth level, that of experience itself. Without an account of such a level, no description of vision can possibly be complete; it is the purpose of the eye, after all, to enable an organism to *see*. Needless to say, we are in no position whatsoever to undertake a scientific analysis of experience.

system **VS**; that of the anatomical system **Anat**; and that of the bio-chemical system **Bio** = $\{b_1, \ldots, b_n\}$. One way in which to depict the hierarchy is by means of labeled brackets [**VS**[**Anat**[**Bio**[$\{b_1, \ldots, b_n\}$]]]], this indicating that every group of right elements belongs to the next leftmost category.[42]

The effect of the labeling may be conveyed as well by a series of rules

$$VS \rightarrow Anat,$$
$$Anat \rightarrow Bio$$
$$Bio \rightarrow b_1, \ldots, b_n.$$

And the effect conveyed by the rules conveyed in turn by an equation

$$VS = Anat + Bio + b_1, \ldots, b_n,$$

or more generally by a polynomial

$$A = f_1 + f_2 + \cdots + f_k,$$

where **A** is any biological category including f_i as subcategories and the + sign indicates not straightforward addition but logical alternation. The polynomial in turn goes over to a system of equations in n variables

$$A_1 = f_1(A_1, \ldots, A_n)$$

.

.

.

$$A_n = f_n(A_1, \ldots, A_n).$$

Each step in this scheme is not only plausible, but natural as well. Since time immemorial, mathematicians have found it useful to repre-

---

42. A labeled bracket is a *static* description of a biological structure. Like every other biological organ, the eye is designed to achieve a certain function—namely, sight; at the level of biological realism, one would expect to find very complicated transformations obtaining among various labeled brackets, so that even the simplest visual act must be described by a sequence $\Phi_1 \rightarrow \Phi_2 \rightarrow \ldots \rightarrow \Phi_n$, where each Φ is a labeled bracket. The description of such a sequence is utterly beyond our analytical abilities. Michael Behe gives a fine sense of the complexity of a small part of the biochemistry of vision in *Darwin's Black Box*, but while we know something of the biochemistry, and something, too, of the algorithmic properties of the visual system at a very abstract level, this through the work of David Marr and his collaborators, what we lack is any understanding of their dynamic interaction.

sent systems by means of simultaneous equations; it is their canonical instrument for the representation of complexity, its employment a sign that some process cannot be represented in the straightforward terms in which a single function undertakes to map one thing onto another. The equations at hand, it is true, are defined in terms of words, and not numbers, but words are mathematical objects in their own right, the algebraic machinery that proceeds smoothly over systems of numerical equations proceeding smoothly here as well.

Like many biological structures, the context-free languages, and so the push down storage automata, admit of definition in terms of a system of equations, a natural circle now forming, one moving from general irreducible complexity to systems of equations to context free languages and back again to systems of equations. There is a mediating operation that has so far gone unremarked, one that is needed to close the circle, and that is the specification of the context free languages in terms of a grammar. The definitions are straightforward, and by now a part of a general academic culture. A context free grammar $G = (V, A, P, S)$ consists of a finite alphabet V; a subset A of V designating terminal symbols; a finite set of productions $P \subset (V—A)$, and a distinguished element $S \in (V—A)$, one functioning as an axiom. The grammar works by means of derivations or rewriting rules.

Consider, as an example, a grammar consisting of one symbol F, and two terminal symbols a and b. The axiom is

$$F \rightarrow aFb.$$

Productions are all of the form

$$F \rightarrow aabb.$$

The language L associated with these rules consist of all and only the terminal strings, expressions of the form $\{a^n b^n\}$, evidence of the deep and surprising power of the notation.

Now let X be a variable ranging over symbols in $(V—A)$, and let P_x be the right hand sides of productions having X on their left side. The set of productions can obviously be rewritten as

$$X \rightarrow P_{x}.$$

This gives rise to a system of equations, one associated with a particular grammar—**G**, say. A solution to the system is given by a family L_x of subsets of A, such that

$$L_x = P_x(L),$$

where

$$P_x(L) = \cup \, p(L),$$
$$p \in P_x$$

and where $p(L)$ is the product of the language obtained by replacing in p each occurrence of a variable Y by the language L_Y.

A far-reaching theorem now follows, one that draws close a conceptual circle.

THEOREM (*Schützenberger*) *Let* **G** *be a context-free grammar. The family* $L = (L_x)$, *where* $L_x = L_G(X)$, *is the least solution of its associated set of equations.*

This theorem rests on certain algebraic machinery; theorem and machinery have the effect of ratifying certain natural forms of representation and so indirectly of ratifying certain intuitions. A context-free grammar is a way of expressing a system of equations and vice versa, the vice versa collapsing the distinction between grammar and equations entirely so that both devices lend themselves to the description of structures in which there is nesting, domination, hierarchy, pattern and symmetry.

IT IS PERHAPS TIME for a retrospective. Two hierarchies have made an appearance, those comprised of automata, and those comprised of their accessible structures—the regular events, in the case of finite state devices, the context-free languages, in the case of push down storage automata. Structures and automata alike give every indication of vanishing into a purely algebraic void; but while the algebra has the effect of vacating the familiar from a class of concepts, it also has the effect of throwing a sharp clean hard light on the essentials. Defined over words, equations are the algebraic middle men in what has passed. This takes some getting used to, but familiarity breeds assent, the more so since the universe of words seems strangely to mimic the universe of numbers, rather like those parallel universes in which one may observe a character identical to oneself in all respects save one. In elementary analysis, the solution to a system of equations is specifiable in terms of power series. Here, too, but with the difference that the underlying variables do not commute.

The introduction of formal power series marks a third abstractive level—things, devices and structures to equations, equations to power series. It preserves, this third step, distinctions in place down the line. Regular and accessible events coincide in the case of finite state automata. So, too, their associated power series. This is the burden of a remarkably lovely and far reaching theorem by M. P. Schützenberger. Power series recognizable by a finite state automata are rational and vice versa

$$\mathcal{A}^{\text{rat}} \langle \Sigma^* \rangle = \mathcal{A}^{\text{rec}} \langle \Sigma^* \rangle.$$

Does the same hold for push down storage automata as well? Not in terms of rational structures, of course; that would defeat the distinction it has taken so many definitions to enforce. Beyond the rational power series, there are those that algebraic, the extraordinarily illuminating pattern *almost* repeating itself once again. With reservations signified by that guarded "almost" in mind, there is the altogether remarkable

$$\mathcal{A}^{\text{alg}} \langle \Sigma^* \rangle = \mathcal{A}^{\text{rec}} \langle \Sigma^* \rangle,$$

where Σ^* is the free monoid of a context-free language.

These purely algebraic identities have a wickedly alluring gleam: they glitter and they glare. A compression of identity is at work, the husks of finite state and push down storage automata dropping away to reveal an algebraic incarnation. This is always an exciting process to watch, but however compelling mathematical pupation, my attention is inexorably drawn elsewhere. Things that stay the same as other things are changing are invariant; reification yields invariants. A subgroup S of a group G in which right and left cosets coincide is an invariant; the dimension of a finite dimensional vector space is another.

Now the line dividing finite state and push down storage automata corresponds in a very intuitive way to the division between natural and artificial processes, and between systems of descriptions that apply to material events and systems of descriptions that apply to biological events. If this seems too rich a claim for many tastes, I have a lite version in mind as well. The line between finite state and push down storage automata corresponds to a division between two concepts of complexity, or between degrees of complexity as one polyvalent concept displays itself. Whatever the content to the distinction, it is less important than the fact that the distinction is refulgent; and if Factor X is good enough for evolutionary theorists casting about for content to the concept of fitness, Factor X can again be pressed into service here, as *whatever* it is that the distinction defines.

Whatever it *is*, the distinction is provided with an invariant, one set of cases expressed by the rational power series, the other, by the algebraic power series.

I would say that these are invariants first of matter and second of life; but I am content to claim only that the invariants serve to characterize Factor X, leaving matters for the moment at that.

AND YET A TROUBLING point remains. My argument has washed on alien shores. Automata and algorithms are by their nature scientifically anomalous instruments.[43] They do not express laws of nature in the sense made familiar by mathematical physics. They play an interesting but poorly understood role in the local economy of explanation. The idea that psychological processes are inherently computational is hardly a model of sturdy common sense. But the identification of Darwinian devices with finite state automata was *my* idea, evidence, if any were needed, that however unwholesome the algorithmic category, I for one have not scrupled at its employment. The proverbial camel is now within

---

43. In the late 1950s, analytic philosophers often wondered whether contextual definitions (in the sense of Carnap) filled an unused space between definitions and laws of nature. Their discussion was inadvertently prophetic. The question needs to be asked about algorithms instead.

the tent, peering indignantly out. Finite state automata cannot express
a certain order of complexity. Very well. Let us go to the next best thing,
and that is the devices that can do the job. Is there a reason that Darwinian theories should not be cast directly in push down terms, the baleful
conclusions of my own argument short lived because short circuited?

It is a reasonable question, one that touches on a number of dark
places in the philosophy of science. Push down storage automata mark
the place where what a computational device does is determined in part
by what it remembers, memory functioning as an ancilla to design and
so playing the role in such devices that it plays elsewhere. It is evidence
of design that must itself be explained, if Gödel's question is ever to be
answered, and plainly if the device doing the explanation itself requires
an explanation of precisely the same sort, we are in the position of a man
bailing water from the front end of his boat and depositing it in the rear.

The unhealthy effect of something so simple as a stack on the larger
Darwinian project emerges when specific cases come under scrutiny.
Consider structures of the form $\{a^m\}$; plainly they are accessible to an A
of old

the result of its action a set of paths, one increasing in length a, aa, aaa,
$aaaa$, ..., a^m.

With the details left in decent obscurity, it is easy to imagine a Darwinian mechanism $\mathcal{DM}$ coming in its behavior to coincide with the paths
specified by the automaton, fitness a function of ever lengthening paths

$$f(a) < f(aa) < f(aaa) < f(aaaa) < \ldots < f(a^m).$$

Consider, on the other hand, structures of the form $\{a^n b^n\}$. These
lie in the Great Beyond that a finite state automaton cannot reach and
does not know. Suppose that $n = 3$ so that $\{a^n b^n\}$ has the form $\{aaabbb\}$.
Symbols are presented to a machine on a tape, one that functions purely
as a device of presentation

The machine turns now to a second tape, this one blank

It is this second tape that functions as the useful embodiment of an imaginary stack. Reading the first tape from left to right, the machine divides its action, printing the symbol c in its stack every time it spots the symbol a. By the time the machine has finished with the a's, the stack is half full

Ah, but on encountering b, the automaton reverses its operations, erasing instead of writing c. When the stack is again empty, the machine has done its work. There is in this design something devilishly clever, the stack managing to achieve the effect of counting without counting at all, action over the stack having first swept in n counters and then swept out m counters and so leaving the stack empty if and only if $n = m$.

But if a simple push down storage automaton is a marvel of ingenuity it seems hardly a model for those natural processes in which complexity is constructed by accretion. The problem is the difficult one of distributing fitness to intermediate steps. In the case of structures of the form $\{a^m\}$, increasing path length is a convenient proxy for fitness. One can imagine natural selection for whatever reason relentlessly favoring length. Structures of the form $\{a^n b^n\}$ are different, the attempt to distribute fitness over intermediate cases leading to contradiction.

Thus $f(c^n) < f(c^m)$, if $n < m$, and thus again

$$f(c) < f(cc) < f(ccc) \ldots,$$

whence

$$f(a^n) > f(a^n b^{m-2}) > f(a^n b^{m-1}) > f(a^n b^m).$$

But equally,

$$f(a^n b^{m-1}) < f(a^n b^m),$$

if f is partwise positive, assessment over the stack and assessment over the output failing to cohere.

Two impulses are plainly at work and plainly in conflict. The difficulty is that increasing path length has nothing whatsoever to do with any measure of fitness, however loose. The example of $\{a^m\}$ is a lurid distraction. Structures of the form *aaabb* are not slightly more symmetrical than structures of the form *aaab*. They are not symmetrical at all. Symmetry arises only when the structure has been completed, a vivid example of irreducible complexity, and one, incidentally, that has nothing to do with function.

Cases such as this are by no means artificial; they are a staple in the literature of evolutionary theory; they occur whenever target sentences or words are supposedly discovered by means of an evolutionary search.[44] If the cases are a staple, so, too, the defective strategies for their analysis. Almost always, they involve, those defective strategies, a discovery procedure in which a description of the desired object is illicitly smuggled into the search procedure itself. If symmetry is at issue, it plainly does little good to resolve contradictory impulses by arguing that $f(aaabb) > f(aaab)$ in virtue of the fact that *aaabb* is *closer* to *aaabbb* than *aaab*. A Darwinian mechanism having access to a definition of distance could no longer be a Darwinian mechanism. Such a device is *designed* to achieve certain ends, that element of carefully contrived intelligence that was to disappear from Darwinian theories reappearing now in models of those theories, rather like a spirit indignantly refusing to stay exorcised. An old, a painful dilemma.[45]

The point is informal, but it is also quite general. Finite state automata occupy a position in the uncharted region between natural laws and complex algorithms. Darwinian theories have always commended themselves to biologists as a bridge between the physical and the living; and insofar as the finite state automata and the Darwinian mechanisms coincide, so, too, the finite state automata. But in a curious reprise of the third man argument, intellectual difficulties have divided themselves

---

44. The obvious example is Richard Dawkins's *The Blind Watchmaker*; but even sophisticated authors often fail to see the problem. See R. Keen and J. Spain, *Computer Simulation in Biology* (John Wiley, 1992), for details.

45. See D. Berlinski, *Black Mischief: Language, Life, Logic & Luck* (Harcourt Brace: 1986), Section V for further details.

without diminishing themselves, the line between the physical and living now *reappearing* as the distinction between finite state and push down storage automata, evidence, if any were needed, that the distinction is itself irrefragable.[46]

SCIENTISTS AND PHILOSOPHERS OFTEN imagine that they are in the business of liberating men and women from their illusions: it is their common boast, although whether men and women find those illusions burdensome or are grateful for the liberation is another matter entirely. The idea that there is an absolute character to the distinction between life and matter and so between biology and physics is among the more stubbornly held illusions, perhaps because it is not an illusion at all, but simply the plain, the ineradicable truth. For reasons that are anything but clear, scientists willing to champion materialism as nothing less than the plain evidence of their senses often treat its denial as a logical impossibility, a contradiction in prospect to which they are singularly indifferent.

In careless conversation, Gödel asked for an account of biological complexity, but however careless the conversation, his question has an intellectual force than transcends its circumstances, if only because nothing suggests, or even plausibly hints, that it has in any way been satisfactorily answered. We have no reason to assume that complexity is conserved, but no reason to doubt it either, and whenever issues are pressed beyond the anecdotal, it quickly emerges that the conservation of complexity functions as a limit, one vacating schemes for the generation of complexity, at least when they are made precise.

In one very large respect, however, the discussion has had something of a toy-like quality. It has been largely static, a matter of adjusting certain structures to certain devices without much concern for much beyond their fit. Biology is preeminently the discipline in which living creatures pursue living ends, their lives marked by continually evolving

---

46. One is reminded of those superb lines by Gottfriend Benn: *Leben ist Bruckenschlagen über Ströme die vergehn.*

trajectories that carry them from what they wish and desire and need to what they achieve, attempt and gain. We know little about the analysis of such activities; beyond the superficial, we know nothing at all. And yet any serious study of complexity must deal with *functional* complexity, an assessment of biological organisms by reference to the tasks that they undertake.[47] Is sight a more complicated task for an organism than digestion, or thermoregulation? On what basis the determination, and what parameters are involved? How should the maintenance of long-range biological rhythms be compared with activities taking place on fractional time scales? What is the correct or even the pertinent mathematical space for the representation of the characteristic triplet of desire, undertaking and satisfaction?

Questions such as these are easy to come by and free for the asking. They tend to mask in a flurry the extent to which the relationship between structure and function is analytically obscure on even the simplest level. Does a biological structure, for example, obey a principle of compositionality, so that what the whole of what the structure does is a function of what its parts do? I have in general no idea. Nor does anyone else.

Theoretical computer science is a discipline confronting problems that are cognate if not common. "In programming linguistics," Peter Mosses writes, "as in the study of natural languages, 'syntax' is distinguished from 'semantics.'" He continues:

> The syntax of a programming language is concerned only with the *structure* of programs: whether programs are legal; the connections and relations between the symbols and phrases that occur in them. Semantics deals with what legal programs *mean*: the behavior they produce when executed by computers.[48]

This sense of semantics already goes far beyond the sense common in philosophy or even formal logic and, indeed, seems reminiscent in an

---

47. In his final interview in *La Récherche*, this was a point to which M. P. Schützenberger recurred again and again. The interview, together with comments from critics, is available on the Web.

48. Peter Mosses, "Denotational Semantics," in *Handbook of Theoretical Computer Science*, *op. cit.*, p. 577.

odd way of Wittgenstein's unhappy dictum that the meaning of a word lies with its use. To a limited extent, it makes sense to speak of a theory of computer behavior, one in which structure and function receive some measure of coordination. The larger project of amplifying that distinction so that it comes to cover biological structure and function remains incomplete, and for all anyone knows, incompleteable.

But until that distinction is clarified, the distinction between structural and functional complexity will remain obscure. And so long as that distinction remains obscure, what evidence can be forthcoming that any mechanism is in any sense an adequate answer to Gödel's question?[49]

49. I have in writing this paper benefited greatly from my correspondence with Arthur Cody. We have discussed every issue at length, agreeing in almost all respects, disagreeing in but a few. Our hope is to write at greater length on this topic.

WAS THERE A BIG BANG?

SCIENCE IS A CONGERIES OF GREAT QUESTS, AND COSMOLOGY IS THE grandest of the great quests. Taking as its province the universe as a whole, cosmology addresses the old, the ineradicable questions about space and time, nature and destiny. It is not a subject for the tame or the timid.

For the first half of the twentieth century, cosmology remained a discipline apart, as a clutch of talented but otherworldly physicists peeped inconclusively at a universe they could barely see: Albert Einstein, of course; the Dutch mathematician Willem de Sitter; the Belgian abbot Georges Lemaître; and the extraordinary Russian mathematician and meteorologist Aleksandr Friedmann, destined to die young, or so the story has it, from exposure to the elements suffered while soaring above Moscow in a weather balloon.

When in 1917 Einstein published his first estimates of the size and shape of the universe, telescopes could not penetrate the heavens beyond the Milky Way. Like a sailor endeavoring to measure the depth of the sea from the shore, astronomers lacked the means to probe the heavens further or to probe them in detail.

This has now changed. Information pours from the night skies, terrestrial telescopes hissing and clicking as they rotate to survey distant galaxies. Somewhere in space, the realigned Hubble telescope peers into the unpolluted depths. Physicists have pictures of the backside of beyond, and they appear to have overheard the cosmic cackle that accompanied the very crack of time, as nothingness gave way to light. The cosmologists have come into their own, handling the universe with an easy familiarity and writing book after book in which they explain in exuberant detail how the great things were done.

WHATEVER THE DREAMS WE dream, the existence of the universe has always seemed a riddle beyond reason, if only because our imagination is forever suspended between ideas of creation and timelessness. Many ancient myths depict the universe as the effect of some Great Cause. In the Babylonian epic *Enuma Elish*, existence is attributed to congress between "primordial Apsu, the begetter" and "Tiamat, she who bore them." Although the connection of cosmic and sexual energies is both familiar and disturbing, it is not congress between gods that is crucial to the myth but the idea that the universe came into being as the result of a Great Cause. And this idea is conveyed by the opening of Genesis as well: "In the beginning, God created the heavens and the earth."

These are words that express an authentically universal concept, one familiar to every culture. But while the concept of creation is common, it is also incoherent. "I venture to ask," the third-century Chinese sage Kuo Hsiang ventured to ask, "whether the Creator is or is not. If He is not, how can He create things? And if He is, then (being one of these things) He is incapable (without self-creation) of creating the mass of bodily forms." That this argument is simple is no reason to think it wrong.

If not creation by a Great Cause, what, then, of timelessness: a universe proceeding sedately from everlasting to everlasting? In Maori myth, the world parents who bring the cosmos into being arise themselves from *po*, a kind of antecedent gruel or stuff, so that the universe appears as an episode in an infinitely extended drama. Some variant of this idea is also universal, a place in every culture where the weary mind takes refuge. Yet if timelessness offers an escape from the paradoxes of creation, the escape can easily seem an evasion. An everlasting universe is itself an object requiring explanation: if it is unprofitable to ask *when* it arose, one nonetheless wonders why the damn thing should be there at all.

It is the remarkable claim of contemporary cosmology to have broken, and broken decisively, the restless movement of the mind as it passes from theory to theory and from myth to myth. "Incomplete though it may be," one physicist has written, "the scope of modern scientific understanding of the cosmos is truly dazzling." This is not hyperbole. It is an assessment widely shared among physicists, and thus the standard by which their claims must be judged.

The universe, cosmologists now affirm, came into existence as the expression of an explosion, the cosmos shaking itself into existence from the bang of an initial singularity. It is tempting to think of the event in humanly comprehensible terms—a *gigantic* explosion or a *stupendous* eruption, as if, popcorn in hand, we were watching the show from far away. But this is absurd. The Big Bang was not an event taking place at a time or in a place. Space and time were themselves created by the Big Bang, the measure along with the measured.

As far as most physicists are concerned, the Big Bang is now a part of the structure of serene indubitability created by modern physics, an event undeniable as the volcanic explosion at Krakatoa. From time to time, it is true, the astrophysical journals report the failure of observation to confirm the grand design. It hardly matters. The physicists have not only persuaded themselves of the merits of Big Bang cosmology, they have persuaded everyone else as well. The Big Bang has come to signify virtually a universal creed; men and women who know nothing of cosmology are convinced that the rumble of creation lies within reach of their collective memory.

LOOKING AT A FEW shards of pottery on the desert floor, the archeologist is capable of conjuring up the hanging gardens of the past, the smell of myrrh and honey in the air. His is an act of intellectual reconstruction, one made poignant by the fact that the civilization from which the artifacts spring lies forever beyond the reach of anything but remembrance and the imagination. Cosmology on the grand scale is another form of archeology; the history of the cosmos reveals itself in layers, like the strata of an ancient city.

The world of human artifacts makes sense against the assumption of a continuous human culture. The universe is something else: an old, eerie place with no continuous culture available to enable us to make sense of what we see. It is the hypothesis that the universe is *expanding* that has given cosmologists a unique degree of confidence as they climb down the cliffs of time.

A universe that is expanding is a universe with a clear path into the past. If things are now far apart, they must at one point have been close together; and if things were once close together, they must at one point have been *hotter* than they are now, the contraction of space acting to compress its constituents like a vise, and so increase their energy. The retreat into the past ends at an initial singularity, a state in which material particles are at *no* distance from each other and the temperature, density, and curvature of the universe are infinite.

The cosmic archeologist may now be observed crawling back up the cliffs of time he so recently descended. During the first 10^{-43} seconds after the Big Bang—10^{-43} is one over a ten followed by 42 zeros—both matter and radiation fill the void. A reign of fluid interchange obtains, with particles of matter and antimatter exchanging identities. As the primitive goo of the cosmos—what the physicist George Gamow called *ylem*, the primordial stuff—continues to expand, it continues to cool. Neutrinos, photons, electrons, positrons, neutrons, and protons agitate themselves throughout space. With the temperature dropping, the neutrinos decamp for parts unknown.

At roughly one-and-one-half seconds after the Big Bang, protons and neutrons lose the ability to exchange identities, and the ratio of neutrons to protons in the universe freezes itself at one to six. Three-and-one-half seconds later, the equilibrium between electrons and positrons collapses, and the positrons follow the neutrinos into the void.

Three minutes pass. The era of nucleosynthesis begins thereafter. Those neutrons that during the freeze-out found themselves bound to the world's vagrant protons now take up an identity as a form of helium.

Other elements wait patiently for the stars to be born so that they may be cooked in their interiors.

The universe continues to expand, pulse, glow, throb, and moan for 400,000 years more, passing insensibly from a place where radiation predominates to an arena where matter has taken charge and is in command. The temperature is now 4,000 degrees Kelvin. The great era of *recombination* is at hand, a burst of cosmic creativity recorded in the walls of time. Free electrons and protons form hydrogen. The interaction between matter and radiation changes dramatically.

Until recombination, photons found themselves trapped within a cosmic pinball machine, ricocheting from one free charged particle to another, the cosmos frustratingly opaque because frustratingly dense. But hydrogen binds the cosmic debris, and for the first time, light streams from one side of creation to the other. The early universe fills with low-temperature blackbody radiation, the stuff destined to be observed fifteen billion years later in Princeton, New Jersey, as cosmic background radiation (CBR).

The separation of light and matter allows the galaxies to form, gravity binding the drifting dust in space. At last, the universe fills with matter, the stars settling into the sky, the far-flung suns radiating energy, the galaxies spreading themselves throughout the heavens. On the earth that has been newly made, living things shamble out of the warm oceans, the cosmic archeologist himself finally clambering over the lip of time to survey the scene and take notes on all that has occurred.

Such is the standard version of hot Big Bang cosmology—"hot" in contrast to scenarios in which the universe is cold, and "Big Bang" in contrast to various steady-state cosmologies in which nothing ever begins and nothing ever quite ends. It may seem that this archeological scenario leaves unanswered the question of *how* the show started and merely describes the consequences of some Great Cause that it cannot specify and does not comprehend. But really the question of how the show started answers itself: before the Big Bang there was nothing. *Darkness was upon the face of the deep.*

NOTWITHSTANDING THE INVESTMENT MADE by the scientific community and the general public in contemporary cosmology, a suspicion lingers that matters do not sum up as they should. Cosmologists write as if they are quite certain of the Big Bang, yet, within the last decade, they have found it necessary to augment the standard view by means of various new theories. These schemes are meant to solve problems that cosmologists were never at pains to acknowledge, so that today they are somewhat in the position of a physician reporting both that his patient has not been ill and that he has been successfully revived.

The details are instructive. It is often said, for example, that the physicists Arno Penzias and Robert Wilson *observed* the remnants of the Big Bang when in 1962 they detected, by means of a hum in their equipment, a signal in the night sky they could not explain. This is not quite correct. What Penzias and Wilson observed was simply the same irritating and ineradicable noise that has been a feature of every electrical appliance I have ever owned. What theoreticians like Robert Dicke *inferred* was something else: a connection between that cackle and the cosmic background radiation released into the universe after the era of recombination.

The cosmic hum is real enough, and so, too, is the fact that the universe is bathed in background radiation. But the era of recombination is a shimmer by the doors of theory, something known indirectly and only by means of its effects. And therein lies a puzzle. Although Big Bang cosmology does predict that the universe should be bathed in a milky film of radiation, it makes no predictions about the uniformity of its temperature. Yet, looking at the sky in every direction, cosmologists have discovered that the CBR has the same temperature, to an accuracy of one part in 100,000.

Why should this be so? CBR filled the universe some 400,000 years after the Big Bang; if its temperature thereafter is utterly and entirely the same, some physical agency must have brought this about. But by the time of recombination, the Big Bang had blown up the universe to a diameter of 90,000,000 light years. A physical signal—light beam, say-sent hustling into the cosmos at Time Zero would, a mere 400,000 years later, be hustling still; by far the greater part of the universe would be untouched by its radiance, and so uninfluenced by the news that it carried. Since, by Einstein's theory of special relativity, nothing can travel faster than light itself, it follows that no physical agency would have had time enough to establish the homogeneity of the CBR, which appears in Big Bang cosmology as an arbitrary feature of the early universe, something that must be assumed and is not explained.

Theories of *inflation* now make a useful appearance. Their animating idea represents a contribution to cosmology from particle physics— a rare example of intellectual lend-lease, and evidence that disciplines dealing with the smallest of objects may be relevant to disciplines dealing with the largest. Within what is now known as the Standard Model, the familiar arrangement in which elementary particles are moved by various forces gives way to a mathematically more general scheme in which fields replace both particles and forces as essential theoretical structures. A field is an expanse or expression of space, something like a surface in two dimensions, or the atmosphere in three, or space and time in four. The Standard Model in particle physics consists of a dozen or so fields, exchanging energy with one another and subsidizing particles by means of the energy they contain.

It is the Higgs field that originally came to play a novel role in Big Bang cosmology. Named after Peter Higgs, the Scottish physicist who charmed it into existence, the Higgs field is purely a conjectured object. Its cosmological potential was first noted by a young American physicist, Alan Guth.

Fields are structures that carry latent energy even under conditions in which the space they control is to all intents and purposes empty. The natural and stable state of the Higgs field is one in which its latent energy is at a minimum. Such is its true vacuum state, the word "vacuum" indicating that the field is empty, and the word "true" that the field is in its lowest energy configuration. But under certain physically possible circumstances, the Higgs field can find itself adventitiously trapped in a *false* vacuum state, a condition in which, like a spring, it is loaded with potential energy. It is thus, Guth conjectured, that the Higgs field might have found itself fluttering about the early universe, energetically throbbing and dying to be of use.

The wish is father to the act. The energy within the Higgs field is repulsive: it pushes things apart. When released, it contributes a massive jolt to the process of cosmic expansion already under way. The universe very quickly doubles in size. Space and time stretch themselves out. Particles zoom from one another. If the ordinary course of cosmic expansion is linear, inflationary expansion is exponential, like the gaunt, hollow-eyed guest gobbling the hors d'oeuvres—and everything else— at a previously decorous cocktail party. Only as the Higgs field tumbles down to its true vacuum state does inflation come to a halt, and the ordinary course of the Big Bang resume.

The mechanism of inflation, cosmologists cheerfully admit, is rather like one of those Rube Goldberg contraptions in which a door is made to open by means of a sequence that includes a flashing neon light, an insulting message in a bottle, a prizefighter wearing patent-leather shoes, and a boa constrictor with an aversion to milk. Nonetheless, they add, inflation provides a natural and plausible explanation for the fact that the CBR is uniform in temperature. If the universe under standard Big Bang cosmology is too large to allow a coordinating physical signal to reach every part of the CBR, then one redemptive idea is to cast around for a universe smaller than the standard one. This, inflation provides. Within an inflationary universe, the CBR owes its uniform temperature to the fact that it has been thoroughly mixed. At the end of the era of recombination, the CBR then surges through space like pre-warmed soup.

Inflation is an idea that has gripped the community of cosmologists. Whether it has advanced their scientific agenda is another question. As we have seen, standard Big Bang cosmology requires that features of the early universe such as the temperature of the CBR be set arbitrarily. This has seemed intellectually repugnant to many physicists: the goal of science is to reduce the arbitrariness of description. But inflationary cosmology has arbitrary features of its own, which, displaced from one corner of the theory, have a habit of popping up in another. "The need for fine-tuning of the universe," the physicist David Lindley observed of Guth's proposal, "has been obviated by fine-tuning the Higgs mechanism instead." So it has.

Soon after its introduction, Guth's model of inflation required adjustment. The Higgs field has been replaced by scalar fields, which, as Guth admits, "in many cases serve no function other than the driving of inflation." These fields must be carefully chosen if they are to do their work, a fact that Guth again honestly acknowledges: "Their nature cannot be deduced from known physics, and their detailed properties have to be hypothesized."

In an interesting example of inflationary theory self-applied, inflationary fields have undergone an exponential increase of their own. Beyond mere inflation, the sort of thing that with great heartiness simply blows the universe up, there is chaotic inflation and even "eternal inflation," both of them the creations of the cosmologist Andrei Linde. Almost all cosmologists have a favored scheme; when not advancing their own, they occupy themselves enumerating the deficiencies of the others.

Streaming in from space, light reaches the earth like a river rich in information, the stars in the sky having inscribed strange and secret messages on its undulations. The universe is very large, light has always whispered; the nearest galaxy to our own—Andromeda—is more than two million light years away. But the universe has also seemed relatively static, and this, too, light suggests, the stars appearing where they have always appeared, the familiar dogs and bears and girdled archers of the constellations making their appointed rounds in the sky each night.

More than anything else, it is this impression that Big Bang cosmology rejects. The cool gray universe, current dogma holds, is a place of extraordinary violence, the galaxies receding from one another, the skin of creation stretching at every spot in space, the whole colossal structure blasting apart with terrible force. And this message is inscribed in light as well.

In one of its incarnations, light represents an undulation of the electromagnetic field; its source is the excitable atom itself, with electrons bouncing from one orbit to another and releasing energy as a result. Each atom has a spectral signature, a distinctive electromagnetic frequency. The light that streams in from space thus reveals something about the composition of the galaxies from which it was sent.

In the 1920s, the characteristic signature of hydrogen was detected in various far-flung galaxies. And then an odd discovery was made. Examining a very small sample of twenty or so galaxies, the American astronomer V. M. Slipher observed that the frequency of the hydrogen they sent into space was shifted to the red portion of the spectrum. It was an extraordinary observation, achieved by means of primitive equipment. Using a far more sophisticated telescope, Edwin Hubble made the same discovery in the late 1920s after Slipher had (foolishly) turned his attention elsewhere.

The galactic redshift, Hubble realized, was an exceptionally vivid cosmic clue, a bit of evidence from far away and long ago, and like all clues its value lay in the questions it prompted. Why should galactic light be shifted to the red and not the blue portions of the spectrum? Why, for that matter, should it be shifted at all?

An invigorating stab in the dark now followed.

The pitch of a siren is altered as a police car disappears down the street, the sound waves carrying the noise stretched by the speed of the car itself. This is the familiar Doppler effect. Something similar, Hubble conjectured, might explain the redshift of the galaxies, with the distortions in their spectral signature arising as a reflection of their recessional velocity as they disappeared into the depths.

Observations and inferences resolved themselves into a quantitative relationship. The redshift of a galaxy, cosmologists affirm, and so its recessional velocity, is proportional to its distance and inversely proportional to its apparent brightness or flux. The relationship is known as Hubble's law, even though Hubble himself regarded the facts at his disposal with skepticism.

Hubble's law anchors Big Bang cosmology to the real world. Many astronomers have persuaded themselves that the law represents an observation, almost as if, peering through his telescope, Hubble had noticed the galaxies zooming off into the far distance. This is nonsense. Hubble's law consolidates a number of very plausible intellectual steps. The light streaming in from space is relieved of its secrets by means of ordinary and familiar facts, but even after the facts are admitted into evidence, the relationship among the redshift of the galaxies, their recessional velocity, and their distance represents a complicated inference, an intellectual leap.

THE BIG BANG RESTS on the hypothesis that the universe is expanding, and in the end the plausibility of its claims will depend on whether the universe *is* expanding. Astronomers can indeed point to places in the sky where the redshift of the galaxies appears to be a linear function of their distance. But in astrophysics, as in evolutionary biology, it is failure rather than success that is of significance. The astrophysical literature contains interesting and disturbing evidence that the linear relationship at the heart of Hubble's law by no means describes the facts fully.

At the end of World War II, astronomers discovered places in the sky where charged particles moving in a magnetic field sent out strong signals in the radio portion of the spectrum. Twenty years later, Alan Sandage and Thomas Mathews identified the source of such signals with optically discernible points in space. These are the quasars—*quasi stellar radio sources.*

Quasars have played a singular role in astrophysics. In the mid-1960s, Maarten Schmidt discovered that their spectral lines were shifted massively to the red. If Hubble's law were correct, quasars should be impossibly far away, hurtling themselves into oblivion at the far edge of space and time. But for more than a decade, the American astronomer Halton Arp has drawn the attention of the astronomical community to places in the sky where the expected relationship between redshift and distance simply fails. Embarrassingly enough, many quasars seem bound to nearby galaxies. The results are in plain sight: there on the photographic plate is the smudged record of a galaxy, and there next to it is a quasar, the points of light lined up and looking for all the world as if they were equally luminous.

These observations do not comport with standard Big Bang cosmology. If quasars have very large redshifts, they must (according to Hubble's law) be very far away; if they *seem* nearby, then either they must be fantastically luminous or their redshift has not been derived from their velocity. The tight tidy series of inferences that has gone into Big Bang cosmology, like leverage in commodity trading, works beautifully in reverse, physicists like speculators finding their expectations canceled by the very processes they had hoped to exploit.

Acknowledging the difficulty, some theoreticians have proposed that quasars have been caught in the process of evolution. Others have scrupled at Arp's statistics. Still others have claimed that his samples are too small, although they have claimed this for every sample presented and will no doubt continue to claim this when the samples number in the billions. But whatever the excuses, a great many cosmologists recognize that quasars mark a point where the otherwise silky surface of cosmological evidence encounters a snag.

WITHIN ANY SCIENTIFIC DISCIPLINE, bad news must come in battalions before it is taken seriously. Cosmologists can point to any number of cases in which disconcerting evidence has resolved itself in their favor; a decision to regard the quasars with a watchful indifference is not necessarily irrational. The galaxies are another matter. They are central to Hubble's law; it is within the context of galactic observation that the crucial observational evidence for the Big Bang must be found or forged.

The battalions now begin to fill. The American mathematician I. E. Segal and his associates have studied the evidence for galactic recessional velocity over the course of twenty years, with results that are sharply at odds with predictions of Big Bang cosmology. Segal is a distinguished, indeed a great mathematician, one of the creators of modern function theory and a member of the National Academy of Sciences. He has incurred the indignation of the astrophysical community by suggesting broadly that their standards of statistical rigor would shame a sociologist. Big Bang cosmology, he writes, "owes its acceptance as a physical principle primarily to the uncritical and premature representation of [the redshift-distance relationship] as an empirical fact…. Observed discrepancies… have been resolved by a pyramid of exculpatory assumptions, which are inherently incapable of noncircular substantiation."

These are strong words of remonstration, but they are not implausible. Having constructed an elaborate scientific orthodoxy, cosmologists have acquired a vested interest in its defense. The astrophysicists J. G. Hoessell, J. E. Gunn, and T. X. Thuan, for example, report with satisfaction that within the structures described by G. O. Abell's *Catalog of Bright Cluster Galaxies* (1958), prediction and observation cohere perfectly to support Hubble's law. Abell's catalog is a standard astronomical resource, used by cosmologists everywhere—but it is useless as evidence for Hubble's law. "In determining whether a cluster meets selection criterion," Abell affirms, "it was assumed that their redshifts were proportional to their distance." If this is what Abell *assumed*, there is little point in asking what conclusions he *derived*.

The fact that the evidence in favor of Hubble's law may be biased does not mean that it is untrue; bias may suggest nothing more than a methodological flaw. But Segal is persuaded that when the evidence is soberly considered, it *does* contravene accepted doctrine, statistical sloppiness functioning, as it so often does, simply to conceal the facts.

A statistical inference is compelling only if the samples upon which it rests are objectively compelling. Objectivity, in turn, requires that the process of sampling be both reasonably complete and unbiased. Segal and his colleagues have taken pains to study samples that within the limits of observation are both. Their most recent study contains a detailed parallel analysis of Hubble's law across four wave bands, one that essentially surveys all stellar objects within each band. The analysis is based on new data drawn from the G. de Vaucoleurs survey of bright cluster galaxies, which includes more than 10,000 galaxies. Hubble's own analysis, it is worthwhile to recall, was limited to twenty galaxies.

The results of their analysis are disturbing. The linear relationship that Hubble saw, Segal and his collaborators cannot see and have not found. Rather, the relationship between redshift and flux or apparent brightness that they have studied in a large number of complete samples satisfies a quadratic law, the redshift varying as the square of apparent brightness. "By normal standards of scientific due process," Segal writes, "the results of [Big Bang] cosmology are illusory."

Cosmologists have dismissed Segal's claims with a great snort of indignation. But the discrepancy from Big Bang cosmology that they reveal is hardly trivial. Like evolutionary biologists, cosmologists are often persuaded that they are in command of a structure intellectually powerful enough to accommodate gross discrepancies in the data. This is a dangerous and deluded attitude. Hubble's law embodies a general hypothesis of Big Bang cosmology—namely, that the universe is expanding—and while the law cannot be established by observation, observation *can* establish that it may be false. A statistically responsible body of contravening evidence has revealed something more than an incidental defect. Indifference to its implications amounts to a decision to place Big Bang cosmology beyond rational inquiry.

WHATEVER THE FACTS MAY be, the Big Bang is also an event informed by the majesty of a great physical theory. Einstein published the equations for general relativity in 1915, and more than 80 years later, general relativity remains the only theoretical instrument remotely adequate to the representation of the universe as a whole.

General relativity is first and fundamentally an account of gravity, the force that pulls ballerinas to the ground and that fixes the planets in elliptical orbits around the sun. At the beginning of the scientific era, Isaac Newton described a universe in which space and time are absolute.

The measured beating of a great clock is heard, and it is heard everywhere at once. Particles move within the unchanging vault of space. Material objects attract one another with a force proportional to their mass and inversely proportional to the square of the distance between them.

A metaphysical reorganization is required before Newton's caterpillar can emerge as Einstein's butterfly. The elements of general relativity are physical *processes*, a word signifying something that starts at one time and at one place and that ends at another time and another place, and so crawls along a continuum whose intrinsic structure has four dimensions. Within the arena of these physical processes, the solid structures of the Newtonian universe undergo a dissolution. The great vault of space and the uniformly beating heart dwindle and then disappear: *this* universe is one in which space and time have fused themselves into a single entity, and *its* heart is an ever-changing but reciprocating relationship between space-time and matter. Material objects direct the space and time that surround them to curve, much as a bowling ball deforms the mattress on which it rests; the curvature of spacetime determines the path undertaken by physical processes, much as an ant crawling on that mattress must travel a curved path to get where it is going.

Newtonian gravity acts at a distance and as a force, the very bowels of the earth reaching to enfold an object and pull it down. But while Newton was able quantitatively to describe how gravity acts, he was unable to say why it acts at all, the aching attraction of matter for matter having no other explanation than the fact that it is so. General relativity provides an explanation of gravity in terms of the curvature of space and time. No forces are involved, and none is invoked, but gravity nonetheless emerges in this universe as a natural expression of the way the cosmos is constructed.

Freely moving objects, Einstein assumed, follow a path covering the shortest distance between points in space and time. Within the ambit of a large material object, the shortest distance between such points is curved. Ballerinas accelerate toward the center of the earth after being thrown upward by their partners because acceleration is required by the geometry in which they are embedded. In this fashion, gravity disappears as a force but remains as a fact.

If the analysis of gravity is at the center of general relativity, the intellectual tools responsible for its analysis—the equations that describe the ever-changing relationship between curvature and material objects—are responsible as well for its local character. For many years, the most precise and most interesting tests of the theory were conducted within the narrow confines of our own solar system. Cosmology, however, is a *global* study, one in which the universe itself is the object of contemplation, and not any of its parts. The conveyance from the local structure of the universe (the sun, the solar system) to the universe as a whole must be negotiated by a daring series of inferences.

In describing matter on a cosmic scale, cosmologists strip the stars and planets, the great galaxies and the bright bursting supernovae, of their uniqueness as places and things and replace them with an imaginary distribution: the matter of the universe is depicted as a great but uniform and homogeneous cloud covering the cosmos equitably in all its secret places. Cosmologists make this assumption because they must. There is no way to deal with the universe object by object; the equations would be inscrutable, impossible to solve. But however useful the assumption of homogeneity may be mathematically, it is false in the straightforward sense that the distribution of matter in the universe is not homogeneous at all.

Having simplified the contents of the universe, the cosmologist must take care as well, and for the same reason, to strip from the matter that remains any suggestion of particularity or preference in place. The universe, he must assume, is isotropic. It has no center whatsoever, no place toward which things tend, and no special direction or axis of coordination. The thing looks much the same wherever it is observed.

The twin assumptions that the universe is homogeneous and isotropic are not ancillary but indispensable to the hypothesis of an expanding universe; without them, no conclusion can mathematically be forthcoming. Together, these two assumptions are like the figured bass needed to chant what in Tibetan is known as *Monlam Chemno*, the great prayer to the cosmos.

AN EQUATION DRAWS THE noose of an identity between two or more items. The field equations of general relativity draw that noose between curvature (the metric structure of the universe) and matter (its stress-energy tensor). But it is one thing to specify an equation, and quite another to solve it.

The mathematician in him having taken command, Einstein endeavored in 1917 to provide cosmological solutions for the field equations of his own theory. He struggled with increasing vexation. The models he was able to derive indicated that the radius of the universe was either expanding or contracting; it was a conclusion that offended his aesthetic sensibilities. By adding a parameter to his equations—the so-called cosmological constant—he was in the end able to discover a static solution, one that revealed a universe finite in extent, but unbounded, like the surface of a sphere. This static solution has a habit of dropping from the view of cosmologists, who routinely aver that Einstein's theory of general relativity uniquely specifies an expanding universe. Not so.

Einstein had hoped that the equations of general relativity would determine a single world model, or cosmic blueprint. In this he was destined to be disappointed. Months after he discovered one solution of the field equations, Willem de Sitter discovered another. In de Sitter's universe, there is no matter whatsoever, the place looking rather like a dance hall in which the music can be heard but no dancers seen, radiation filling the empty spaces and ricocheting from one end of creation to another.

In the 1920s, both Aleksandr Friedmann and Georges Lemaître discovered the solutions to the field equations that have dominated cosmology ever since, their work coming to amalgamate itself into a single denomination as Friedmann-Lemaître (FL) cosmology. Gone from their models is the cosmological constant (although it is resurrected in various inflation scenarios), and as a result the universe breathes voluptuously, its radius expanding or contracting with time.

FL cosmology does not assign to the universe a unique geometrical identity, or specify its fate forever; general relativity is mathematically compatible with a number of different physical scenarios. Like the surface of a sphere, the universe may well be closed, the whole thing falling back on itself at the end of time. A certain symmetry prevails, the life of things and all the drama of creation caught between two singularities as the universe traces a trajectory in which its initial effervescent explosion is followed by a subsequent enervating contraction.

Or yet again, the universe may well be open, space and time forever gushing into the void but with ever-decreasing intensity, like an athlete panting in shallow breaths. Such a universe is purely a mathematician's world, one seen only by the exercise of certain obscure mental muscles, and regarded by physicists (and everyone else) with glum distaste.

Or, finally, the universe may be one that occupies the Euclidean space of high-school textbooks and intuition alike, balanced precariously but balanced forever on a knife's edge between expansion and contraction.

If its predictive capacities seem unstable, FL cosmology has other peculiarities as well. Whatever the specific form its solutions take, they are alike in assigning dynamic properties to the universe as a whole. The classical distinction between the eternal vault of space and time and its entirely perishable contents has disappeared. The universe in FL cosmology is *itself* bound to the wheel of being, with space and time no more permanent than water and air. Light blazes, the show commences, and like some magnificent but mysterious organism, the universe expands with an exuberant rush of energy and floods nothingness with the seeds of being. In this fundamental respect, FL cosmology breaks both with tradition and with common sense.

THE INTERPRETATION OF A physical theory partakes of a dark art, one in which mathematical concepts are ceded dominion over the physical world. In practicing this art, the mathematician, like the necromancer that he is, is always liable to the temptation of confusing the structures over which he presides with things in the real world.

On the assumption—on the *assumption*—that the universe is expanding, it is irresistibly tempting to run time backward until the far-flung debris of the cosmos collects itself back into a smaller and smaller area. It seems evident, though, that this process of contraction and collapse may be continued only so far. An apple may be divided in halves, and then thirds, and then quarters, but even though the mathematical sequence of one half, one third, one fourth, and so on contains infinitely many terms, and converges ultimately to zero, the apple itself may be divided only finitely many times.

This straightforward point has been the source of grave confusion. "The universe," the astronomer Joseph Silk writes, "began at time zero in a state of infinite density." It is there that (by definition) a *singularity* may be found. "Of course," Silk adds, "the phrase 'a state of infinite density' is completely unacceptable as a physical description of the universe.... An infinitely dense universe [is] where the laws of physics, and even space and time, break down."

These are not words that inspire confidence. Does the phrase "a state of infinite density" describe a physical state of affairs or not? If it does, the description is uninformative by virtue of being "completely unacceptable." If it does not, the description is uninformative by virtue of being completely irrelevant. But if the description is either unacceptable or irrelevant, what reason is there to believe that the universe began in an initial singularity? Absent an initial singularity, what reason is there to believe that the universe began at all?

When prominent cosmologists tie themselves in knots, charity tends to assign the blame to the medium in which they are navigating—books for a general audience—rather than the message they are conveying. But when it comes to the singularities, the knots form in *every* medium, evidence that the message is at fault and not the other way around. Cosmologists often claim that the mathematicians among them have demonstrated what they themselves may be unable clearly to express.

In a passage that is typical, the astrophysicist Kip Thorne writes that "[Stephen] Hawking and [Roger] Penrose in 1970 proved—without any idealizing assumptions—that our universe must have had a space-time singularity at the beginning of its Big Bang expansion." But while it is true that Hawking and Penrose proved something, what they demonstrated remains within the gerbil wheel of mathematics; any additional inference requires a connection that the mathematician is not in a position to provide.

The concept of a singularity belongs *essentially* to mathematics. Singularities are not experimentally accessible objects. They cannot be weighed, measured, assessed, replicated, balanced, or seen by any modality of the senses. Within certain mathematical contexts, the concept has real content. An ordinary curve goes up one side of the blackboard and down the other; it *changes* its direction at a singular point. There are singularities within the calculus, and singularities in complex function theory where imaginary numbers loiter, and singularities in the space of smooth maps. There are singularities within general relativity as well, but the term covers a variety of cases, and the singularities within general relativity are distinctly odd.

In most mathematical theories a natural distinction is drawn between a figure and its *background*: a curve arcs within the broader ambit of an enveloping space, a mapping is easily distinguished from the spaces it connects. Typically, it is the figure that admits of a singularity: the curve changes its direction or the mapping breaks down, while the background stays the same. But in general relativity, it is the background that suffers a singularity, the very fabric of space and time giving way with a rip as curvature zooms off to infinity and space and time contort themselves. For the purposes of describing such singularities, the usual mathematical techniques are unavailing.

That having been said, here is what Hawking and Penrose brought under the control of a mathematical demonstration. The setting is FL cosmology and *only* FL cosmology. There are three kinds of universe to consider, and innumerably many species within each type. Those that are open and forever gushing into the void are called hyperbolic. Within almost all of those hyperbolic universes, almost all processes begin at a point in the past. Within the two types of universe that remain, there is bound to be at least one process that has begun somewhere in the past.

Despite the tics—"almost all," "at least one"—the Penrose-Hawking theorems do indeed demonstrate that some universes begin in an initial singularity. But the light thrown by the Penrose-Hawking theorems flickers over a mathematical theory and so a mathematical universe. The universe that we inhabit is a physical system. Nothing but grief can come of confusing the one for the other. FL cosmology requires the existence of space-time singularities, but there is nothing in the Penrose-Hawking theorems to suggest that a space-time singularity corresponds to an explosion, or marks the beginning of an expansion, or describes an accessible portion of space and time, or connects itself to any physical state of affairs whatsoever. Mathematical concepts achieve physical significance only when the theories in which they are embedded are confirmed by experience. If a space-time singularity is not a physical event, no such confirmation can logically be forthcoming. With the argument rolled backward, it follows that if these mathematical theories are not confirmed by experience, then neither have they achieved any physical significance.

It is Einstein who expressed the most reasonable and deeply thought views on this matter. "One may not therefore assume the validity of the [field] equation for very high density of the field and of matter," he remarked, "and one may not conclude that the beginning of the expansion must mean a singularity in the mathematical sense. All we have to realize is that the equations may not be continued over such regions."

The sharp, clean, bracing light that the Big Bang was to have thrown on the very origins of space and time lapses when it is most needed. The relevant equations of general relativity fall silent at precisely the moment we most wish they would speak.

LIKE SO MANY HAUNTING human stories, the scientific story of the Big Bang is circular in the progression of its ideas and circular thus in its deepest nature. Cosmologists have routinely assumed that the universe is expanding because they have been persuaded of FL cosmology; and they have been persuaded of FL cosmology because they have routinely assumed that the universe is expanding. The pattern would be intellectually convenient if it were intellectually compelling.

If the evidence in favor of Big Bang cosmology is more suspect than generally imagined, its defects are far stronger than generally credited. Whatever else it may be, the universe is a bright, noisy, energetic place. There are monstrously large galaxies in the skies and countlessly many suns burning with fierce thermonuclear fires. Black holes are said to loiter here and there, sucking in matter and light and releasing it slowly in the form of radiation. Whence the energy for the show, the place where accounts are settled? The principles of nineteenth-century physics require that, in one way or another, accounts *must* be settled. Energy is neither created nor destroyed.

Hot Big Bang cosmology appears to be in violation of the first law of thermodynamics. The global energy needed to run the universe has come from nowhere, and to nowhere it apparently goes as the universe loses energy by cooling itself.

This contravention of thermodynamics expresses, in physical form, a general philosophical anxiety. Having brought space and time into existence, along with everything else, the Big Bang itself remains outside any causal scheme. The creation of the universe remains unexplained by any force, field, power, potency, influence, or instrumentality known to physics—or to man. The whole vast imposing structure organizes itself from absolutely nothing.

This is not simply difficult to grasp. It is incomprehensible.

Physicists, no less than anyone else, are uneasy with the idea that the universe simply popped into existence, with space and time "suddenly switching themselves on." The image of a light switch comes from Paul Davies, who uses it to express a miracle without quite recognizing that it embodies a contradiction. A universe that has *suddenly* switched itself on has accomplished something within time; and yet the Big Bang is supposed to have brought space and time into existence.

Having entered a dark logical defile, physicists often find it difficult to withdraw. Thus, Alan Guth writes in pleased astonishment that the universe really did arise from "essentially... nothing at all": as it happens, a false vacuum patch "10^{-26} centimeters in diameter" and "10^{-32} solar masses." It would appear, then, that "essentially nothing" has both spatial extension and mass. While these facts may strike Guth as inconspicuous, others may suspect that nothingness, like death, is not a matter that admits of degrees.

The attempt to discover some primordial stuff that can be described both as nothing and as something recalls the Maori contemplating the manifold mysteries of *po*. This apparently gives Stephen Hawking pause. "To ask what happened before the universe began," he has written, "is like asking for a point on the Earth at 91 degrees north latitude." We are on the inside of the great sphere of space and time, and while we can see to the boundaries, there is nothing beyond to see if only because there is nothing beyond. "Instead of talking about the universe being created, and maybe coming to an end," Hawking writes, "one should just say: the universe is."

Now this is a conclusion to which mystics have always given their assent; but having concluded that the universe just "is," cosmologists, one might think, would wish to know *why* it is. The question that Hawking wishes to evade disappears as a question in physics only to reappear as a question in philosophy; we find ourselves traveling in all the old familiar circles.

STANDING AT THE GATE of modern time, Isaac Newton forged the curious social pact by which rational men and women have lived ever since. The description of the physical world would be vouchsafed to a particular institution, that of mathematical physics; and it was to the physicists and not the priests, soothsayers, poets, politicians, novelists, generals, mystics, artists, astrologers, warlocks, wizards, or enchanters that society would look for judgments about the nature of the physical world. If knowledge is power, the physicists have, by this arrangement, been given an enormous privilege. But a social arrangement is among other things a contract: something is given, but something is expected as well. In exchange for their privilege, the physicists were to provide an account of the physical world at once penetrating, general, persuasive, and true.

Until recently, the great physicists have been scrupulous about honoring the terms of their contract. They have attempted with dignity to respect the distinction between what is known and what is not. Even quantum electrodynamics, the most successful theory ever devised, was described honestly by its founder, Richard Feynman, as resting on a number of unwholesome mathematical tricks.

This scrupulousness has lately been compromised. The result has been the calculated or careless erasure of the line separating disciplined physical inquiry from speculative metaphysics. Contemporary cosmologists feel free to say anything that pops into their heads. Unhappy examples are everywhere: absurd schemes to model time on the basis of the complex numbers, as in Stephen Hawking's *A Brief History of Time*; bizarre and ugly contraptions for cosmic inflation; universes multiplying beyond the reach of observation; white holes, black holes, worm holes, and naked singularities; theories of every stripe and variety, all of them uncorrected by any criticism beyond the trivial.

The physicists carry on endlessly because they can. Just recently, for example, Lee Smolin, a cosmologist at the University of Pennsylvania, has offered a Darwinian interpretation of cosmology, a theory of "cosmological natural selection." On Smolin's view, the Big Bang happened within a black hole; new universes are bubbling up all the time, each emerging from its own black hole and each provided with its own set of physical laws, so that the very concept of a law of nature is shown to be a part of the mutability of things.

There is, needless to say, no evidence whatsoever in favor of this preposterous theory. The universes that are bubbling up are unobservable. So, too, are the universes that have bubbled up and those that will bubble up in the future. Smolin's theories cannot be confirmed by experience. Or by anything else. What law of nature could reveal that the laws of nature are contingent? Yet the fact that when Smolin's theory is self-applied it self-destructs has not prevented physicists like Alan Guth, Roger Penrose, and Martin Rees from circumspectly applauding the effort nonetheless.

A scientific crisis has historically been the excuse to which scientists have appealed for the exculpation of damaged doctrines. Smolin is no exception. "We are living," he writes, "through a period of scientific crisis." Ordinary men and women may well scruple at the idea that cosmology is in crisis because cosmologists, deep down, have run out of interesting things to say, but in his general suspicions Smolin is no doubt correct. What we are discovering is that many areas of the universe are apparently protected from our scrutiny, like sensitive files sealed from view by powerful encryption codes. However painful, the discovery should hardly be unexpected. Beyond every act of understanding, there is an abyss. Like Darwin's theory of evolution, Big Bang cosmology has undergone that curious social process in which a scientific theory is promoted to a secular myth. The two theories serve as points of certainty in an intellectual culture that is otherwise disposed to give the benefit of the doubt to doubt itself. It is within the mirror of these myths that we have come to see ourselves. But if the promotion of theory into myth satisfies one human agenda, it violates another. Myths are quite typically false, and science is concerned with truth. Human beings, it would seem, may make scientific theories or they may make myths, but with respect to the same aspects of experience, they cannot quite do both.

2001

WHAT BRINGS
A WORLD INTO BEING?

Since their inception in the seventeenth century, the modern sciences have been given over to a majestic vision: there is nothing in nature but atoms and the void. This is hardly a new thought, of course; in the ancient world, it received its most memorable expression in Lucretius' *On the Nature of Things*. But it has been given contemporary resonance in theories—like general relativity and quantum mechanics—of terrifying (and inexplicable) power. If brought to a successful conclusion, the trajectory of this search would yield a single theory that would subsume all other theories and, in its scope and purity, would be our only necessary intellectual edifice.

In science, as in politics, the imperial destiny drives hard. If the effort to subordinate all aspects of experience to a single set of *laws* has often proved inconclusive, the scientific enterprise has also been involved in the search for universal *ideas*. One such idea is information.

Like energy, indeed, information has become ubiquitous as a commodity and, like energy, inescapable as an idea. The thesis that the human mind is nothing more than an information-processing device is now widely regarded as a fact. "Viewed at the most abstract level," the science writer George Johnson remarked recently in the *New York Times*, "both brains and computers operate *the same way* by translating phenomena—sounds, images, and so forth—into a code that can be stored and manipulated" (emphasis added). More generally, the evolutionary biologist Richard Dawkins has argued that life is itself fundamentally a river of information, an idea that has in large part also motivated the successful effort to decipher the human genome. Information is even said to encompass the elementary particles. "All the quarks and electrons in the cosmic wilds," Johnson writes, "are exchanging information each time they interact."

These assertions convey a current of intellectual optimism that it would be foolish to dismiss. Surely an idea capable of engaging so many distinct experiences must be immensely attractive. But it seems only yesterday that other compelling ideas urged their claims: chaos and nonlinear dynamics, catastrophe theory, game theory, evolutionary entropy, and various notions of complexity and self-organization.

The history of science resembles a collection of ghosts remembering that once they too were gods. With respect to information, a note of caution may well be in order if only because a note of caution is always in order.

If INFORMATION CASTS A cold white light on the workings of the mind in general, it should certainly shed a little on the workings of language in particular. *Moby-Dick*.

The words and sentences of Herman Melville's, to take a suggestive example, have the power to bring a world into being. The beginning of the process is in plain sight. There are words on the printed page, and they make up a discrete, one-dimensional, linear progression. Discrete—there are no words between words (as there are fractions between fractions); one-dimensional—each word might well be specified by a single number; linear—as far as words go, it is one thing after another. The end of the process is in sight as well: a richly organized, continuous, three- (or four-) dimensional universe. Although that universe is imaginary, it is recognizably contiguous to our own.

Bringing a world into being is an act of creation. But bringing a world into being is also an activity that suggests, from the point of view of the sciences, that immemorial progression in which causes evoke various effects: connections achieved between material objects, or between the grand mathematical abstractions necessary to explain their behavior.

And therein lies a problem. Words are, indeed, material objects, or linked as abstractions to material objects. And as material objects, they have an inherent power to influence other material objects. But no informal account of what words do as material objects seems quite sufficient to explain what they do in provoking certain experiences and so in creating certain worlds.

In the case of *Moby-Dick*, the chemical composition of words on the printed page, their refractive index, their weight, their mass, and ultimately their nature as a swarm of elementary particles—all this surely plays *some* role in getting the reader sympathetically to see Captain Ahab and imaginatively suffer his fate. The relevant causal pathways pass from the printed ink to our eyes, a river of light then serving to staple the shape of various words to our tingling retinal nerves; thereafter our nervous system obligingly passes on those shapes in the form of various complicated electrical signals. This is completely a physical process, one that begins with physical causes and ends with physical effects.

And yet the experience of reading begins where those physical effects end. It is, after all, an experience, and the world that it reveals is imaginary. If purely physical causes are capable of creating imaginary worlds, it is not by means of any modality known to the physical sciences.

Just how *do* one set of discrete objects, subject to the constraints of a single dimension, give rise to a universe organized in completely different ways and according to completely different principles?

It is here that information makes its entrance. The human brain, the linguist Steven Pinker has argued in *How the Mind Works*, is a physical object existing among other physical objects. Ordinary causes in the world at large evoke their ordinary effects within the brain's complicated folds and creases. But the brain is, *also*, an information-processing device, an instrument designed by evolution for higher things.

It is the brain's capacity to process information that, writes Pinker, allows human beings to "see, think, feel, choose, and act." Reading is a special case of seeing, one in which information radiates from the printed page and thereafter transforms itself variously into various worlds.

So much for what information does—clearly, almost everything of interest. But what is it, and how does it manage to do what it does? Pinker's definition, although informal, is brisk and to the point. Information, he writes, "is a correlation between two things that is produced by a lawful process." Circles in a tree stump carry information about the tree's age; lines in the human face carry information about the injuries of time.

Words on the page also contain or express information, and as carriers of information they convey the stuff from one place to another, piggybacked, as it were, on a stream of physical causes and their effects.

Why not? The digital computer is a device that brilliantly compels a variety of discrete artifacts to scuttle along various causal pathways, ultimately exploiting pulsed signals in order to get one thing to act upon another. But in addition to their physical properties, the symbols flawlessly manipulated by a digital computer are capable of carrying and so conveying information, transforming one information-rich stream, such as a database of proper names, into another information-rich stream, such as those same names arranged in alphabetical order.

The human mind does as much, Pinker argues; indeed, what it does, it does in the same way. Just as the computer transforms one information stream into another, the human mind transforms one source of information—words on the printed page—into another—a world in which whalers pursue whales and the fog lowers itself ominously over the spreading sea.

Thus Pinker; thus almost everyone.

The theory that gives the concept of information almost all of its content was created by the late mathematician Claude Shannon in 1948 and 1949. In it, the rich variety of human intercourse dwindles and disappears, replaced by an idealized system in which an information *source* sends signals to an information *sink* by means of a communication channel (such as a telephone line).

Communication, Shannon realized, gains traction on the real world by means of the firing pistons of tension and release. From far away, where the system has its source, messages are selected and then sent, one after the other—perhaps by means of binary digits. In the simplest possible set-up, symbols are limited to a single digit: *1*, say. A binary digit may occupy one of two states (*on* or *off*). We who are tensed at the system's sink are uncertain whether *1* will erupt into phosphorescent life or the screen will remain blank. Let us assume that each outcome is equally likely. The signal is sent—and then received. Uncertainty collapses into blessed relief, the binary digit *1* emerging in a swarm of pixels. The exercise has conveyed one unit, or bit, of information. And with the definition of a unit in place, information has been added to the list of properties that are interesting because they are *measurable*.

The development of Shannon's theory proceeds toward certain deep theorems about coding channels, noise, and error-reduction. But the details pertinent to this discussion proceed in another direction altogether, where they promptly encounter a roadblock.

"Frequently," Shannon observes, "messages have *meaning*: that is, they refer to or are correlated according to some system with certain physical or conceptual entities." Indeed. Witness *Moby-Dick*, which is *about* a large white whale. For Shannon, however, these "semantic aspects of communication" take place in some other room, not the one where his theory holds court; they are, he writes with a touch of asperity, "irrelevant to the engineering problem." The significance of communication lies only with the fact that "the actual message is one selected from a set of possible messages"—*this* signal, and not some other.

Shannon's strictures are crucial. They have, however, frequently proved difficult to grasp. Thus, in explaining Shannon's theory, Richard Dawkins writes that "the sentence, 'It rained in Oxford every day this week,' carries relatively little information, because the receiver is not surprised by it. On the other hand, 'It rained in the Sahara every day this week,' would be a message with high information content, well worth paying extra to send."

But this is to confuse a signal with what it signifies. Whether I am surprised by the sentence, "It rained in the Sahara desert every day this week," depends only on my assessment of the source sending the signal. Shannon's theory makes no judgments whatsoever about the subjects treated by various signals and so establishes no connection whatsoever to events in the real world. It is entirely possible that whatever the weather in Oxford *or* the Sahara may be, a given source might send both sentences with equal probability. In that case, they would convey precisely the same information.

The roadblock now comes into view. Under ordinary circumstances, reading serves the end of placing one man's thoughts in contact with another man's mind. On being told that whales are not fish, Melville's readers have learned something about whales and so about fish. Their uncertainty, and so their intellectual tension, has its antecedent roots in facts about the world beyond the symbols they habitually encounter. For most English speakers, the Japanese translation of *Moby-Dick*, although conveying precisely the same information as the English version, remains unreadable and thus unavailing as a guide to the universe created by the book in English.

What we who have conceived an interest in reading have required is some idea of how the words and sentences of *Moby-Dick* compel a world into creation. And about this, Shannon's theory says nothing.

For readers, it is the connections that are crucial, for it is those connections themselves—the specific correlations between the words in Melville's novel and the world of large fish and demented whalers—that function as the load-bearing structures. Just how, then, are such connections established?

Apparently they just are.

IF, IN reading, every reader embodies a paradox, it is a paradox that in living he exemplifies as well. "Next to the brain," George Johnson remarks, "the most obvious biological information-processor is the genetic machinery of the cell."

The essential narrative is by now familiar. All living creatures divide themselves into their material constituents and an animating system of instruction and information. The plan is in effect wherever life is in command: both the reader and the bacterial cell are expressions of an ancestral text, their brief appearance on the stage serving in the grand scheme of things simply to convey its throbbing voice from one generation to another.

Within the compass of the cell itself, there are two molecular classes: the proteins, and the nucleic acids (DNA and RNA). Proteins have a precise three-dimensional shape, and resemble tight tensed knots. Their essential structure is nonetheless linear; when denatured and then stretched, the complicated jumble of a functional protein gracefully reveals a single filament, a kind of strand, punctuated by various amino acids, one after another.

DNA, on the other hand, is a double-stranded molecule, the two strands turned as a helix. Within the cell, DNA is wound in spools and so has its own complicated three-dimensional shape; but like the proteins, it also has an essentially linear nature. The elementary constituents of DNA are the four nucleotides, abbreviated as A, C, G, and T. The two strands of DNA are fastened to one another by means of struts, almost as if the strands were separate halves of a single ladder, and the struts gain purchase on these strands by virtue of the fact that certain nucleotides are attracted to one another by means of chemical affinities.

The structure of DNA as a double helix endows one molecule with two secrets. In replicating itself, the cell cleaves its double-stranded DNA. Each strand then reconstitutes itself by means of the same chemical affinities that held together the original strands. When replication has been concluded, there are two double-stranded DNA molecules where formerly there was only one, thus allowing life on the cellular level to pass from one generation to the next.

But if DNA is inherently capable of reproducing itself, it is also inherently capable of conveying the linear order of its nucleotides to the cell's amino acids. In these respects, DNA functions as a template or pattern. The mechanism is astonishingly complex, requiring intermediaries and a host of specialized enzymes to act in concert. But whatever the details, the central dogma of molecular biology is straight as an arrow. The order of nucleotides within DNA is read by the cell and then expressed in its proteins.

Read by the cell? Apparently so. The metaphor is inescapable, and so hardly a metaphor. As the DNA is read, proteins form in its wake, charged with carrying on the turbulent affairs of the cell itself. It was an imaginary reader, nose deep in Melville's great novel, who suggested the distinction between what words do as material causes and what they achieve as symbols. The same distinction recurs in biology. Like words upon the printed page, DNA functions in any number of causal pathways, the tic of its triplets inducing certain biochemical changes and suppressing others.

And this prompts what lawyers call a leading question. We quite know what DNA is: it is a macromolecule and so a material object. We quite know what it achieves: apparently everything. Are the two sides of this equation in balance?

The cell is, after all, a living system. It partakes of all the mysteries of life. The bacterium *escherichia coli*, for example, contains roughly 2,000 separate proteins, and every one of them is mad with purpose and busy beyond belief. Eucaryotic cells, which contain a nucleus, are more complicated still. Chemicals cross the cell membrane on a tight schedule, consult with other chemicals, undertake their work, and are then capped in cylinders, degraded and unceremoniously ejected from the cell. Dozens of separate biochemical systems act independently, their coordination finely orchestrated by various signaling systems. Enzymes prompt chemical reactions to commence and, work done, cause them to stop as well. The cell moves forward in time, functional in its nature, continuous in its operations.

Explaining all this by appealing to the causal powers of a single molecule involves a disturbing division of attention, rather as if a cathedral were seen suddenly to rise from the head of a carrot. Nonetheless, many biologists, on seeing the carrot, are persuaded that they *can* discern the steps leading to the cathedral. Their claim is often presented as a fact in the textbooks. The difficulty is just that, while the carrot—DNA, when all is said and done—remains in plain sight, subsequent steps leading to the cathedral would seem either to empty in a computational wilderness or to gutter out in an endless series of inconclusive causal pathways.

FIRST, THE COMPUTATIONAL WILDERNESS. Proteins appear in living systems in a variety of three-dimensional shapes. Their configuration is crucial to their function and so to the role they play in the cell. The beginning of a causal process is once again in plain sight—the linear order expressed by a protein's amino acids. And so, too, is the end—a specific three-dimensional shape. It is the mechanism in the middle that is baffling.

Within the cell, most proteins fold themselves into their proper configuration within seconds. Folding commences as the protein itself is being formed, the head of an amino-acid chain apparently knowing its own tail. Some proteins fold entirely on their own; others require molecular chaperones to block certain intermediate configurations and encourage others. Just how a protein manages to organize itself in space, using only the sequence of its own amino acids, remains a mystery, perhaps the deepest in computational biology.

Mathematicians and computer scientists have endeavored to develop powerful algorithms in order to predict the three-dimensional configuration of a given protein. The most successful of these algorithms gobble the computer's time and waste prodigally its power. To little effect. Protein-folding remains a mystery.

Just recently, IBM announced the formation of a new division, intended to supply computational assistance to the biological community. A supercomputer named Blue Gene is under development. Operating at processing speeds 100 times faster than existing supercomputers, the monster will be dedicated largely to the problem of protein folding.

The size of the project is a nice measure of the depth of our ignorance. The slime mold has been slithering since time immemorial, its proteins folding themselves for just as long. No one believes that the slime mold accomplishes this by means of supercomputing firepower. The cell is not obviously an algorithm, and a simulation, needless to say, is not obviously an explanation. Whatever else the cell may be doing, it is not using Monte Carlo methods or consulting genetic algorithms in order to fold its proteins into their proper shape. The requisite steps are chemical. No other causal modality is available to the cell.

If these chemical steps were understood, simulations would be easy to execute. The scope of the research efforts devoted to simulation suggest that the opposite is the case: simulations are difficult to achieve, and the requisite chemical steps are poorly understood.

If computations are for the moment intractable, every analysis of the relevant causal pathways is for the moment inconclusive.

As they are unfolding, proteins trigger an "unfolding protein response," one that alerts an "intracellular signaling system" of things to come. It is this system that in turn "senses" when unfolded proteins accumulate. The signal sent, the signaling system responds by activating the transcription of still other genes that provide assistance to the protein struggling to find its correct three-dimensional shape. Each step in the causal analysis suggests another to come.

But no matter the causal pathways initiated by DNA, some *overall* feature of living systems seems stubbornly to lie beyond their reach. Signaling systems must themselves be regulated, their activities timed. If unfolding proteins require chaperones, these must make their appearance in the proper place; their formation requires energy, and so, too, do their degradation and ejection from the cell. Like the organism of which it is a part, the cell has striking global properties. It is *alive*.

Our own experience with complex dynamical systems, such as armies in action (or integrated microchips), suggests that in this regard command and coordination are crucial. The cell requires what one biologist has called a "supreme controlling and coordinating power." But if there is such a supreme system, biologists have not found it. The analysis of living systems is, to be sure, a science still in its infancy. My point, however, is otherwise, and it is general.

Considered strictly as a material object, DNA falls under the descriptive powers of biochemistry, its causal pathways bounded by chemical principles. Chemical actions are combinatorial in nature, and *local* in their effect. Chemicals affect chemicals within the cell by means of various weak affinities. There is no action at a distance. The various chemical affinities are essentially arrangements in which molecules exchange their parts irenically or like seaweed fronds drift close and then hold fast.

But command, control, and coordination, if achieved by the cell, would represent a phenomenon incompatible with its chemical activities. A "supreme controlling and coordinating power" would require a device receiving signals from every part of the cell and sending its own universally understood signals in turn. It would require, as well, a universal clock, one that keeps time globally, and a universal memory, one that operates throughout the cell. There is no trace of these items within the cell.

Absent these items, it follows that the cell quite plainly has the ability to organize itself *from itself*, its constituents bringing order out of chaos on their own, like a very intricate ballet achieved without a choreographer. And what holds for the cell must hold as well for the creatures of which cells are a part. One biologist has chosen to explain a mystery by describing it as a fact. "Organisms," he writes, "from daisies to humans, are naturally endowed with a remarkable property, an ability to *make themselves.*"

Naturally endowed?

Just recently, the biologist Evelyn Fox Keller has tentatively endorsed this view. The system of control and coordination that animates the cell, she observes in *The Century of the Gene*, "consists of, and lives in, the interactive complex made up of genomic structures and the vast network of cellular machinery in which those structures are embedded." This may well be so. It is also unprecedented in our experience.

We have no insight into such systems. No mathematical theory predicts their existence or explains their properties. How, then, *do* a variety of purely local chemical reactions manage to achieve an overall and global mode of functioning?

INFORMATION NOW MAKES ITS second appearance as an analytic tool. DNA is a molecule—that much is certain. But it is also, molecular biologists often affirm, a library, a blueprint, a code, a program, or an algorithm, and as such it is quivering with information that it is just dying to be put to good use. As a molecule, DNA does what molecules do; but in its secondary incarnation as *something else*, DNA achieves command of the cell and controls its development.

A dialogue first encountered on the level of matter (DNA as a molecule and nothing more) now reappears on the level of metaphor (DNA as an information source). Once again we know what DNA is like, and we know what it does: apparently everything. And the question recurs: are the sides of this equation in balance?

Unfortunately, we do not know and cannot tell.

Richard Dawkins illustrates what is at issue by means of a thought experiment. "We have an intuitive sense," he writes,

> that a lobster, say, is more complex (more "advanced," some might even say more "highly evolved") than another animal, perhaps a millipede. Can we *measure* something in order to confirm or deny our intuition? Without literally turning it into bits, we can make an approximate estimation of the *information contents* [emphasis added] of the two bodies as follows. Imagine writing a book describing the lobster. Now write another book describing the millipede down to the same level of detail. Divide the word-count in one book by the word-count in the other, and you have an approximate estimate of the relative information content of lobster and millipede.

These statements have the happy effect of enforcing an impression of quantitative discipline on what until now have been a series of disorderly concepts. Things are being measured, and that is always a good sign. The comparison of one book to another makes sense, of course. Books are made up of words the way computer programs are made up of binary digits, and words and binary digits may both be counted.

It is the connection outward, from these books or programs to the creatures they describe, that remains problematic. What level of detail is required? In the case of a lobster, a very short book comprising the words "Yo, a lobster" is clearly not what Dawkins has in mind. But adding detail to a description—and thus length—is an exercise without end; descriptions by their very nature form an infinitely descending series.

Information is entirely a static concept, and we know of no laws of nature that would tie it to other quantitative properties. Still, if we cannot answer the question precisely, then perhaps it might be answered partially by saying that we have reached the right level of descriptive detail when the information in the book—that is, the lobster's DNA—is roughly of the same order of magnitude as the information latent in everything that a lobster is and does. This would at least tell us that the job at hand—constructing a lobster—is doable insofar as information plays a role in getting *any*thing done.

Some biologists, including John Maynard Smith, have indeed argued that the information latent in a lobster's DNA must be commensurate with the information latent in the lobster itself. How otherwise could the lobster get on with the business at hand? But this easy response assumes precisely what is at issue, namely that it is *by means of information* that the lobster gets going in the first place. Skeptics such as ourselves require a direct measurement, a comparison between the information resident in the lobster's DNA and the information resident in the lobster itself. Nothing less will do.

DNA is a linear string. So far, so good. And strings are well-defined objects. There is thus no problem in principle in measuring their information. It is there for the asking and reckoned in bits. But what on earth is one to count in the case of the lobster? A lobster is not discrete; it is not made of linear symbols; and it occupies three or four dimensions and not one. Two measurements are thus needed, but only one is obviously forthcoming.

The unhappy fact is that we have in general no noncircular way of specifying the information in any three- (or four-) dimensional object except by an appeal to the information by which it is generated. But this appeal makes literal sense only when strings or items like strings are in concourse. The request for a direct comparison between what the lobster has to go on—its DNA—and what it is—a living lobster—ends with only one measurement in place, the other left dangling like a mountain climber's rope.

We are thus returned to our original question: how do symbols—words, strings, DNA—bring a world into being?

Apparently they just do.

ONE MIGHT HOPE THAT in one discipline, at least, the situation might be different. Within the austere confines of mathematical physics, where a few pregnant symbols command the flux of space and time, information as an idea might come into its own at last.

The laws of physics have a peculiar role to play in the economy of the sciences, one that goes beyond anything observed in psychology or biology. They lie at the bottom of the grand scheme, comprising principles that are not only fundamental but irreducible. They must provide an explanation for the behavior of matter in all of its modes, and so they must explain the *emergence* as well as the organization of material objects. If not, then plainly they would not explain the behavior of matter in all of its modes, and, in particular, they would not explain its *existence*.

This requirement has initiated a curious contemporary exercise. Current cosmology suggests that the universe began with a big bang, erupting from nothing whatsoever 15 billion years ago. Plainly, the creation of something from nothing cannot be explained in terms of the behavior of material objects. This circumstance has prompted some physicists to assign a causative role to the laws of physics *themselves*.

The inference, indeed, is inescapable. For what else is there? "It is hard to resist the impression," writes the physicist Paul Davies, "of something—some influence capable of transcending space-time and the confinements of relativistic causality—possessing an overview of the entire cosmos at the instant of its creation, and manipulating all the causally disconnected parts to go bang with almost exactly the same vigor at the same time."

More than one philosopher has drawn a correlative conclusion: that, in this regard, the fundamental laws of physics enjoy attributes traditionally assigned to a deity. They are, in the words of Mary Hesse, "universal and eternal, comprehensive without exception (omnipotent), independent of knowledge (absolute), and encompassing all possible knowledge (omniscient)."

If *this* is so, the fundamental laws of physics cannot themselves be construed in material terms. They lie beyond the system of causal influences that they explain. And in this sense, the information resident in those causal laws is richer—it is more abundant—than the information resident in the universe itself. Having composed one book describing the universe to the last detail, a physicist, on subtracting that book from the fundamental laws of physics, would rest with a positive remainder, the additional information being whatever is needed to bring the universe into existence.

WE ARE NOW AT the very limits of the plausible. Contemporary cosmology is a subject as speculative as scholastic theology, and physicists who find themselves irresistibly drawn to the very largest of its intellectual issues are ruefully aware that they have disengaged themselves from any evidential tether, however loose. Nevertheless, these flights of fancy serve a very useful purpose. In the image of the laws of nature zestfully wrestling a universe into existence, one sees a peculiarly naked form of information—naked because it has been severed from every possibility of a material connection. Stripped of its connection to a world that does not yet exist, the information latent in the laws of physics is nonetheless capable of *doing* something, by bringing the universe into being.

A novel brings a world into creation; a complicated molecule an organism. But these are the low taverns of thought. It is only when information is assigned the power to bring something into existence from *nothing whatsoever* that its essentially magical nature is revealed. And contemplating magic on this scale prompts a final question. Just how did the information latent in the fundamental laws of physics unfold itself to become a world?

Apparently it just did.

WHERE PHYSICS
AND POLITICS MEET

EDWARD TELLER HAS UNDERTAKEN, AT THE AGE OF 93, TO TELL THE story of his life. In conducting an exercise of this sort, most men find much to admire, but little to censure in themselves. An autobiography thus tends to be an exercise in double deception: the reader deceived by the author, the author by himself. Teller's *Memoirs: A Twentieth Century Journey in Science and Politics* (Perseus, 2001) does not constitute a notable exception to the genre's rules. His work done, he finds it good. The reader may come in the end to agree with him—interesting evidence that in compiling an autobiography a man may reveal the truth without ever telling it completely.

Teller was born in Budapest in 1907, during the long autumn of the Austro-Hungarian empire. The autumn over, there followed the First World War, civil insurrection, and dictatorships of the left and the right. From the first, he found himself devoted to science, and among the sciences, to mathematical physics. He was by nature a man persuaded of his own pleasantness and consequently a romantic in his personal life and a sentimentalist in politics. Until 1931, he admits, he "was ignorant of most of the facts about political affairs." The Nazis appalled him, if only because they appalled almost everyone; Stalin and the Soviet Union, on the other hand, provoked a suspension of his judgment. He listened with patience to physicists such as Lev Landau, persuaded of the Communist cause and prepared to perish for its sake. It was not until he read Arthur Koestler's *Darkness at Noon* in 1943 that his understanding improved. Thereafter, it became very much improved.

Like many other physicists, including the young J. Robert Oppenheimer, Teller was profoundly influenced—he was enraptured—by the creation of quantum mechanics, the strangest and most powerful of physical theories. In this he was fortunate but not lucky. He studied with Werner Heisenberg at Göttingen and then with Niels Bohr in Copenhagen. But by the early 1930s, the greater part of the great work had already been done, classical physics shattered. Teller was able to witness the last revolution in mathematical physics, but he had arrived too late to participate at its birth.

Teller left Europe for the United States in 1935 and promptly found a generous sinecure at George Washington University. No more than a dozen physicists quite understood the new quantum mechanics, and Teller was one of them. No doubt, he added a certain European cachet to a department otherwise known for its industrious mediocrity. It is nonetheless not easy to imagine him addressing undergraduates or attending faculty meetings. He was, by his own account, a large, a rumpled personality, at once brooding and expansive, the engine of his ambition firing without pause.

By 1939, the world had conspired to make more room for him than George Washington University afforded. Developments in physics had made it clear that the relationship between matter and energy was something more than the theoretical artifact Einstein had predicted in 1905.

Enrico Fermi had spotted the telltale signs of nuclear fission in experiments conducted in 1932 (although Fermi unaccountably had failed to register what he had seen). Six years later, the German chemists Otto Hahn and Fritz Strassmann carried out similar experiments, bombarding uranium with a stream of neutrons. With the help of Carl Friedrich von Weizsäcker, they drew the appropriate conclusions. They had witnessed nuclear fission. Their analysis was strengthened and confirmed a year later by Lise Meitner.

The consequences were plain to every one of the émigré physicists. A bomb could be built. The only question was whether it would be built by Nazi Germany or the United States. In 1939 Teller and Leó Szilárd traveled to Einstein's house on Long Island in order to elicit his support for the development of an atomic weapon, carrying a letter Szilárd had drafted. Einstein had long been an indecisive pacifist, willing to condemn but not to reject the use of violence in international affairs. He signed the letter without significant change. Szilárd forwarded the letter to President Roosevelt.

THE AMERICAN SCIENTIFIC AND military establishment required four years to construct an atomic weapon, their success an extraordinary feat of disciplined engineering. J. Robert Oppenheimer was recruited from the University of California at Berkeley to lead a team assembled at Los Alamos. Oppenheimer had been known as a reserved and somewhat arrogant professor, quick to ridicule his intellectual inferiors. But he undertook the transformation of his personality and against great odds succeeded in disciplining the immensely fractious and self-important group of scientists he had assembled. It was in all respects a remarkable performance. Teller's account of Oppenheimer's success is the more impressive inasmuch as Teller never manages to conceal the fact that he disliked the man because he felt inadequate in his presence.

The project the physicists had set themselves possessed an undeniable moral urgency. Germany had retained a cadre of talented physicists, most notably Werner Heisenberg, who had chosen to remain in Germany by invoking the unassailable but irrelevant logical truth that "if my brother steals a silver spoon he is still my brother." It was natural to suppose that he was lending his support to the development of a German bomb—as, in fact, he was. And Germany embodied a brilliant and advanced technological society. An American scientist concerned that he might be performing the devil's work at Los Alamos had only to imagine an atomic weapon in Hitler's hands to resolve his doubts.

For reasons that are still unclear, the Nazi state failed to develop an atomic weapon. German physicists were sequestered in Britain after the war, and their conversations secretly recorded. The recordings reveal a surprising level of technical incompetence, with Heisenberg offering his colleagues an analysis of the relevant details that was in error by an order of magnitude. Teller suggests Heisenberg's incompetence was feigned. This is a thesis as implausible as it is generous. Heisenberg was a notorious arithmetic bungler, and it is more likely that his carelessness, rather than his character, kept the bomb from German hands.

Although Teller worked on the development of the atomic bomb, sometimes with diligence and often with resentment, his real interests were in the development of thermonuclear weapons. Enrico Fermi had raised the possibility of such weapons with Teller in 1941, observing that an atomic weapon might be used to trigger a thermonuclear reaction. Two quite different processes are at work. Atomic weapons divide heavier atoms into lighter elements. The change between states is released as energy. Thermonuclear weapons fuse lighter atoms into heavier elements. And the change between states is again released as energy. Thermonuclear devices have a potentially unlimited yield; if needed, they can destroy the planet; if employed, no doubt they will.

Teller thus found himself in an odd position. His colleagues were endeavoring to surmount the immense technical difficulties involved in the construction of an atomic weapon—difficulties that involved both detailed theoretical calculations and practical problems of metallurgy. And all the while, Teller was urging that he be allowed to undertake a still more demanding project, one that presumed that the problem at hand had already been settled. In retrospect, his advocacy seems both rational and farsighted, but at the time, it may have seemed otherwise, rather as if a man who has not mastered the bicycle were demanding access to a race car.

TELLER'S INTEREST IN THERMONUCLEAR weapons was in part an expression of his particular talents as a physicist. He was not a great theoretician like his teachers, nor a master of detail like Hans Bethe. He had no gift for experiment. His particular talent lay in his ability to undertake intellectual gestures that were somewhat larger and more audacious than his colleagues. A hint was all he needed. Thereafter, his imagination was consumed. His interest in thermonuclear weapons represented largely an intelligent man's obsessive curiosity about a number of technically challenging problems.

It is on this level that Teller's story is personal. But like the other physicists at Los Alamos, Teller was fate's servant as well as her master, and his determination to construct thermonuclear weapons, although prompted by an entirely private scientific calculus, also served several purely political ends.

The destruction of Hiroshima and Nagasaki in 1945 disturbed many of the physicists who had made it possible. They imagined a world in which nuclear weapons were constructed by hands less capable than their own. In a widely circulated remark, Oppenheimer compared potential nuclear antagonists to two scorpions in a bottle. The alternative, he thought, must be the international control of atomic weapons. Enrico Fermi, I. I. Rabi, and Hans Bethe agreed. Physicists reposed their hopes for world peace in international schemes such as the Acheson-Lilienthal report, which formed the basis for the Baruch Plan. Oppenheimer's scorpions consumed their imagination. Few of the physicists observed that as an honor system is unworkable among thieves, disarmament schemes are unworkable among states.

IN ANY EVENT, THE Soviet Union detonated its own atomic bomb in 1948. Those physicists who had been concerned to place atomic weapons beyond the control of the United States now devoted their efforts to placing thermonuclear weapons beyond its reach. In their minority report to the General Advisory Committee of the Atomic Energy Commission, Enrico Fermi and I. I. Rabi condemned a hydrogen weapon as a "necessarily evil thing considered in any light." The majority report (signed by James B. Conant, Hartley Rowe, Cyril Smith, L. A. DuBridge, Oliver Buckley, and J. Robert Oppenheimer) was hardly less emphatic, if slightly more precise: "The extreme dangers to mankind inherent in this proposal wholly outweigh any military advantage that could come from this development."

Teller, on the other hand, argued that the United States must have such weapons simply because the Soviet Union would have them. Physicists in the Soviet Union argued the same in reverse. These arguments enjoyed an enviable simplicity, and if they were the expression of competitive envy in international affairs, it has not often been observed that envy plays a lesser role among states than it does among men.

President Truman decided the issue in favor of Teller. Work on the hydrogen bomb commenced in earnest in 1950. It is by no means clear from Teller's *Memoirs* just who played the decisive role. Teller assigns himself credit for the essential idea that fusion could be induced by compression of a bomb's core, and he assigns the physicist Richard Garwin credit for the detailed design that made his idea workable. The mathematician Stanislaw Ulam played, on this account, an altogether minor role. Mathematical gossip generally has it the other way around, with Ulam rousing himself from his habitual torpor—he was reputed to enjoy the gift of laziness—just long enough to confide the right idea to his cleaning woman. The details can no longer be verified. In any event, the bomb was built, and the world armed itself. A balance of terror was the result, and as one might have expected, the world was suitably terrified. The editors of the *Bulletin of the Atomic Scientists* placed a "doomsday clock" on their journal's cover. It was forever set close to midnight.

Curiously enough, midnight did not come. The balance of terror that has prevailed from 1954 until the present day has proved remarkably stable. Nations that acquired thermonuclear weapons found it prudent not to use them. If the aim of arms control was the avoidance of nuclear war, arms proliferation provided the outcome that arms control had only promised. A stable balance of terror is hardly an ideal condition for the human race, but it appeared to be the only accessible condition. What had been learned could not be unlearned.

ALTHOUGH TELLER HAS HAD a long and illustrious career as a warrior priest, he remains in the public mind a dark and disturbing figure. During the 1950s, he participated in an epic bureaucratic battle against J. Robert Oppenheimer, who had acquired a position of great influence on the Atomic Energy Commission. It was a battle that destroyed both men: Oppenheimer because he lost, Teller because he won.

Heavy-bearded and dark, Teller seemed to many to convey an air of moral malignity. He was popularly known as the "father of the H-bomb," and if he objected to the paternal association, he did not widely advert to the fact. Oppenheimer seemed saintly in comparison, projecting—in disagreeably many interviews—a sense of guilt that he had taken pains to conceal, and nothing to employ, while exercising power. In meeting President Truman after the destruction of Hiroshima and Nagasaki, Oppenheimer announced "he had blood on his hands." Physicists who were uncomfortably aware that events were no longer in their control were pleased to be reminded that once it had been in their hands.

Yet it was Teller, and not Oppenheimer, who had urged that atomic weapons be demonstrated to the Japanese before their military employment. And it was Oppenheimer, and not Teller, who rejected this counsel, arguing that the physicists who had built the bomb should have nothing to say about the circumstances of its use.

If with respect to this issue the popular impression and the historical record are at odds, they are also at odds when it comes to other issues. Robert Oppenheimer was a man of considerable personal negligence. During the 1930s he had acquired a circle of left-wing friends large enough to form a cohort. He had conducted the Los Alamos project with due regard for security, but the fact remains that the Russians were able to construct an atomic weapon in 1948 because they had seen the details of its design in 1945. Oppenheimer received all of the glory for the American bomb; inevitably, he received some of the blame for the Russian bomb as well.

Oppenheimer had initially opposed the hydrogen bomb on the grounds of its uselessness. But no one proposing the development of thermonuclear weapons suggested using them. By 1950 or 1951, Oppenheimer came reluctantly to see what Teller had already seen: A weapon may be politically or diplomatically useful even if militarily useless. In this Oppenheimer saw the light, but he saw it too late. With the advent of Senator McCarthy and the changed climate in Washington, his days in power were numbered.

He did not help himself by his behavior. He had performed brilliantly during World War II, but as he came to reflect on his role, he became progressively more soulful. He suffered visibly. He seemed to suggest that the physicists had corrupted themselves. These attitudes were at odds with his unflagging wish to remain in power, an enterprise not often guided by considerations of epistemological sin.

The American defense establishment lost confidence in J. Robert Oppenheimer some time after he appeared to lose confidence in himself. President Eisenhower appointed Lewis Strauss as the new chairman of the Atomic Energy Commission in 1953. Strauss was hardly capable of addressing Oppenheimer as a physicist. He did not need to. He needed only to enlist the support of a prominent physicist. He found Edward Teller.

The commission determined to strip Oppenheimer of his security clearance, an act that would effectively end his career. Between April and May of 1954, its Personnel Security Board held hearings on Oppenheimer's fitness to retain access to secret documents. Oppenheimer's left-wing associations were thoroughly rehearsed. Whatever they may in their heart of hearts believed, almost all of the great physicists testified on his behalf.

Teller maintains that until almost the last minute he was prepared to join them. But just before his own testimony, Roger Robb, the counsel for the Atomic Energy Commission, shrewdly showed him sections of Oppenheimer's testimony. At issue was Oppenheimer's friendship with Haakon Chevalier, a professor of French at Berkeley.

OPPENHEIMER'S TESTIMONY WAS BOTH implausible and self-serving. He had been approached for information at Los Alamos in August 1943. And in his statements to security officers at the time, he had lied about the identity of his interlocutor. Now he was prepared to affirm that it had been Haakon Chevalier. In explaining how he came to dissemble so copiously, Oppenheimer could think of no better explanation than that he "was an idiot."

And thereafter a shaken Teller felt compelled to share his doubts with the investigating committee. In words that would haunt him, he said, "I would like to see the vital interests of this country in hands which I understand better and therefore trust more." He counseled against extending Oppenheimer's security clearance, a recommendation Oppenheimer's enemies accepted with alacrity.

If Oppenheimer had suffered previously because of the weight of his power, he now suffered grievously because of the effect of its absence. He withdrew to his directorship at the Institute for Advanced Study at Princeton. He grew very thin. He gave occasional public lectures that were admired to the extent that they were not understood. A lesser man would have suffered more; a greater man, less.

Teller in his turn was cut by the community of physicists. Oppenheimer's wound was to his soul; Teller's to his vanity. He had, it was widely believed, testified against Oppenheimer to punish him for not supporting more vigorously the development of the hydrogen bomb. The testimony that he reproduces in his autobiography suggests that this may be so. It is unyielding in its insinuation.

If Teller never completely recovered the respect of his colleagues, it was an affront that did nothing to impede his career as a warrior priest. The men responsible for the American military establishment were not generally known for their fine moral delicacies. During Reagan's presidency, it was Teller who was instrumental in developing support for the Strategic Defense Initiative. Among the photographs Teller includes in his *Memoirs*, there is one of him standing beside Reagan, the president beaming enigmatically, Teller impossibly old, stooped, and mottled, but still obviously alert and cunning.

The arms race between the United States and the Soviet Union continued for more than 40 years, each side spending billions refining weapons both agreed could not be used. The Soviet Union is no more, and the major European powers are at peace with one another. If this is not Teller's legacy, it is at least a credit to his influence.

The Cold War's balance of terror has become magnificently obsolete, and states now at peace occasionally think they might dispense with the weapons which made that peace possible. But no state trusts that the last state to disarm would not use its singular power to overawe the rest. And so the world remains armed.

And this, too, is a part of Edward Teller's legacy.

2002

GOD, MAN AND PHYSICS

Gᴏᴅ's ᴇxɪsᴛᴇɴᴄᴇ ɪs ɴᴏᴛ ʀᴇǫᴜɪʀᴇᴅ ʙʏ ᴛʜᴇ ᴘʀᴇᴍɪsᴇs ᴏғ ǫᴜᴀɴᴛᴜᴍ mechanics or general relativity, the great theories of twentieth-century physics—but then again, it is not contravened by their conclusions either. What else can we do but watch and wait?

The agnostic straddle. It is hardly a posture calculated to set the blood racing. In the early 1970s Jacques Monod and Steven Weinberg thus declared themselves in favor of atheism, each man eager to communicate his discovery that the universe is without plan or purpose. Any number of philosophers have embraced their platform, often clambering onto it by brute force. Were God to exist, Thomas Nagel remarked, he would not only be surprised, but disappointed.

A great many ordinary men and women have found both atheism and agnosticism dispiriting—evidence, perhaps, of their remarkable capacity for intellectual ingratitude. The fact remains that the intellectual's pendulum has swung along rather a tight little arc for much of the twentieth-century: atheism, the agnostic straddle, atheism, the agnostic straddle.

The revival of natural theology in the past 25 years has enabled that pendulum to achieve an unexpected amplitude, its tip moving beyond atheism and the agnostic straddle to something like religious awe, if not religious faith.

It has been largely the consolidation of theoretical cosmology that has powered the upward swing. Edwin Hubble's discovery that the universe seemed to be expanding in every direction electrified the community of cosmologists in the late 1920s, and cosmologists were again electrified when it became clear that these facts followed from Einstein's general theory of relativity. Thereafter, their excitement diminished, if only because the idea that the universe was expanding suggested inexorably that it was expanding from an origin of some sort, a *big bang*, as the astronomer Fred Hoyle sniffed contemptuously.

In 1963 Arno Penzias and Robert Wilson inadvertently noticed the background microwave radiation predicted by Big Bang cosmology; when Robert Dicke confirmed the significance of their observation, competing steady-state theories of creation descended at once into desuetude. And thereafter a speculative story became a credible secular myth.

But if credible, the myth was also incomplete. The universe, cosmologists affirmed, erupted into existence 15 billion years ago. Details were available, some going back to the first three minutes of creation. Well and good. But the metaphoric assimilation of the Big Bang to the general run of eruptions conveyed an entirely misleading sense of similarity. The eruption of Mount Vesuvius took place in space and time; the Big Bang marks the spot at which time and space taper to a singularity and then vanish altogether.

It follows that the universe came into existence from nothing whatsoever, and for no good reason that anyone could discern, least of all cosmologists. Even the most ardent village atheist became uneasily aware that Big Bang cosmology and the opening chapters of the Book of Genesis shared a family resemblance too obvious profitably to be denied.

Thereafter, natural theology, long thought dead of inanition, began appearing at any number of colloquia in mathematical physics, often welcomed by the same physicists who had recently been heard reading its funeral obsequies aloud. In *The God Hypothesis: Discovering Design in our "Just Right" Goldilocks Universe* (Rowman & Littlefield, 2001) Michael A. Corey is concerned to convey their news without worrying overmuch about the details. His message is simple. There is a God, a figure at once omnipotent, omniscient, eternal, and necessary. Science has established his existence.

How very embarrassing that this should have been overlooked.

AT THE VERY HEART of revived natural theology are what the physicist Brandon Carter called "anthropic coincidences." Certain structural features of the universe, Carter argued, seemed finely tuned to permit the emergence of life. This is a declaration, to be sure, that suggests far more than it asserts. *Structural features? Finely tuned? Permit?* When the metaphors are squeezed dry, what more is at issue beyond the observation that life is a contingent affair? This is not a thesis in dispute.

Still, it often happens that commonplace observations, when sharpened, prompt questions that they had long concealed. The laws of physics draw a connection between the nature of certain material objects and their behavior. Falling from a great height, an astrophysicist no less than an airplane accelerates toward the center of the earth. Newton's law of gravitational attraction provides an account of this tendency in terms of mass and distance (or heft and separation). In order to gain traction on the real world, the law requires a fixed constant, a number that remains unchanged as mass and distance vary. Such is Newton's universal gravitational constant.

There are many comparable constants throughout mathematical physics, and they appear to have no very obvious mathematical properties. They are what they are. But if arbitrary, they are also crucial. Were they to vary from the values that they have, this happy universe—such is the claim—would be too small or too large or too gaseous or otherwise too flaccid to sustain life. And these are circumstances that, if true, plainly require an explanation.

Carter was a capable physicist; instead of being chuckled over and dismissed by a handful of specialists, the paper that he wrote in 1974 was widely read, Fred Hoyle, Freeman Dyson, Martin Rees, Stephen Hawking, Paul Davies, Steven Weinberg, Robert Jastrow, and John Gribbin all contributing to the general chatter. Very few physicists took the inferential trail to its conclusion in faith; what is notable is that any of them took the trail at all.

The astronomer Fred Hoyle is a case in point, his atheism in the end corrected by his pleased astonishment at his own existence. Living systems are based on carbon, he observed, and carbon is formed within stars by a process of nucleosynthesis. (The theory of nucleosynthesis is, indeed, partly his creation.) Two helium atoms fuse to form a beryllium intermediate, which then fuses again with another helium atom to form carbon. The process is unstable because beryllium intermediates are short-lived.

In 1953 Edwin Salpeter discovered that the resonance between helium and intermediate beryllium atoms, like the relation between an opera singer and the glass she shatters, is precisely tuned to facilitate beryllium production. Hoyle then discovered a second nuclear resonance, this one acting between beryllium and helium, and finely tuned as well.

Without carbon, no life. And without specific nuclear resonance levels, no carbon. And yet there *he* was, Hoyle affirmed, carbon based to the core. Nature, he said in a remark widely quoted, seems to be "a put-up job."

Inferences now have a tendency to go off like a string of firecrackers, some of them wet. Hoyle had himself discovered the scenario that made carbon synthesis possible. He thus assigned to what he called a "Super-calculating Intellect" powers that resembled his own. Mindful, perhaps, of the ancient wisdom that God alone knows who God is, he did not go further. Corey is, on the other hand, quite certain that Hoyle's Supercalculating Intellect is, in fact, a transcendental deity—*the* Deity, to afford Him a promotion in punctuation.

And Corey is certain, moreover, that he quite knows His motives. The Deity, in setting nuclear resonance levels, undertook his affairs "*in order* to create carbon-based life forms."

Did He indeed? It is by no means obvious. For all we know, the Deity's concern may have lain with the pleasurable intricacies of nucleosynthesis, the emergence of life proving, like so many other things, an inadvertent consequence of his tinkering. For that matter, what sense does it make to invoke the Deity's long-term *goals*, when it is His existence that is at issue? If nothing else, natural theology would seem to be a trickier business than physicists may have imagined.

As it happens, the gravamen of Corey's argument lies less with what the Deity may have had in mind and more with the obstacles He presumably needed to overcome. "The cumulative effect of this fine tuning," Corey argues, "is that, against all the odds, carbon was able to be manufactured in sufficient quantities inside stellar interiors to make our lives possible." That is the heart of the matter: *against all the odds*. And the obvious question that follows: Just how do we know this?

Corey does not address the question specifically, but he offers an answer nonetheless. It is, in fact, the answer Hoyle provides as well. They both suppose that something like an imaginary lottery (or roulette wheel) governs the distribution of values to the nuclear resonance levels of beryllium or helium. The wheel is spun. And thereafter the *right* resonance levels appear. The odds now reflect the pattern familiar in any probabilistic process—one specified outcome weighed against all the rest. If nuclear resonance levels are, in fact, unique, their emergence on the scene would have the satisfying aspect of a miracle.

It is a miracle, of course, whose luster is apt to dim considerably if other nuclear resonance levels might have done the job and thus won the lottery. And this is precisely what we do not know. The nuclear resonance levels specified by Hoyle are *sufficient* for the production of carbon. The evidence is all around us. It is entirely less clear that they are *necessary* as well. Corey and Hoyle make the argument that they are necessary because, if changed slightly, nucleosynthesis would stop. "Overall, it is safe to say"—Corey is speaking, Hoyle nodding—"that given the utter precision displayed by these nuclear resonances with respect to the synthesis of carbon, not even *one* of them could have been *slightly* different without destroying their precious carbon yield." This is true, but inconclusive. Mountain peaks are isolated but not unique. Corey and

Hoyle may well be right in their conclusions. It is their argument that does not inspire confidence.

The trouble is not merely a matter of the logical niceties. Revived natural theology has staked its claims on probability. There is nothing amiss in this. Like the rest of us, physicists calculate the odds when they cannot calculate anything better. The model to which they appeal may be an *imaginary* lottery, roulette wheel, or even a flipped coin, but imaginary is the governing word. Whatever the model, it corresponds to no plausible *physical* mechanism. The situation is very different in molecular biology, which is one reason criticism of neo-Darwinism very often has biting power. When biologists speculate on the origins of life, they have in mind a scenario in which various chemicals slosh around randomly in some clearly defined physical medium. What does the sloshing with respect to nuclear resonance?

Or with respect to anything else? Current dogma suggests that many of the constants of mathematical physics were fixed from the first, and so constitute a part of the initial conditions of the Big Bang. Corey does not demur; it is a conclusion that he endorses. What then is left of the anthropic claim that the fundamental constants have the value that they do despite "all odds"? In the beginning there was no time, no place, no lottery at all.

Mathematical physics currently trades in four fundamental forces: gravity, electromagnetism, and the strong and weak forces governing the nucleus and radioactive decay. In general relativity and quantum mechanics, it contains two great but incompatible theories. This is clearly an embarrassment of riches. If possible, unification of these forces and theories is desirable. And not only unification, but unification in the form of a complete and consistent theoretical structure.

Such a theory, thoughtful physicists imagine, might serve to show that the anthropic coincidences are an illusion in that they are not coincidences at all. The point is familiar. Egyptian engineers working under the pharaohs knew that the angles of a triangle sum to more or less one hundred and eighty degrees. The number appears as a free parameter in their theories, something given by experience and experiment. The Greeks, on the other hand, could *prove* what the Egyptians could only calculate. No one would today think to ask why the interior angles of a Euclidean triangle sum to precisely one hundred and eighty degrees. The question is closed because the answer is necessary.

THE GRAND HOPE OF modern mathematical physicists is that something similar will happen in modern mathematical physics. The Standard Model of particle physics contains a great many numerical slots that must be filled in by hand. This is never counted as a satisfaction, but a more powerful physical theory might show how those numerical slots are naturally filled, their particular values determined ultimately by the theory's fundamental principles. If this proves so, the anthropic coincidences will lose their power to vex and confound.

Nonetheless, the creation of a complete and consistent physical theory will not put an end to revived natural theology. Questions once asked about the fundamental constants of mathematical physics are bound to reappear as questions about the nature of its laws. The constants of mathematical physics may make possible the existence of life, but the laws of mathematical physics make possible the existence of *matter*. They have, those laws, an overwhelmingly specific character. Other laws, under which not much exists, are at least imaginable. What explanation can mathematical physics itself provide for the fact that the laws of nature are arranged as they are and that they have the form that they do? It is hardly an unreasonable question.

Steven Weinberg has suggested that a final theory must be logically *isolated* in the sense that any perturbation of its essential features would destroy the theory's coherence. Logical isolation is by no means a clear concept, and it is one of the ironies of modern mathematical physics that the logical properties of the great physical theories are no less mysterious than the physical properties of the universe they are meant to explain. Let us leave the details to those who cherish them.

The tactic is clear enough. The laws of a final theory determine its parameters; its logical structure determines its laws. No further transcendental inference is required, if only because that final theory explains itself.

This is very elegant. It is also entirely unpersuasive. A theory that is logically isolated is not necessarily a theory that is logically unique. Other theories may be possible, some governing imaginary worlds in which light alone exists, others worlds in which there is nothing whatsoever. The world in which we find ourselves is one in which galaxies wink and matter fills the cup of creation. What brings about the happy circumstance that the laws making this possible are precisely the laws making it real? The old familiar circle.

All this leaves us where we so often find ourselves. We are confronted with certain open questions. We do not know the answers, but what is worse, we have no clear idea—no idea whatsoever—of how they might be answered. But perhaps that is where we should be left: in the dark, tortured by confusing hints, intimations of immortality, and a sense that, dear God, we really do not yet understand.

EINSTEIN AND GÖDEL:
FRIENDSHIP BETWEEN EQUALS

A PICTURE TAKEN IN PRINCETON, NEW JERSEY, IN AUGUST 1950 shows Albert Einstein standing next to the Austrian logician Kurt Gödel. Both men are looking at the camera. Einstein is wearing a rumpled shirt and baggy slacks held up by suspenders. His body sags. Gödel, dressed in a white linen suit and wearing owlish spectacles, looks lean and almost elegant, the austerity of his expression softened by an odd sensuality that plays over the lower half of his face. The men are at ease; they are indulging the photographer. Clearly, they are friends.

It is hardly surprising that they should have come to know each other. They were members of the Institute for Advanced Study at Princeton, and their offices were close. As refugees from the Third Reich, they had both felt the harsh breath of history and had in common the rich, throaty German language, a world of words in which the pivot of memory turns on Goethe, not Shakespeare. Although Einstein was a physicist and Gödel a mathematician, they shared an intellectual daring that transcended their disciplines.

Gödel's incompleteness theorems, published in 1931 when he was only 25, had rewritten the ground rules of modern science much as Einstein's theory of relativity had done 15 years before. Elementary arithmetic, Gödel demonstrated, is incomplete and will remain so. Whatever axiomatic system you base your calculations on, there are true statements that lie beyond the system's reach. Adding such statements to the system as further axioms does no good. The enriched system is also incomplete, the infection moving upward by degrees.

Einstein once remarked to Oskar Morgenstern, one of the cofounders of game theory, that he went to the Institute chiefly to walk home with Gödel. (*"Um das Privileg zu haben, mit Gödel zu Fuss nach Hause gehen zu durfen."* There is in the original German a note of gentle deference that cannot quite be translated.) They did so often until Einstein's death in 1955. Yet their scientific affinities grew out of profound personal differences. Einstein was a man of unshakable self-confidence. Gödel retreated before controversy and twice suffered nervous collapses; he was, under the best of circumstances, a valetudinarian, and under the worst, a hypochondriac. When the two men met, in 1933, word of young Gödel's genius had yet to leave the academic cloister, where it was conveyed in whispers. Einstein, on the other hand, was 54, nearing the end of his productive career. Although he retained a sense of impudent playfulness, he had also acquired a marmoreal aspect, transcending fame itself to become one of the century's mythic figures, his plump, sad face known throughout the world.

These differences were inevitably reflected in the nature of the friendship. In a letter written to the biographer Carl Seelig, Einstein's secretary remarked on the "awed hush" that greeted Einstein whenever he appeared at seminars or conferences. Not even the sharp-tongued Wolfgang Pauli, a fellow Nobel Prize winner in physics, could bring himself to treat the great man as if he were mortal. Gödel seemed to share something of this attitude. In letters to his mother, he appeared to take pleasure in affirming that through his friendship with Einstein, he was basking in reflected glory. "I have so far been to his house two or three times," he wrote in 1946. "I believe it rarely happens that he invites anybody to his house."

Still, in the grandeur of their scientific achievements, Einstein and Gödel both stood alone and so must have turned to each another in part because they could turn to no one else. Although the content of their conversations has been lost, we can imagine at least one topic they must have discussed on those long evening walks. In 1948 Gödel turned his attention to Einstein's supreme creation, the general theory of relativity, and succeeded in coaxing a new and flamboyant universe from the alembic of its symbols. He did so by providing an exact solution to the heart of the theory—a field equation that allows one to calculate the force of a gravitational field—and his analysis reflects the distinctive characteristics of all his work. It is original and logically coherent, the argument set out simply but with complete and convincing authority. A sense of superb taste prevails throughout. There is no show.

And it is odd. It is distinctly odd.

THE LEADING IDEA OF general relativity—the fusion of space and time— is not hard to grasp. After all, space and time are fused in ordinary life as well. We locate an event (the assassination of JFK, for example) both in terms of where it took place (Dallas, Texas) and when it took place (roughly 1:30 EST on the afternoon of November 22, 1963). Three numbers suffice to mark the space of Dallas, Texas, on a three-dimensional map: longitude, latitude, and altitude. The place is pinpointed as an event in space-time if another number, the time, is added. And if an event can be defined by four numbers, then a series of events can be defined by a series of such numbers, trailing one another like elephants marching trunk to tail. In general relativity such series are called world lines.

General relativity then forges a far-flung connection between the geometry of space and time and the behavior of objects in motion within space and time. Imagine a marble placed on a mattress. Given a tap, the marble will move in a straight line. But place a bowling ball on the mattress, too, and the marble, given precisely the same tap, will roll down the sagging surface, its path changing from a straight to a curved line. The bowling ball's weight deforms the medium of the mattress, and the deformed medium influences the marble's movement.

Replace the bowling ball and marble with planets, stars, or wheeling galaxies, and the mattress with space-time itself, and a homely metaphor is transformed into the leading principle of a great physical theory. In a universe with no massive objects, there is no deformation of space and time, and the shortest route between two points is a straight line. When matter makes its fateful appearance, the shortest routes will curve. The first and most celebrated confirmation of this theory came in 1919, when astronomers established that the mass of the sun causes a beam of light to curve, as Einstein had predicted.

"For us believing physicists," Einstein once wrote, "the distinction between the past, the present, and the future is only an illusion." It was a melancholy remark, made as Einstein faced death, but it flowed directly from Einstein's special theory of relativity. Imagine a group of observers scattered carelessly throughout the cosmos. Each is able to organize the events of his life into a linear order—a world line of the kind just described. Each is convinced that his life consists of a series of nows, moving moments passing from the past to the present to the future. Special relativity urges a contrary claim. The observers scattered throughout space and time are all convinced their sense of now is universal. Now is, after all, now, is it not? Apparently not. Time passes at a different rate depending on how fast a person is moving: While one hour passes on Earth, only a few seconds might pass on a rocket ship hurtling away from Earth at nearly the speed of light. It is entirely possible that one man's now might be another man's past or future.

Gödel's solution to the field equation vindicated the deepest insight of Einstein's theory, namely that time is relative. But Einstein's theory of relativity suggests only that time does not exist in the conventional sense, not that time exists in no sense whatsoever. Einstein's claim is more subtle. He suggests that *change* is an illusion. Things do not become, they have not been, and they will not be: They simply are. Time is like space; it is precisely like space. In traveling to Singapore, I do not bring Singapore into existence. I reach Singapore, but the city has been there all along. So, too, I reach events in the future by displacing myself in time. I do not bring them into being. And if nothing is brought into being, there is no change.

MOST COSMOLOGISTS NOW AGREE that the universe expanded from a primordial explosion we call the Big Bang. Physicists talk, after all, about the first three minutes. But if this makes sense, it makes sense as well to talk of times after the first three minutes. And if time has an origin, and a uniform measure, then we are again within the bounds of Newton's universal clock, marking time throughout the cosmos. It is everywhere approximately 14 billion years after the Big Bang, and it is that time now.

But a universe proceeding from nothing to nowhere by means of an enthusiastic expansion is only one possibility. There are others. Some interpretations of the field equation are realized in a static but unstable universe, one that simply hangs around for all eternity if it manages to hang around at all. Then again the universe might be rotating in a void, turning serenely like a gigantic pinwheel. In a universe of this sort, each observer sees things as if he were at the center of the spinning, with the galaxies—indeed, the whole universe—rotating about him. And this strange assumption, Gödel demonstrated, satisfies the field equation of general relativity exactly.

A rotating universe is an idea reminiscent of ancient astrologers, who imagined observers clustered on the earth and the celestial sphere turning around them. But in Gödel's conception, the galaxies aren't the only things rotating. Everything else goes along for the ride. The galaxies rotate, and as they do, they drag space and time with them. Just as an expanding universe blows up space and time, a rotating universe turns space and time around in spirals. The same idea is at work but with a different effect. In a rotating universe, for instance, time travel becomes possible. By moving in a large enough circle around an axis, at something approaching the speed of light, an observer might catch his own temporal tail, returning to his starting point at some time earlier than his departure. The requisite paths are known as closed timelike curves.

When Gödel first published his work, the general reaction from other scientists was one of polite curiosity. Einstein was respectful but cautious, suggesting that perhaps Gödel's conclusions would simply be rejected on "physical grounds." (Gödel's solution rules out an expanding universe, which Einstein had by then grudgingly accepted.) Gödel himself never succeeded in making sense of time travel, whatever his solution suggested. Aside from the sorts of paradoxes beloved by science fiction writers—a time traveler who accidentally kills his own grandparents, say—time travel has more subtle, theoretical problems. There is no suggestion in Gödel's work, for instance, that time itself can come to a stop and then reverse course. Yet time travel would entail a journey, and in general relativity, as in real life, every journey takes time. From the traveler's perspective time would move forward minute by minute, even as it leaps backward when he arrives.

There are still deeper points at issue. If time moves in circles, and an observer can return to his own past, it seems to follow that effects might be their own causes. It is one thing to give up on time; it is quite another to give up on cause as a fundamental physical property.

And finally, there is the philosophical point, the one at the heart of Gödel's concerns. Granted, rotating universes may be physically unrealistic. But they are possible, and once seen as possibilities, they cannot be unseen. Within these strange contraptions, time is an illusion. But if time is an illusion in some universes, then the features of time that we take for granted in this particular universe must be accidents of creation, a matter of how matter and its motions are arranged in the world. But a philosophical view leading to this conclusion, Gödel remarked dryly, "can hardly be considered satisfactory." Time is far too deep a concept to arise accidentally.

GÖDEL SPENT THE SECOND half of his life absorbed by philosophy. Despite his experiences in Europe, he believed that "the world is rational." He was an optimist and a theist; and although he thought that "religions are for the most part bad," he insisted that "religion [itself] is not." A deity was at the center of his metaphysics. He entertained speculations about the afterlife, arguing that "the world in which we live is not the only one in which we shall live or have lived." He dismissed the Darwinian theory of evolution and declared flatly that "materialism was false."

He was a mathematical Platonist, arguing with boldness that the human intellect is capable of perceiving pure mathematical abstractions, just as the human senses are capable of grasping material objects.

Still, in the end, Gödel concluded that his efforts had been unavailing. "By his own account," the late logician Hao Wang wrote in *Reflections on Kurt Gödel*, he did not "attain what he was looking for in philosophy." Much the same is true for Einstein. The great unified theory for which he had searched for more than 30 years ultimately eluded him. For much of that time, he had worked in isolation, a younger generation of physicists regarding his obsession as just that, treating him with reverence, of course, but a form of reverence in which the first faint curl of contempt might be seen.

Did Einstein and Gödel discuss such issues? Gödel's journals and biographers do not say. But the depth of their friendship made what diplomats often call frank discussions unnecessary. Gödel was skeptical of Einstein's quest for a unified theory, and Einstein must have regarded Gödel's philosophical investigations with a certain amused detachment. The deep and ineradicable melancholy in Einstein's personality made it impossible for him to regard optimism or theism with anything more than a sense of tolerant skepticism.

Their lives, too, revealed countervailing currents. Einstein sought solace not only in solitude but in a deliberate, carefully contrived release from the ordinary human bonds of family and friends. He divorced his first wife and never once saw his daughter, who was probably put up for adoption. His second marriage, to his cousin Elsa, was hardly an affair of passion.

Like Einstein, Gödel found ordinary social intercourse an immense chore. For most of his adult life, he was happily married to a former dancer in Viennese cabarets. Yet he was notoriously reclusive, working at the Institute for Advanced Study in a darkened room, never attending other scientists' lectures, solitary, obsessed, half-mad, consumed from within by the fires of an intellectual passion so powerful that by the end of his life they seemed quite literally to have consumed his frail flesh entirely. Gödel died, in 1978, of "inanition," in the lapidary words of his death certificate. He had refused to eat.

Long before then, Einstein's theory of general relativity had undergone an efflorescence, its dark, difficult interior yielding many interesting mathematical secrets. A wealth of observational evidence made it possible to test the theory against a cosmic instead of a local background. Most of all, a new generation of physicists fell under Einstein's sway and attended to the strange, unearthly music of his dream of a unified theory. Einstein's vision has proved too powerful to be dismissed. As for Gödel's, it awaits another time to be fully understood. Or perhaps another universe.

LUCKY JIM

A DOCTORATE FROM INDIANA UNIVERSITY IN 1949, THE CAVENdish laboratories at Cambridge University, the discovery of DNA. Thereafter, immortality. James Watson has plainly come to regard his life as a sign of grace.

And with some reason, I suppose. Watson was 23 when in the early 1950s he joined Francis Crick in a scientific partnership. They proposed to discover the secret of life. The odds in their favor were not great. Biologists knew that in perpetuating themselves, living systems must squeeze their identity into what the physicist Erwin Schrödinger had called a code script. Deoxyribonucleic acid, or DNA, was plainly involved. Beyond this, experiments had revealed little and various theories nothing. Maurice Wilkins and Rosalind Franklin had for years studied DNA by means of X-ray crystallography, but it was slow, frustrating, and inconclusive work, rather like deducing the score of a symphony from its echoes in a concert hall. But the matter had come to occupy Linus Pauling, and as far as Watson and Crick were concerned, his presence on the scene was ominous. Pauling possessed an intelligence of almost supernatural vigor. He seemed eager to offer a revelation.

And yet there it was. The patient plodding researchers continued to plod patiently, consuming time but not covering distance; Pauling's infallible intuition failed him as he emerged noisily from the California Institute of Technology in 1951 with a bizarre triple helix in hand. Watson and Crick spotted the truth. DNA was a double helix, its two strands supported by chemical struts, adenine paired with thymine and guanine with cytosine. The molecule's structure at once revealed its secrets. DNA expressed a cryptogram and so contained a message, its four chemical constituents comprising an elementary alphabet. And it penetrated the future by unwinding itself and then separating, its halves recombining to form two double stranded helices where before there was only one.

This was all very elegant. The double helix electrified the emerging discipline of molecular biology. It electrified the world as well, Watson and Crick winning the 1962 Nobel Prize in medicine and physiology.

That they had made a discovery of great importance, no one disputed, least of all Watson. But the real story, he believed, had been pointlessly sanitized. And so, in 1968, he published a memoir of rectification under the title, *The Double Helix: A Personal Account of the Discovery of the Structure of DNA.* The book was a considerable success, the more so since Watson expressed with candor his conviction that scientific research is ruthless, unprincipled, and driven largely by an undignified scramble for fame. Watson's narrative supported his claim. Having appropriated Rosalind Franklin's research results because they were crucial, Watson admitted that he and Crick had denied her the appropriate credit because it was easy.

Watson's book amused the general public and outraged his colleagues, Crick denouncing it as something like "that found in the lower class of women's magazines." E. O. Wilson, the environmental biologist and Watson's colleague at Harvard in the 1960s, was moved to describe Watson as the most unpleasant human being he had ever met—the "Caligula of Science."

Blood dries quickly, Charles de Gaulle observed, and so does outrage. Watson's memoir came to be appreciated as an achievement in brashness. Scientists whom Watson had neglected personally to offend quickly reached the conclusion that in disciplines other than their own, scientific research was every bit as nasty as Watson had indicated. The book is considered a classic.

Now in *Genes, Girls, and Gamow: After the Double Helix* (Knopf, 2002), Watson proposes to take up the story where *The Double Helix* ends. "For better or worse," he writes, "I and my friends were present at the birth of the DNA paradigm—by any standards one of the great moments in the history of science, if not of the human species. In this way we were unique players in a momentous drama. Thus there will be many readers wanting to know better what actually happened in our lives."

This is not so. Watson has been misinformed.

Every life is no doubt precious, but few are interesting. The days follow one another. There is the sound of someone snoring. It rains. Watson's first memoir described a quest, and the quest gave to his narrative its powerful effect of artistic compression. Something was ventured, and something gained. But his second memoir has plainly been compiled from diary notes. It reads as life moves. Watson travels from England to California and back again to England. He is always ill at ease and often maladroit. There are outstanding scientific problems to be solved, but Watson does not solve them. Long walks are taken, often on disagreeable mountain paths. The days accumulate.

Charm is occasionally a substitute for literary skill, but Watson lacks both, and he is inclined to offer candor as a substitute. It is a mistake. "Dick Feynman and I sat next to each other," he writes, and "although we could not say it to others, we felt we might be Caltech's most obvious candidates for future Nobel awards."

A part of Watson's narrative is devoted to reviving the dead, a familiar if often fruitless pastime. George Gamow was an imaginative Russian physicist and a dabbler in molecular biology. Watson loved the man and is devoted to his memory. Gamow enters variously into his memoir, performing card tricks or otherwise engaged in amiable trifles and then shambles off, a larger figure, one hopes, in real than in remembered life.

The burden of Watson's memories lies elsewhere. Now that he is old, Watson is eager to convey the extent to which his scientific achievements were a distraction from his search for pretty young women, a search that was never-ending because never successful. The enchanting young women are forever too busy to see Watson or, having seen him, too busy to see him again. Some depart for foreign ports with what seems a rare urgency. Others enter his life like quick sunbursts and then leave Watson dazzled but disappointed when they cover themselves in clouds.

The experiences that Watson recounts must have been painful to relive, which makes them painful to read. After almost half a century he is still suffering in retrospect the sentiments that only a young man can suffer when someone deeply loved tells him, no, honey, things are just not going to work out. In continuing to attend so earnestly to the ones that got away, Watson appears something of a schnook.

For ten years or so following his great triumph, Watson sought fitfully to enlarge its scope. Living systems divide naturally into two molecular classes. DNA involves command, control, and coordination, and is found in the cell's nucleus. But the proteins comprise the stuff of life, the basic building blocks of every living system. And they are for the most part located in the cell's cytoplasm, an arena in which countlessly many seething chemical activities take place. Whatever the message contained in DNA, it must somehow be conveyed to the proteins. The conveyance was in 1953 a mystery. Watson and Crick were thus in the position of an observer who can see that an architect's plans are being carried out, but cannot determine how his commands are communicated. Crick speculated that a sequence of intermediate molecules must link DNA with various emerging proteins. He was right. It is ribonucleic acid, or RNA, that acts as a second source of information within the cell, and, as the name suggests, RNA is single stranded, containing only one gently floating filament. If DNA is the master, RNA is its messenger, ferrying information from DNA and impressing it on the proteins.

Watson pursued a number of experiments with RNA. He thought diligently, although not obsessively. He exchanged letters with various eminences and visited their laboratories. But lightning did not strike twice. The lightning that missed him struck Crick instead. It struck him so many times, in fact, that he is now widely regarded as a one-man miracle. Watson, on the other hand, became a powerful scientific administrator, first at the Cold Spring Harbor Laboratory and, more recently, as the director of the Human Genome Project.

Still, there remains the great work. It is now fifty years since Watson and Crick initiated "the DNA paradigm," and Watson is understandably eager to claim that in discovering DNA he and Crick had really discovered the secret of life, just as they had hoped to do.

Once again, he is mistaken. Details have accumulated, but the secret remains a secret. DNA was introduced originally as a code script, and it has enjoyed successive incarnations as a blueprint, design, template, or computer program. The metaphors are helpful, but they are misleading as well. A molecule is sent into the future, and directly thereafter an organism appears, bouncing, energetic, and alive. The incredible discrepancy between the beginning and the end of this process suggests that something remarkable has taken place. A molecule is one thing, a living creature another. Molecular biology has traced the story from DNA to the various proteins. Thereafter a mystery begins as the proteins somehow organize themselves to form an organism. Metaphors all lapse just when they are most needed. A computer program cannot, after all, create a computer.

As so often happens in the sciences, molecular biology has resolved its mysteries by magical thinking. Whatever the process, it is DNA, according to official doctrine, that is still crucial, still in charge, an agency capable of achieving every biological effect. Evolutionary biologists now assign to the human genome full responsibility for altruism, date rape, aggression, eating disorders, and a taste for *Mansfield Park*. The truth is we do not know how the genome achieves any effect beyond the molecular. Although more powerful by far than astrology, molecular biology is not appreciably different in kind, the various celestial houses having about as much to do with human affairs as the various genes.

This is hardly a criticism of Watson and Crick's discovery, but it is a fact. Still, if DNA has been assigned magical properties, there is at least one respect in which the designation is merited. Whatever it may be doing in the real world, that elegant molecule bestowed on James Watson a form of immortality.

STEPHEN JAY GOULD, 1942–2002: IN MEMORIAM

This following tribute was published electronically by the Discovery In-stitute on the occasion of Gould's death, May 20, 2002.

—Editor

STEPHEN JAY GOULD WAS THE MOST IMPORTANT PALEONTOLOGIST of his generation, the impact of his life best measured by the wide-spread sense of loss occasioned by his death.

Gould wrote widely on a variety of topics in evolutionary theory, and if he sometimes gave the impression of diluting his accomplishments by dividing his attention, the body of work that resulted seemed to have some of the quirkiness and originality of the subjects he chose to study.

The Structure of Evolutionary Theory, published by Harvard Univer-sity Press just months before his death, represents Gould's attempt to organize his scattered thoughts into a systematic treatise. Despite its length at more than 1400 pages, it fails in this respect, but in the end, it does something more important. It shows a man of sensitive intelligence endeavoring to master his doubts by disclosing them honestly.

Gould's theory of punctuated equilibria, which he advanced with Niles Eldredge in the mid-1970s, emphasized the extent to which Dar-win's theory of evolution fails adequately to account for the most obvious facts about the history of life. Darwin predicted that the fossil record would display a continuous distribution of animal forms. A great many species, by way of contrast, enter the fossil record without antecedents and depart without descendents. Before Gould's work, these facts were widely known, but not widely adverted. Gould succeeded in bringing them to the attention of evolutionary biologists, and often with great rhetorical skill. Their response was often to cast doubt upon the facts when possible, and upon Gould when necessary.

Gould quite understood that a theory in conflict with the facts is a great unhappiness. He thus argued that if evolution proceeds discontinuously, as the natural history of the dead suggests, natural selection must in part act on the level of a species. This is not a doctrine notable for its clarity. If species are treated as biological individuals, critics asked with some asperity, what is the source of their collective variation? And if they are not treated as individuals, what is it that natural selection acts upon? In *The Structure of Evolutionary Theory*, Gould came very close to expressing the obvious but forbidden thought that while variation and natural selection bulk large in evolutionary theology, they should weigh little in evolutionary theory. If this is so, what then remains of Darwinism as a doctrine? It was a question that Gould declined to ask, perhaps because criticism at the hands of his intellectual inferiors made him sensitive to the distinction between fearlessness and folly.

No notice of Gould's death should omit the most important fact about his life. The man was widely loved. The other day in Paris, while waiting in line at a bookstore, I overheard a customer asking for one of Gould's books entitled—he was quite sure—*La vie est belle*. The words in French mean that "life is beautiful." The urgency of his request suggested that these were words he was badly in need of hearing. Gould's book is, of course, entitled *Wonderful Life*, the title alluding to Frank Capra's film, *It's a Wonderful Life*; and the book itself deals with the discovery of the Burgess Shale. The French had mistranslated the title. It hardly mattered. The general impression current in the bookstore was that an American believed that life was beautiful and had written a book to say so.

The exchange did Gould great honor, for no matter what he had intended, this was, indeed, the impression that he had managed to convey.

HAS DARWIN MET HIS MATCH?

THE REVEREND WILLIAM PALEY PUBLISHED *NATURAL THEOLOGY: Or Evidences of the Existence and Attributes of the Deity, Collected from the Appearances of Nature* in 1802, shortly after the French astronomer and mathematician Pierre-Simon Laplace brought out the first two volumes of his *Traité du Mécanique Céleste*. Twenty-three years later, Laplace published the fifth and final volume of his magnificent treatise, and so brought to a close the first and greatest of scientific revolutions. He was thereafter promoted to the scientific pantheon, where, dignified among the immortals, he has reposed ever since. Paley, by contrast, has been the victim of a number of near-death experiences, and like so many men who have survived close calls, he has remained eager to remind the world that he is still alive.

Natural Theology is an argument for the existence of the Christian deity, a defense of the faith; but its centerpiece—the argument from design—has long slipped its doctrinal noose. "When we come to inspect [a] watch," Paley observed, we note that "its several parts are framed and put together for a purpose." It follows that "the watch must have had a maker." By the same token, living systems are very much like human artifacts in their complexity. Their own "marks of design," Paley urged, "are too strong to be got over." It follows again that this "design must have had a designer. The designer must have been a person. That person is God."

This argument retains something of the clarifying effect of a hammer struck twice. The marks of design—bang! The existence of a designer—bang again! But the publication in 1859 of Charles Darwin's *On the*

Origin of Species allowed biologists, who had long appreciated the force of Paley's inference, to escape its conclusions. Coordination and design are required to explain the emergence of human artifacts—waterwheels as well as watches, sun dials as well as sun dresses—but it is an alliance between time and chance that explains the complexity of living creatures. No design is in place in the natural world; no designer is needed.

Darwin's theory of evolution—random variation and natural selection—made it possible, as the contemporary biologist Richard Dawkins has written, to be "an intellectually fulfilled atheist." And if not fulfilled, then certainly at ease.

WITHIN THE PAST TEN years or so, however, such satisfactions as atheism affords have come to seem a little premature. Paley's claims have been pressed anew, as the argument from design, its revival widespread, has appeared in both evolutionary biology and (as we shall see) mathematical physics.

Within biology itself, the argument from design has become the gravamen of the "intelligent design" movement, which has engendered a current of sympathy among philosophers, biologists, and ordinary men and women long skeptical of Darwinian doctrine. The movement's spiritual adviser has been Phillip E. Johnson, a professor of criminal law at the University of California. In 1991, Johnson published a shrewd and widely read critique of Darwinian theory under the title, *Darwin on Trial*. Like Michael Denton's *Evolution: A Theory in Crisis*, published five years earlier, the book had some of the effect of wet dynamite. It sputtered for a while and then exploded.

Johnson's criticisms were not new; they were, in fact, part of a community of criticisms that had been marshaled against Darwin for many years. But Johnson went further than his predecessors. In what now seems a rhetorical masterstroke, he argued that biologists had embraced Darwin and rejected design by virtue of their allegiance to a worldview that was wholly extra-scientific: namely, philosophical naturalism. This is the view that, in Johnson's words, "the entire realm of nature [is] a closed system of material causes and effects, which cannot be influenced by anything from the outside."

Providence provided Johnson with opponents dutifully willing to confirm his diagnosis by exhibiting the disease that it signified. Thus, the literary critic Frederick Crews has argued recently in the *New York Review of Books* that "There *is* a fundamental principle in all science— only material and observable forces can count" (emphasis added). But if there is such a principle, Johnson has rejoined to critics like Crews, it is not to be found as the premise to any of the great physical theories; nor is it evident in their conclusions. It functions as an article of faith.

Although Johnson's book did not elicit the assent of the biological establishment—hardly a group interested in searching self-criticism—it did something more important. It attracted the attention of a number of younger scholars—Michael Behe, William Dembski, Paul Nelson, Jonathan Wells, Stephen Meyer. These were men with very respectable training in biochemistry, molecular biology, genetics, philosophy, and mathematics. They found Johnson's critique bracing in part because it confirmed an allegiance to Christian doctrine, and in part because, with the assumption of naturalism tentatively indicted in philosophy, they saw no reason to exclude the hypothesis of design in biology. In this they were correct. There is no reason.

Their decision to resurrect the design hypothesis represented a daring move on the chessboard, an unexpected flanking attack. Secular intellectuals have always affirmed their adherence to Darwin's theory because they have felt that the alternative is the abyss. The abyss had now acquired a loud, lively, and unembarrassed voice: living creatures are designed, and if the designer's identity is not obvious, it can easily be inferred. In the last years, advocates of intelligent design have put forward their views in books, conference reports, Internet publications, public lectures, scientific debates, and before various school boards—where, in a marvelous display of cheekiness, they have demanded that the public schools teach the controversy that they themselves have engendered.

Still, if the Reverend Paley has by their efforts been revived, he has been nearly dead before. Despite his reappearance on the intellectual scene, it is worth observing that a man who survives being drowned only to find himself in danger of being hanged may well possess a diminishing capacity to evade disaster.

DARWIN'S THEORY OF EVOLUTION and theories of intelligent design are in conceptual conflict. Although they may well both be false, they cannot both be true. Each party to the dispute has thus become persuaded that with the other character in the same room, there is simply not enough oxygen to survive.

It is the fossil record that is a preliminary witness to both theories of evolution and design, the stacked dead endeavoring to convey some discernible signal to their various partisans amid a great deal of background noise. But the signal is often confusing.

Living systems, Darwin maintained, are organized continuously, one species trailing off into the next. But the dead have done little to confirm this thesis. "Most of the fossil record," the paleontologist Robert Carroll has noted, "does not support a strictly gradualistic interpretation." It does not, that is, support the obvious and intended interpretation of Darwin's theory.

This circumstance is properly a source of satisfaction to proponents of intelligent design. In *Icons of Evolution* (2002), Jonathan Wells, who holds a degree in molecular genetics from Berkeley, summarizes the common view that the Cambrian explosion, the seemingly abrupt appearance of diverse and highly developed fauna in the early Paleozoic era, is a great imponderable in the geological record. The various life forms that erupted then look for all the world as if they were simply planted in the ground by some obviously competent designing intelligence.

But all that is on the one hand. On the other hand, if a great many species enter the fossil record fully formed, and then depart unchanged for Valhalla some million years later, the reverse is true as well, most notably in the evolution of the vertebrates. The margin between the reptiles and the mammals appears far more friable today than it did a century ago. Paleontologists now speak of mammal-like reptiles, chiefly the Pelyocosaurs and the Therapsids. Within the amniotes—the broad class comprising the birds, the reptiles, and the mammals—we have the benefit of an unusually full record of transition sequences and can see them proceeding inexorably over the course of 150 million years, with a variety of organisms busily changing their skeleton, gross anatomy, way of life, feeding habits, digestion, locomotion, and even their cellular physiology. These changes are demonstrable through the fossil record to a fine level of detail, the delicate bones of the mammalian ear emerging in stages from the bones of the reptilian jaw. In the late Triassic and early Jurassic, one can almost see the sequences touching, Cynodont to Morganucodon, scales to fur, cold blood to hot.

When the dead are interrogated at length, then, oxygen levels do drop in certain chambers of conflict, but curiously enough, both Darwinian and design theorists seem to be turning blue as a result.

IF THE DEAD ARE enigmatic and unavailing, what of the world of the living: the world in which creatures yawn, stretch, sleep, slither, and scramble for advantage? There, unceasing variety is the rule. Dogs divide themselves into more than 500 breeds, ranging from the very large to the very small; the domestic cat is found underfoot in only 50 or so flavors. The question is why.

Just why should dogs come in so many shapes and sizes, and not cats? Why *in principle?* Why, for that matter, should mimetic design in certain moths and butterflies exceed the observational powers of even the most capable predators, the design toppling over into a form of artistic ecstasy, while camouflage in other species is a matter of a few half-hearted disguises, a splash of stripes, a daub of pigment? Why do certain orchids involve themselves in an incredibly intricate reproductive routine while others, growing nearby, are content to slap a dusting of pollen on their shimmering petals and trust that their precious seed will be carried on the hind limbs of some heedless insect?

Why echolocation in the bats but not the buzzards? Why dancing among the bees but not the butterflies? Webs among the spiders but not the centipedes? Poison among the toads but not the frogs? Pouches among the possums but not the penguins?

And why the peacock's tail?

In the face of such questions, Darwinian theorists may be observed standing in silence. They are looking upward, apparently much occupied in assessing cracks in the ceiling. Beyond saying that that is just the way things are, what *could* they say? Intelligent-design theorists, on the other hand, hear the heavy hammer-beats of Paley's argument: the "marks of design" are too strong to be got over; such design must have a designer.

Yes? But what then follows? What follows *precisely*, once a designer has been acknowledged?

In commenting on the peacock's tail, Phillip Johnson lightheartedly imagined that such preposterous contrivances were just what one might expect from a designer with a sense of whimsy. But why stop at whimsy? Why not inadvertence, anger, mockery, incompetence, color blindness, bad taste, profligacy, or any other psychological disposition that can plausibly be connected to action?

And suppose the peacock's tail were found mounted on the hindquarters of a donkey; or the garden shrew were given the power to speak Dutch. Suppose a world in which cats acquired the arts of ingratiation while dogs declined all further services to man—*Sniff that suitcase? The one that might contain a bomb? I think not*—and whales returned from the sea to take up residence in the rich pasture land from which they originally departed. Suppose the carnival of the animals and the world's natural riot permutated in a thousand different ways. What then?

Why, nothing would change—save for the fact that it would be the donkey and not the peacock that prompted an inference to design. *His* choice, design theorists might say. But if His choice, why made, and according to which principles?

Design theorists may now be observed standing in companionable silence alongside Darwinian biologists. They, too, seem to be gazing upward, studying the same queer little cracks in the ceiling.

IN 1994, DAN E. Nilsson and Suzanne Pelger published a paper in the *Proceedings of the Royal Society* entitled, "A Pessimistic Estimate of the Time Required for an Eye to Evolve."[50] By "pessimistic," they meant an estimate that, if anything, exaggerated the length of time required for the eye's evolution. Even so, their conclusions were remarkable. "A light-sensitive patch," they wrote, "will gradually turn into a focused-lens eye" in only a few hundred thousand years.

Darwin had himself been troubled by the existence of the mammalian eye, whose evolution by random mutation and natural selection has always seemed difficult to imagine. Nilsson and Pelger's paper provided a welcome redemptive note. A few hundred thousand years and the job would be done. Authors have waited longer for their royalty checks.

As Nilsson and Pelger's paper gained currency, it amassed content it did not actually possess. Biologists who failed to read what Nilsson and Pelger had written—the great majority, apparently—assumed that they had constructed a computer simulation of the eye's evolution, a program that could frog-march those light-sensitive cells all the way to a functioning eye using nothing more than random variation and natural selection.[51] This would have been an impressive and important achievement, a vivid demonstration that Darwinian principles can create simulated biological artifacts.

But no such demonstration has been achieved, and none is in prospect. Nilsson and Pelger's computer simulation is a myth. In a private communication, Nilsson has indicated to me that the requisite simula-

---

50. Editor's note: In the original article, Pelger's name was misspelled as "Pilger." Berlinski apologized for that error in "A Scientific Scandal," and it is corrected in this version.

51. The physicist Matt Young offers this inadvertently rich account of his own inability to read the literature: "Creationists used to argue... that there was not enough time for an eye to develop. A computer simulation by Dan Nilsson and Suzanne Pelger gave the lie to that claim: Nilsson and Pelger estimated conservatively that 500,000 years was enough time."

tion is in preparation; his assurances are a part of that large and generous family of promises of which "your check is in the mail" may be the outstanding example.

What Nilsson and Pelger in fact described was the evolution not of an eye but of an eye*ball*, and they described it using ordinary back-of-the envelope calculations. Far from demonstrating the emergence of a complicated biological structure, what they succeeded in showing was simply that an imaginary population of light-sensitive cells could be flogged relentlessly up a simple adaptive peak, a point never at issue because never in doubt.

Despite a good deal of research conducted over the last twenty years, the mammalian visual system is still poorly understood, and in large measure not understood at all. The eye acts as a focusing lens and as a transducer, changing visual signals to electrical ones. Within the brain and nervous system, complicated algorithms must come into play before such signals may be interpreted. And no theory has anything whatsoever of interest to say about the fact that the visual system terminates its activities in a visual *experience*, an episode of consciousness. We cannot characterize the most obvious fact about sight—that it involves seeing something.

These are again circumstances that properly afford a measure of satisfaction to members of the intelligent-design community. But in what respect is our understanding improved by assuming that the visual system is the result of intelligent design? Unless very specific religious hypotheses are invoked, neither the identity nor the nature of the designer is known. The principles that he employs are a mystery, and the objects of his design are not well understood. Certain questions now reappear with unyielding insistence. Could a designer whose nature we cannot fathom, using principles we cannot specify, construct a system we cannot characterize?

If the question is unyielding, so, too, is its answer: who knows?

"If," Darwin wrote, "it could be demonstrated that any complex organ existed which could not possibly have been formed by numerous, successive, slight modifications, my theory would absolutely break down." These are lapidary words. They suggest a man prepared to subject his ideas to the sternest possible test. And they issue in an exemplary challenge: *show* me that complex organ or organism.

Darwin's challenge is as easy to state as it is difficult to meet—an example, perhaps, of the old boy's skill in adaptive self-protection. How could the requisite demonstration be conducted? There is, on the one hand, the organ or organism as it now exists. And there is, on the other hand, its history, the path taking it from the past to the present. But the evolutionary history of a great many species has been lost, the butterfly, the beetle, and the bat all emerging from time's endless fog as butterfly, beetle, and bat. To show that these organisms "could not *possibly* have been formed" (emphasis added) by a Darwinian mechanism demands a complicated argument, one that begins with their observable properties and then strikes negatively at *every* possible path by which they might have been created by "numerous, successive, slight modifications."

In *Darwin's Black Box*, published in 1995, the biochemist Michael Behe identified such an observable property—what he called "irreducible complexity"—and proposed precisely such a negative argument. Darwin's challenge having been met to Behe's satisfaction, logic then played its familiar role: if there is no Darwinian path to certain biological structures, they must have emerged by design.

Beyond being complex, just what is an *irreducibly* complex system? It is, Behe writes, "a single system composed of several well-matched interacting parts that contribute to the basic function, wherein the removal of any one of the parts causes the system to effectively cease functioning." Irreducible complexity applies to both biological and nonbiological systems. As an example of the latter, Behe offers the mousetrap. Its parts—hammer, spring, holding bar, platform, catch—interact; they each contribute to the trap's function, which is to catch or kill mice; and if any part of the trap is removed, the trap fails properly to function.

Behe's definition of irreducible complexity specifies a concept while simultaneously providing a means for its detection. That is its great strength. Is this organ, organism, system, or creature irreducibly complex? Remove a part. The system's subsequent failure is a sign of its irreducible complexity. The first steps in forging a scientific concept—make it clear, make it operational—have been taken.

Of course, a mousetrap is a human artifact, and the argument that Behe makes in cooked-up cases needs to be ported to biological systems. Here, the structures that Behe considers are biochemical: their parts consist of molecules. (To his credit, Behe offers a powerful and convincing case that Darwinian theories that fail to encompass the details of biochemistry are exercises in self-deception.) He considers in turn the bacterial flagellum, which is a tiny but horribly efficient rotary motor found in very primitive bacteria, the blood-clotting cascade, systems of molecular transport, and the immunodefense system—and he considers these systems on the level of their biochemical constituents.

As these examples reveal, and as readers of technical journals already know, biochemical systems display coordination, regulation, flexibility, delicacy of effect, unheard-of precision, efficiency, and great complexity. Behe's astonishment at the vivid and miraculous world thus disclosed is hardly a matter of hyperbole. It reflects the living truth. And one further point, the obvious one: the systems that Behe discusses are irreducibly complex. Strip just one protein from the blood-clotting system, whether by design or by disease, and the normal cascade leading from a cut to a clot fails, the poor creature evacuating its life along with its blood.

IT IS PRECISELY SUCH irreducibly complex systems, Behe writes, that meet Darwin's challenge. His argument now becomes conceptual rather than empirical. Some things cannot be done, and the "cannot" in this case has the full force of logical impossibility:

An irreducibly complex system cannot be produced directly... by slight successive modifications of precursor systems, because any precursor to an irreducibly complex system that is missing a part is by definition non-functional.

Just as an arbitrary angle cannot be trisected by means of a ruler and compass, so an irreducibly complex system cannot be produced directly by "numerous, successive, slight modifications" of its precursor systems. A mousetrap—or any other irreducibly complex system—cannot be assembled from its precursors if those precursors are missing a part of the original system. "Any precursor to an irreducibly complex system that is missing a part is by definition nonfunctional."

This is true. And it is clear. But it is not conclusive.

Behe's argument succeeds in canceling one path from the past to the present: the familiar, assembly line model in which a complex artifact like a mousetrap is assembled piece by piece, the base going first, the hammer, spring, holding bar, and catch attached in stages. But this mode of construction, which is common enough in the case of virtually every human artifact, is never seen in the biological world. That is in itself cause for suspicion. But the real objection is logical. If Behe's argument were to meet Darwin's challenge, it would have to cancel not one but *every* possible Darwinian path between the precursors of an irreducibly complex system and the system itself. He would thus have to show that irreducibly complex systems arise by means of an assembly line and *only* by such means. And that is plainly not so—even in the case of artifacts.

A three-legged stool, to chose a modest example, is irreducibly complex. Remove a leg and the stool topples. But a three-legged stool may be constructed by "numerous, successive, slight modifications" of a solid cylindrical block of wood, modifications that serve to hollow out three arches, leaving only the stool's three legs and seat upon completion. The stool is irreducibly complex; its precursors are not. And the path to complexity just sketched falls entirely within Darwin's ambit of small changes, each conferring some selective advantage. As those arches are carved, weight is reduced and stability increased.

A gap has now opened between Behe's definition and the conclusion he means to draw. With a few deft maneuvers, the gap may be further enlarged. An irreducibly complex system must fail if a part is removed; well and good. But from this it hardly follows that an irreducibly complex system must fail if its parts are *modified*, or if new parts are substituted for old.

The gap is large enough to have accommodated by now a number of Behe's detractors. After the publication of his book, biologists wasted no time in demonstrating how the common mousetrap might be modified in successive stages by an adroit substitution of parts, the platform disappearing in favor of the ground itself or a table top, and so on to the hammer, spring, holding bar, and catch. They then imagined this backward tinkering going forward by evolutionary means, the mousetrap getting better and better as its parts became more and more refined. For a time, the Internet vibrated with imaginary mousetraps of varying degrees of ingenuity, with a number of biologists losing themselves entirely in detailed designs, their diagrams drawn to scale and an aspect of obsessive lunacy stealing over their work.

Still: mousetraps are one thing, three-legged stools another, and living systems yet another. Biologists who had mastered the mousetrap were rather less successful in explaining the origins of the *biological* systems Behe had described in patient detail. When he himself searched the existing literature for plausible models, Behe had found little of interest and nothing of value. In reviewing his book, biologists asserted that the problem he had identified did not exist, and would anyway soon be remedied. A number of biologists have in fact proposed biochemical pathways leading to the formation of complex biochemical systems, the Internet again vibrating with their proposals. But Behe has in turn criticized their suggestions, arguing generally that where the pathways are Darwinian, they do not succeed, and where they succeed, they are not Darwinian. In this, it seems to me, he has been entirely convincing. The origins of irreducibly complex biological systems remain what they have always been, and that is an utter mystery.

But these rattling exchanges of small-arms fire represent little more than tactical exercises, the usual (and usually hidden) pop and counter-pop of controversy within biology itself. Behe's concerns were with what military theorists call grand strategy, and his ambition was to justify an inference to design by means of a *logical* resolution of Darwin's challenge.

What Behe has shown is that, with respect to certain biochemical systems, Darwinian biologists do not have a clue. What he *needed* to show was something stronger. He needed to show that they do not have a prayer.

As I intimated early on, the argument from design has been revived not only within biology but also within mathematical physics. Surveying the possibilities for scientific explanation, Jacques Monod, the eminent French molecular biologist, concluded rather grimly that the universe is ruled by necessity and by chance; there is nothing else. Monod made this claim in an elegant volume entitled *Chance and Necessity* (1970), a work that has come to play an important if often unacknowledged role in contemporary debate over these matters.

The laws of physics control the behavior of matter. It is there that necessity rules, its iron fist unyielding. Necessity explains the movement of the sun, the moon, and the planets; it explains the dancing play of atoms and molecules; it governs the very skin of creation. But chance—luck—is also a great governing force. It is chance that accounts for the origin of life, and chance again that has governed the emergence of human beings, with their complicated languages, their insatiable desires, and their doomed sense of curiosity.

But is this all there is? Do necessity and chance exhaust the explanatory options?

The biochemistry of living systems is based on carbon. The periodic table, which begins with hydrogen, and then straggles up and down the chart, covers more than 100 atoms. Among them, only carbon can bond with other atoms in four different directions, and so only carbon has the flexibility to create ever larger and more complicated organic structures.

As it happens, there is plenty of the stuff around, a state of affairs no less perplexing than the fact that apples unsupported fall downward. Why should this be so?

In 1946, the astronomer Fred Hoyle published the first of his pioneering studies, *The Synthesis of the Elements from Hydrogen*. The rich and complex panoply of chemical elements that is characteristic of the universe, Hoyle argued, must have been forged by a sequential process, one starting with hydrogen and then continuing step by step until the construction of carbon, the universe enlarging itself in stages. The process could not be chemical in any ordinary sense; chemistry leaves the interior of the atom untouched. The creation of matter must thus have been handled within the interior of the stars.

If hydrogen is the first step in the chain, deuterium is the next, the fused product of two hydrogen atoms and a vital link in the formation of carbon. The fusion that fashions deuterium depends crucially on the magnitude of the strong nuclear force. (There are four fundamental forces in nature: gravity, electromagnetism, and the strong and weak nuclear forces.) Like all forces, the strong nuclear force is expressed as a number. The value of that number is critical. Were it weaker than it is, hydrogen atoms would have found themselves unable to fuse; were it stronger, the stars would long ago have burned themselves out.

None of this took place. The strong nuclear force has the value that it does.

Although the next step is relatively simple—deuterium atoms fuse together to form helium—thereafter the story grows complicated. The laws of physics would normally prevent helium atoms from spontaneously fusing to form anything more interesting than helium. But two vagrant helium atoms meeting in the interior of a star *can* fuse together to form what is called a beryllium intermediate. It is an intricate celestial dance, and one that is highly unstable because beryllium intermediates are very short-lived.

In 1953, Edwin Salpeter discovered that the nuclear resonance between helium atoms and intermediate-beryllium atoms is precisely tuned to facilitate the creation of beryllium. (Nuclear resonance is the vibration produced when the frequency of the absorbing nucleus is identical to the frequency of the emitting nucleus.) The process is wonderfully elegant and entirely improbable. What would otherwise have been a process sputtering into the abyss has now been promoted to a process taking helium into a new element.

But there is yet no carbon—and here Hoyle entertained his most daring conjecture. Quite before any evidence was available, he predicted the existence of a *second* nuclear-resonance sounding directly between beryllium and helium, one that would in its turn enable the great stellar furnaces to produce carbon in abundance. This prediction was verified. The steps involved in the construction of carbon lay revealed—and so, too, the path to life.

Hoyle was deeply troubled by the specific nuclear-resonance levels that he had discovered. That they expressed physical properties of material objects was not in doubt. But what explained the appearance of those physical properties in the great causal chain stretching from the Big Bang to the emergence of life? On this point, the laws of physics and the vagaries of chance would both seem unavailing. "A commonsense interpretation of the facts," Hoyle observed, "suggests that a superintellect has monkeyed with physics, as well as with chemistry and biology, and that there are no blind forces worth speaking about in nature."

THE FACTS ARE NOT in dispute. In both molecular biology and mathematical physics, things seem to have happened *in defiance of all the odds*. It is the analysis of these six ultimately crucial words that has occupied William Dembski, a young mathematician of impeccable academic training and a specialist in probability theory.

Dembski is the author of *The Design Inference* (1998) and *No Free Lunch* (2001). These are serious and difficult works. They merit respect. In much that he has written, Dembski's proximate target has been Darwin's theory of evolution, although, like so many of Darwin's critics, Dembski has also discovered that Darwin is a most elusive target, bobbing and weaving around the ring and every now and then pausing to deliver a smart blow of his own. But if Darwin is Dembski's proximate target, his real target all along has been Jacques Monod, and his aim has been to demonstrate that necessity and chance do not exhaust the options for scientific explanation. There is design as well.

Other members of the design-theory community have seen design emerging as Darwinian theory weakens. Dembski is interested in restoring design to the community of *scientific properties*—like mass, force, or momentum—and thus endowing the argument from design with a direct form of legitimacy.

Dembski begins with a very plausible general premise: events that can be explained neither by the laws of physics nor by chance must be explained by an appeal to the intervention of an intelligent agent.

It is surely plain that many events cannot be explained in terms of the laws of mathematical physics, or any other laws for that matter. The precise way in which Shakespeare arranged 113 words to fashion his 18th sonnet ("Shall I compare thee to a summer's day..."), owes nothing to any system of physical laws. Shakespeare might, after all, have written, "I shall compare thee to a summer's day," losing poetical effectiveness but violating no physical constraints.

It is just as plain that many events cannot be explained by chance. The odds are just too small. "Phenomena with very small probabilities," the French mathematician Emile Borel remarked, "do not occur." They do not occur, that is, by chance.

Borel's thesis, if not a tautology, is certainly a truism. And yet Dembski argues that Borel's maxim, taken as it stands, is false. "Plenty of highly improbable events," he writes in *The Design Inference*, "happen by chance all the time."

> The precise sequence of heads and tails in a long sequence of coin tosses, the precise configuration of darts from throwing darts at a dart board, and the precise seating arrangement of people at a cinema are all highly improbable events that, apart from any further information, can properly be explained by chance.

Truisms thus seem to be in conflict. But of the two, it is Borel's maxim, Dembski concludes, that requires revision. Certain kinds of improbable events happen "all the time." Others do not. In the case of those that do not, an additional property has to be taken into account—the "further information" that Dembski mentions. They are *specified*.

A specification is something like a pattern, a description, or an identifying tag; it represents a selection from among alternatives, and so involves a division of the stream of possibilities. Dembski offers an example drawn from archery. If an archer is given a large target, and encouraged simply to let fly, then, no matter where his arrows land, they will have reached an improbable place and so have realized an unlikely event. Chance is still in charge. But if a target—say, a familiar bull's-eye—has been fixed, then a halter has been placed on the riot of possibilities. The fixed target serves as a specification, picking out one set of places to the exclusion of all others, and the archer hitting this target repeatedly has defeated chance. He has done so by exhibiting skill and attention: marks of design, intimations of intelligence.

In Scrabble, any particular long sequence of letters forming a nonsensical word—PZQATRV—is improbable; but, conforming to no pattern, it is not specified. A short sequence of letters spelling out a particular English word ("dog," say) is specified (since it conforms to a pattern), but not improbable; it might have appeared on the Scrabble board by chance. But Shakespeare's 18th sonnet is both specified and improbable. It conforms to a pattern, and it is unlikely to have appeared by chance.

These observations suggest to Dembski a reformulation of Borel's maxim: it is specified phenomena with very small probabilities that do not occur by chance.

Having offered Borel a face-saving detail, Dembski is now in the position to offer Paley a life-saving deal. Specified improbability is the very mark of design: the trace left in matter of a cognitive act. It is specified improbability that characterizes thimbles and thermostats and watches, bone-handled knives, computer programs, steam shovels, symphonic suites, lace lingerie and lipstick, Vicuna coats, Etruscan vases, sand-castles, epic poems, French epigrams, German novels, mathematical theorems, paintings done in oil and a sign found in Lubbock, Texas, spelling out the words *Good Eats*.

And it is specified improbability, again, that characterizes DNA and the various proteins, the bacterial flagellum, and the blood-clotting cascade. It is in fact ubiquitous, this property, a kind of steady note sounding throughout living systems, from the Tyrannosaurus roaring in the Jurassic right down to the cellular telomeres currently vibrating under its influence. If the specified improbabilities in human artifacts suggest inexorably a human designer, the specified improbabilities in biological artifacts suggest inexorably a biological designer as well.

Nor is there any need to stop at the margins of the biological world. Nuclear-resonance levels are both improbable and specified—improbable because nuclear-resonance levels could have had other values, and specified because these nuclear-resonance levels and no others are needed in order for life to exist. The inference to a designer proceeds apace; it gathers force. It is by these means that Paley has been offered yet another lease on life.

BUT THE HINGE OF doubt begins to creak. Eliminating chance through small probabilities—these words are the sub-title of *The Design Inference*—is an enterprise that cannot begin if probabilities cannot be objectively assigned to various events. It is not always easy to see how this might be done. It is one thing to assign probabilities to a lottery, or a horse race; but just how might one go about assigning probabilities to, for example, the creation of *Las Meninas*, Velasquez's great painting of the Spanish royal court—the first step in establishing its claim to design? Probabilities require a world of alternatives. There are either no alternatives to *Las Meninas* because it is unique, or infinitely many because it resembles every other splotch of paint on canvas.

What probability, for that matter, should we assign to the emergence of a chicken? Given an egg, the chicken's appearance is almost inevitable; absent that egg, impossible. Which is it to be? What is the probability that Luxembourg might succeed in conquering Brazil? Or that Chinese officials might declare Basque the new language of China? Or that the queen of England might have been constructed of ice instead of flesh, her body kept at constant temperature by means of a far-reaching Masonic conspiracy?

It is not easy to say. And that, of course, is the problem.

If it is difficult to assign probabilities to a great many things, events, processes, or circumstances, it is still more difficult to determine whether they conform to a specification, the bridge to the bridge to agency. A specification is a human gesture. It may be offered or withheld, delayed or deferred; it may be precise or incomplete, partial or unique. Paley's watch may thus be specified in terms of its timekeeping properties, but it may also be specified in terms of the arrangement or number of its parts; and its parts may be specified in horological, mechanical, molecular, atomic, or even sub-atomic terms. Each specification introduces a different calculation of probabilities. Paley's watch may be improbable under one specification, probable under another.

And whatever the specification of specification, it is in any event hardly clear that intentional acts must be specified in the first place. A random walk, a careless gesture, the significant sigh, the soul-shattering sniffle—to what specification do they conform? The word *basta* overheard in a dark Italian alley, a woman's voice conveying all the urgency of human intention—but a *specification*? A specification is, after all, a way of distinguishing among possibilities. What possibilities are at issue in these cases? And what choices from among a sample of possibilities do the events represent?

These examples could be multiplied at will. They suggest that there are plenty of intentional acts—acts reflecting a choice—that neither require nor conform to a specification. If that is so, then specified improbability is not necessary to trigger an inference to design. If it is not necessary, there then remains the question of whether it is sufficient.

According to Dembski, if and only if there is both specification and improbability can chance be eliminated in favor of design; otherwise, not. No one, for example, would attribute the origin of Shakespeare's 18th sonnet to chance. That is true: the sonnet is indeed highly improbable. But what does *specified* improbability have to do with it? If 113 words are randomly rearranged, every possible arrangement of them has the same probability of occurrence, whether the arrangement represents Shakespeare's limpid lines or sheer gibberish. No one would attribute *any* particular organization of Shakespeare's words, whether specified or not, to chance. And for just the reason Borel offered: phenomena with very small probabilities do not occur. A specification does nothing to alter the odds. It is improbability that does the heavy lifting; the specifications have gone along for the ride.

Once this is seen, Dembski's reformulation of Borel's principle must be discreetly withdrawn; it rests on a mistake. It is true that "plenty of highly improbable events happen by chance," but false that they happen *all the time*. They happen, in fact, precisely as many times as one might expect, given their probabilities.

Adieu to specifications and adieu to Dembski's initial premise. If specified improbabilities are not necessary to trigger the elimination of chance, and if the elimination of chance is not sufficient to trigger an inference to design, it follows that specified improbabilities are not sufficient, either. We are left with unlikely events and the sense of puzzlement that they inevitably provoke. In both molecular biology and mathematical physics, things seem to have happened in defiance of all the odds. And we do not know why.

IF DESIGN IS TO be a scientific category, it must answer, as Dembski (following Paley) quite understands, to a scientific property. And if design leaves marks in the natural world, biologists and physicists dissatisfied with naturalism as a doctrine are committed to identifying those marks. But the curious thing is that the more these mysterious marks of design are sought, the more elusive they seem to be.

Nineteenth-century biologists often imagined that living systems were perfumed by an *élan vital*, which French military theorists proudly compared to the mysterious *élan* animating their soldiers. Although intellectual fashions have changed over the course of one hundred years, the *élan vital* has not disappeared. It has simply been renamed. Citing the evolutionary biologist George Williams, Phillip Johnson has argued that living systems comprise two domains, the material and the codical. The material domain is exhausted by certain material objects, chiefly DNA and the various proteins. The codical domain consists of something more. Many scientists have taken to identifying this "something more" with the concept of information, but since the usage of that term is so loose we might as well refer to it as the ineffable inimitable.

Both Johnson and Williams agree that what I am calling the ineffable inimitable "is fundamentally distinct from the chemical medium in which it is recorded." Thus Williams argues that Cervantes's great novel *Don Quixote* just *is* the ineffable inimitable, "most often coded as a pattern of ink on paper" but also capable of existing "in many other media." Less interested in metaphysical niceties, Johnson draws a more straightforward conclusion from Williams's remarks: if matter and the ineffable inimitable are truly incommensurable, "then Darwinism cannot be true as a theory." Random variation and natural selection are physical properties of material objects. Explaining the origin of the ineffable inimitable by an appeal to the material properties of living systems is rather like explaining the origin of *Don Quixote* by an appeal to the physical properties of ink and paper. Not to be outdone, Richard Dawkins has remarked that he has long admired the idea of information—the ineffable inimitable—and made it the foundation of his research on many occasions.[52]

George Williams is a prominent evolutionary biologist; and Phillip Johnson his critic. What the ineffable inimitable is, neither man is prepared precisely to say.

Over the years, other scholars have been more forthcoming. The ineffable inimitable has been identified with a fluid, an aura, a spirit, an

---

52. Dawkins's remarks, as well as those of Johnson and Williams, appear in *Intelligent Design Creationism and its Critics* (2001), edited by Robert T. Pennock.

entelechy, a soul, a field, and a force; it has been described in terms of entropy, energy, complexity, Kolmogorov complexity, organization, self-organization, hierarchical self-organization, organized hierarchies, catastrophe theory, graph theory, automata theory, and various theories of computation and control. Irreducible complexity and specified improbability are the latest incarnations of the ineffable inimitable.

The search for the ineffable inimitable continues, if only because it answers to a human need. But not every urgent human need is destined to be met, and not every anxiety of the heart has its causes in reality itself. Experience might suggest another counsel. The search for the ineffable inimitable is fruitless.

AN INQUIRY OF THIS sort must inevitably end by disappointing every party to the dispute while exhilarating none. Members of the intelligent-design community have been stalwart in their attack on Darwin's theory of evolution. This has been their strength. Mindful of the old-fashioned wisdom that you can't beat something with nothing, they have proposed design as an alternative, without ever quite realizing the extent to which Darwinism and design share very similar weaknesses.

Darwin's theory of evolution makes use of a fantastic extrapolation in which the mechanism of random variation and natural selection, responsible for a number of trivial local effects, is read into the great global record of life itself, so that the development of antibiotic resistance in bacteria becomes the model for the development of such complex structures as the mammalian eye, or the immunodefense system. As design theorists have noted, this is rather like arguing that since kangaroos can hop, they can, given time enough and chance, learn to fly.

The same faith in the same flaw—the fallacy of extrapolation, to give it a name—runs through design theories as well. Given the human ability to fashion objects and create artifacts, design theorists argue that some similar process is at work in the construction of biological structures. Is it? What justifies the assumption that a process that accounts for the construction of a pocket watch, a sundress, or a nation-state might also account for the emergence of the blood-clotting cascade or the immunodefense system? It is entirely possible, after all, that complicated biological structures lie quite beyond the possibility of design, no matter the designer and regardless of the principles employed. Faith in a designed universe might well be rather like faith in a planned economy, a doctrinal commitment that cannot survive a confrontation with experience.

It has likewise been the ambition of the design-theory community to interpose design as a category that might occupy the contested territory between necessity and chance. No philosophical argument renders this ambition absurd; and no scientific theory suggests that it is impossible. But if theories of design cannot be ruled out of court by facile invocations of philosophical naturalism, design theorists have for their part tended to underestimate the enduring intellectual force behind Monod's claim that the categories of chance and necessity are mutually exclusive and jointly exhaustive.

WHAT INTERESTED HIM THE most, Albert Einstein once remarked, was whether the deity had any choice in the creation of the universe. Einstein is very often presumed to have had a religious sensibility; this remark reveals its profoundly austere nature. For Einstein's comment suggests two, and only two, possibilities: a deity with no choice, or a deity with a world to choose and then to make.

If the deity's actions in creating the world are necessary, it is the principles that explain the *nature* of this world that are of interest and compel attention. In the end, I suspect, those principles will be matters of pure logic, simple and compelling, and if ever they are discovered, human beings will share with the deity pleasure in their contemplation. But this is a vision that subordinates the designer to the design, one in which the designer is bound to the same wheel of inexorability that binds the rest of us.

If, on the other hand, the universe represents a deity's free creation, it is the principles that explain his *choice* that are of interest and compel attention. A universe in which necessity does not control creation must have presented itself to the deity as an immense ocean of possibilities, our own universe an island amid other possible islands, the whole archipelago arising from unfathomable nothingness. And those divine possibilities, so far as we can tell, must have been all equally likely.

God alone knows what God is thinking; as suffering men and women have long known, His inscrutability is one of His less attractive features. But if all options are equally likely, and no motives known, we are in the position of observers contemplating a vast, cosmic lottery, one whose outcome is neither favored by the odds nor specified by its designer. It is chance that now returns as the default hypothesis, if only because it is the only hypothesis that is completely consistent with our ignorance.

By this means, it might well seem, Monod's resilient double distinction, chance and necessity, returns to haunt the intellectual scene, with design, having been admitted as a possibility, once more in danger of collapsing along with the contested space between the two categories.

And Paley, poor Paley? Dead at last, or at least not very vigorously alive.

2003

DARWINISM VERSUS INTELLIGENT DESIGN: DAVID BERLINSKI & CRITICS

In December 2002, Commentary *magazine published David Berlinski's essay "Has Darwin Met His Match?"; page 286 in this volume. The article, like its predecessor "The Deniable Darwin," created a stir and elicited another round of letters from scientists and philosophers. Below are the letters, and at the bottom is Dr. Berlinski's reply. Originally published in March 2003 in* Commentary.

—EDITOR

Paul R. Gross
University of Virginia, Charlottesville, Virginia

CONGRATULATIONS ARE IN ORDER. In his latest *Commentary* essay on "Darwinism"—as it is often called by those who do not know much evolutionary biology—David Berlinski seems to have reversed himself. He is now critical of the effort to rehabilitate the ancient "argument from design," these days holiday-wrapped as "intelligent-design theory." This change of mind is all the more praiseworthy because Mr. Berlinski is not only the author of "The Deniable Darwin" (*Commentary*, June 1996) but, according to his current author's note, closely associated with the Discovery Institute, the conservative Christian think tank that serves as the primary promoter of "intelligent design."

The manner of Mr. Berlinski's dismissal is less creditable, however. His argument might be paraphrased as follows: intelligent design has failed, evolutionary biology has failed, and therefore nobody has a plausible scientific explanation for the diversity of life on earth. This is absurd.

The reasons Mr. Berlinski gives for the failure of intelligent-design theory have all been given before—in the professional literature, in the introductory biology courses of all decent colleges, in half a dozen very recent specialist books, on fifty flourishing scientific websites. However well intelligent-design apologetics is doing as politics, there is nothing new about its abject failure as science.

On the other hand, none of the reasons Mr. Berlinski hints at for rejecting "Darwinism" stand up. They are the familiar creationist pabulum, supposedly demonstrating the grand flaws and gaps in evolutionary biology. But these arguments, too, have been addressed in the professional literature, in tens of thousands of papers, and in dozens of excellent, bestselling popular science books—and they have been soundly refuted. Unfortunately, a non-biologist reader of "Has Darwin Met His Match?," innocent of this vast body of knowledge, will have no notion of its content or even, perhaps, of its existence, and may therefore take Mr. Berlinski's assertions as true as well as deep, which they are not.

Mr. Berlinski says that the argument from design has been taken up again in evolutionary biology and mathematical physics. This, if true, would give it some scientific seriousness. But it is not true. There is not one publication in recent biological journals, out of the tens of thousands of articles published annually—a huge subset of them devoted specifically to evolution—that undertakes to rehabilitate the argument from design. None of the intelligent-design "theorists" mentioned in his essay has ever published the claim in an original article in a regular, refereed biological journal. Nor, of course, has Mr. Berlinski himself done so.

As for William Dembski's lucubrations on chance and probability—which even Mr. Berlinski now finds flawed—I know of no professional publications (other than Dembski's own books and his frenetic responses to critics) that treat them as of interest for mathematical physics.

There are, however, a dozen point-by-point refutations of those claims, many of them available on the Internet, by well-known physicists, philosophers, mathematicians, and biologists.

Mark Perakh
Bonsall, California

WHAT IS PERHAPS MOST amusing about David Berlinski's article is his apparent change of mind on the subject of intelligent design. Having supplied rave blurbs to the books of such prominent advocates of this "theory" as William Dembski and Michael Behe, he now casts doubt on the concepts they promote. What explains this new view? The advocates of intelligent design are anxious to be taken seriously as scientists. From their standpoint, it may seem to be a step forward that Mr. Berlinski gives their ideas a status equal to that of Darwinism, even if only by casting doubts on both equally.

The list of unsubstantiated assertions in Mr. Berlinski's paper is long. Though enviably eloquent, it adds nothing to the debate other than to try to put the failed hypothesis of intelligent design on a par with genuine science.

Jason Rosenhouse
Kansas State University, Manhattan, Kansas

DAVID BERLINSKI'S ARGUMENTS AGAINST intelligent-design theory suffer only from a lack of originality; critics have been making the same points for years. Still, he gives a good picture of why most scientists find this field unpromising.

Mr. Berlinski's ongoing antipathy to mainstream biology, by contrast, is based on a caricature. Blubbering about gaps in the fossil record cannot change the fact that, with millions of fossils collected and classified, not one is out of place from a Darwinian standpoint. Also, Darwinism requires continuity at the level of the genotype, not the phenotype.

And while it is true that information is independent of the medium containing it, there is nothing mysterious about the idea that changing the medium can alter the information. Genes do mutate, thereby coding for different proteins from before, and new functions are sometimes observed to arise as a result. The source of the code is indeed mysterious, a fact that would be troubling if Darwinism were a theory about the origins of life. Since it is not, Mr. Berlinski's hand-wringing on the subject is inappropriate.

The fruits of evolutionary theory are disseminated in thousands of research articles in dozens of journals every year. Obviously, the people charged with the responsibility of entering their labs and solving problems find it useful. Numerous complex systems have been studied and the major steps of their evolution revealed. Where data are copious, they are all in accord with Darwinian expectations; where mysteries remain, the problem is a lack of data, not a lack of theoretical robustness.

In response to all this, intelligent-design theorists fold their arms and shake their heads. That is their right. But for all their bloated claims and hyperbolic rhetoric, they have made no contribution toward solving an actual biological problem. For that matter, neither has David Berlinski.

Clay Shirky
Brooklyn, New York

DAVID BERLINSKI EXPENDS A lot of energy trying to make evolutionary theory look like religion. He mentions Darwin often and gestures toward a set of beliefs called "Darwinism," as if Darwin were some sort of high priest and Darwinism a sect. But biologists do not practice "Darwinism" any more than physicists practice "Einsteinism." For biologists, *The Origin of Species* is a historical work, not a how-to manual.

Unlike Berlinski's "Darwinism," evolutionary theory is a field of vigorous and current debate. Darwin had no idea how heredity worked—it took fifty years for R. A. Fisher and his colleagues to relate Darwin's work to Mendel's. Darwin likewise had no explanation for altruism—it took a hundred years and the work of William Hamilton to develop a plausible hypothesis. And there are still myriad open questions today.

Mr. Berlinski does not mention Fisher or Hamilton or any of the other thousands of biologists exploring these open questions because he is anxious to present evolutionary theory as a fixed philosophy rather than a dynamic science. Indeed, the only publishing biologist he mentions is Jonathan Wells, the Unification minister whose prayers and conversations with the Rev. Sun Myung Moon convinced him, in Wells's own words, "that I should devote my life to destroying Darwinism," which he reports as his spur for studying biology.

But no matter how vivid his theological motivations may be, Wells (or indeed anyone in the intelligent-design camp) will eventually have to offer an alternative explanation of the variety we see among living things that does not rely on the key evolutionary idea of descent with modification. Furthermore, that explanation will have to be tested, and those tests will have to be scientific, not religious.

Here even Mr. Berlinski gets off the intelligent-design bus, because he can see where it is heading. If the only way to defeat a scientific hypothesis is with another scientific hypothesis, then we will never get to what he seemingly wants: a world where science stops trying to explain things. He likes intelligent design for criticizing evolutionary theory, but he dislikes it because it is itself too much like science, and may end up having to make testable assertions in the domain of what he calls "the ineffable inimitable."

Because he cannot get his hands on the steering wheel, Mr. Berlinski reaches for the brakes, asserting that a large domain of interest is or should be permanently exempt from scientific inquiry. This has been a standard plea of the religious for the last several centuries, and Mr. Berlinski's formulation—that the search for "the ineffable inimitable" is fruitless—is a classic of the genre. History has not been kind to those who predict an end to scientific progress, and declare God to be the sole possible explanation for the remaining mysteries.

S. L. Baccus
Houston, Texas

INTELLIGENT-DESIGN THEORY CAN BE summed up, as best I can deter-mine, by two propositions: there is a creator and evolutionary theory is false. Its advocates do not believe the creator is sophisticated enough to have created our universe in such a complex way. It appears they would prefer a magician, waving a wand and shouting magical words of creation.

Nor should one overlook that the creator, as defined by Mr. Berlinski and his associates at the Discovery Institute's Center for Science & Culture (CSC), is the Christian God. The group's ultimate goal appears to be getting Christian beliefs taught in our schools. In his book, *The Wedge of Truth: Splitting the Foundations of Naturalism* (2000), Phillip E. Johnson writes,

> The proper metaphysical basis for science is not naturalism or mate-rialism but the fact that the creator of the cosmos not only created an intelligible universe but also created the powers of reasoning which en-able us to conduct scientific investigations…. [T]he materialist story thrives only as long as it does not confront the biblical story directly… So it is of the greatest importance that we ask the question: "Has God *done* something to give us a start in the right direction, or has He left us alone and on our own?" When we have reached that point in our ques-tioning, we will inevitably encounter the person of Jesus Christ, the one who has been declared the incarnate Word of God, and through whom all things came into existence.

In the mission statement on its website, the CSC bends over back-ward to avoid appearing as the Christian creationist group that it is. It is only in written material and speeches for Christian audiences that its representatives admit their real beliefs and goals.

Morton Rosoff
Yonkers, New York

LIKE CREATIONISTS, DAVID BERLINSKI targets natural selection in his critique of Darwinian theory, oblivious to other mechanisms of evolution that Darwin never knew about: pseudogenes, genetic drift, symbiosis, chromosomal rearrangements, interbreeding, developmental proteins, etc.

"Creation science," a slick variation of creationism, seeks to transport philosophical and theological ideas into a scientific program to investigate intelligent design, but offers no empirical evidence, no models, no verifiable predictions, no possibility of correction or elaboration. It is not intended to improve our knowledge or extend scientific horizons but to support religious faith.

The writers whose work Mr. Berlinski describes propose a version of the "anthropic principle": that the universe and its fundamental parameters must be such as to admit the creation of observers. The tautological foundation of this idea is that observers find themselves in universes that allow them to observe. It is only a short step from here to intelligent design, the advocates of which have expressed theological inclinations.

It is true that science has its own philosophical or faith-like underpinnings. They consist of methodological tools like Occam's Razor (don't invent unnecessary entities to explain something), falsification (can a hypothesis, in principle, be falsified?), and balance (extraordinary claims need extraordinary evidence). These devices have served science with great success for 300 years.

Matt Young
Colorado School of Mines, Golden, Colorado

CONSIDERING THE GREAT NUMBER of biologists whom David Berlinski charges with failing to read Dan E. Nilsson and Suzanne Pelger's paper on the evolution of the mammalian eye, I am flattered that he singled me, a physicist, out for criticism. It is Mr. Berlinski, however, not the biologists or I, who suffers, as he writes, from "an inability to read the literature."

In describing the paper by Nilsson and Pelger (not Pilger, as Mr. Berlinski misidentifies her), I wrote that they had performed a computer simulation of the development of the eye. I did not write, as Mr. Berlinski suggests, that they used nothing more than random variation and natural selection, and I know of no reference that says they did.

Mr. Berlinski's complaint that they described an eyeball, not an eye, is typical of those who tilt at neo-Darwinism. If Nilsson and Pelger had simulated the development of the light-sensitive patch with which they started, Mr. Berlinski would have asked where the original cells came from, and so on back to primordial ooze. Criticizing what we know about biology by harping on what we do not know is like criticizing a sturdy, durable, and functional concrete wall because there are chinks in it.

The paper by Nilsson and Pelger is a sophisticated simulation that even includes quantum noise; it is not, contrary to Mr. Berlinski's assertion, a back-of-the-envelope calculation. It begins with a flat, light-sensitive patch, which they allow to become concave in increments of 1 percent, calculating the visual acuity along the way. When some other mechanism will improve acuity faster, they allow, at various stages, the formation of a graded-index lens and an iris, and then optimize the focus. Unless Nilsson and Pelger performed the calculations in closed form or by hand, theirs was, as I wrote, a "computer simulation." Computer-*aided* simulation might have been a slightly better description, but not enough to justify Mr. Berlinski's sarcasm at my expense.

More important, had Mr. Berlinski read more carefully, he would have recognized that Nilsson and Pelger's accomplishment was not to "flog" a collection of cells up an adaptive peak, "a point," he writes, "never at issue because never in doubt." Rather, they showed that the required time was geologically short—a point very much at issue among Mr. Berlinski and his neocreationist colleagues.

Tony Doyle
New York City

ALTHOUGH DAVID BERLINSKI OFFERS sharp criticisms of the theory of intelligent design, he is inclined to pull punches. First, he leaves unchallenged Phillip E. Johnson's claim that naturalism is "wholly extrascientific" and thus no more justified than the theological account offered by those who would maintain that the organic world presents compelling evidence of design. This is misleading. Naturalists simply contend that, for any event or phenomenon that needs explaining, we should seek only physical causes.

This is not dogma. We have a robust idea of physical causation, supported by countless examples; time and again, the search for physical causes and only physical causes has paid off. By contrast, we do not have a clue about how the theist's nonphysical "causation" is supposed to work. The appeal to nonphysical causes, to which intelligent-design theory is committed, amounts to an appeal to ignorance.

Mr. Berlinski rejects William Dembski's version of the argument from design, but he is well disposed toward Dembski's "very plausible general premise":

> [E]vents that can be explained neither by the laws of physics nor by chance must be explained by an appeal to the intervention of an intelligent agent. It is surely plain that many events cannot be explained in terms of the laws of mathematical physics, or any other laws for that matter. The precise way in which Shakespeare arranged 113 words to fashion his 18th sonnet ("Shall I compare thee to a summer's day…") owes nothing to any system of physical laws.

Not so fast. Shakespeare presumably moved his hand while composing those immortal lines. Hand movements are physical, so they have fully physical explanations, like any other physical event. It follows that they are strictly governed by the laws of physics; hence they are physically explicable. Of course this sort of explanation might not interest us much. We might prefer to know something about the person who inspired the lines or whether the poet penned the verse in July swelter. But this does not mean that his composition owed "nothing to any system of physical laws" or that we need to appeal to a divine "intelligent agent" to explain his behavior. The only eligible intelligent agent was Shakespeare himself, physical from head to toe.

Besides, suppose for the sake of argument that we cannot in principle explain physically how poetic inspiration produces a sonnet. Just how is a non-natural or supernatural agent supposed to help us? Can the theist explain how God is capable of creating self-conscious beings who occasionally write gorgeous love poems? Doesn't positing a designer or agent here just multiply mysteries?

Chris Beall
Lafayette, Colorado

DAVID BERLINSKI LEAVES UNTOUCHED the major contradiction at the heart of the intelligent-design argument. Intelligent-design advocates want to have it both ways. On the one hand, they claim that wherever we find a very complex and ingeniously arranged organ, we have evidence of both design and a designer. From the design we can clearly infer the designer's purpose—to make an organ that serves its owner well. Astonishing engineering to an obvious purpose surely tells us that we are looking at the work of a purposeful designer.

But the world in which organisms really live consists, primarily, of other organisms. And some organisms have organs that are very complex and ingeniously arranged so as to do the work of killing or parasitizing other organisms. Their potential victims have equally complex organs whose purpose is clear—to evade, fight off, or endure the insults enabled by their enemies' ingenious designs. What is the purpose of a designer working out clever mechanisms for organisms to work so hard to take and kill, keep and survive?

At this point, design advocates retreat into mysticism. The designer, it turns out, works in mysterious ways, and his purposes are beyond us. But the problem with the mysterious designer should be obvious. It is from his purposefulness, not just his skill, that we should infer his existence, yet when his work is examined at a scale greater than a single organism, he appears to be working at cross-purposes.

The remaining mystery is that these contradictions, which logically weaken the claims of design advocates, seem instead to strengthen their conviction. This is cleared up by a simple observation: when belief grows as evidence wanes, we can be certain we are not dealing with science and uncertain knowledge, but with religion and certain faith. While there is nothing wrong with proclamations of faith, we are under no obligation to consider them serious contributions to our attempt to understand how the world really works.

George C. Williams
South Setauket, New York

THOUGH I DO THINK the Reverend William Paley's approach a good one—for information about the Creator, examine creation—David Berlinski seriously understates God's less attractive features. It is an impressive universe out there, so that calling God *almighty* is appropriate. Calling Him *good* I will briefly deal with below. Calling Him *wise* was Paley's emphasis—as it is of recent anti-Darwin critics like Phillip E. Johnson—and is what I want to discuss here.

There are two ways of evaluating functional designs: according to their general plan or according to the quantitative precision of their parts. The second approach gives impressive positive results. We have come to understand the functional precision of such eye measurements as the distance from the lens to the retina, the dimensions and shape of the lens, etc. The eye is of nearly optimum dimensions.

As for general design features, consider two: our upside-down retinas and the number of our eyes. The retinal orientation is the result of its embryonic origin and positioning in a tiny and nearly transparent ancestor of all vertebrates. When descendants of this early ancestor got big enough for the upside-down retina to be disadvantageous, the problem was partly solved by making the blood vessels and other services between the retina and the lens as thin and transparent as possible. As for having just two eyes: it orients them optimally in relation to such conflicting ecological needs as distance perception and breadth of coverage. We have excellent distance perception because our arboreal ancestors maximized their ability to judge the distance to the next tree branch. Unfortunately, having merely two forward-pointing eyes is seriously deficient in seeing what may be sneaking up behind.

All our adaptations and those of other organisms are similarly compromised by historical constraints. Adaptive changes can come only by natural selection altering quantitative variables such as size and shape (or numbers when they are large, like scale rows of fishes).

As I argue in my book *Plan and Purpose in Nature* (1996), Paley was right that a scientific treatment of the creation can turn theology into a rigorous science. As noted above, God can make elaborate biological adaptations by trial-and-error selection, but it never occurs to Him to alter the basic plan. So He is stupid! Is He good? The answer is obvious from any objective evaluation of His biological creation. Go out into the woods and note the ratios of success to failure, of happiness to fear and pain. As Thomas Huxley recognized in his 1893 lecture "Evolution and Ethics," the Creator is clearly evil.

Karl Wessel
Rancho Palos Verdes, California

IN HIS CRITIQUE OF philosophical naturalism, David Berlinski fails to see that restricting scientific research to material causes and effects must follow as the immediate logical consequence of a position he himself defends in his article: that the intentions of supernatural beings cannot provide the basis of any falsifiable, or even testable, theories. Indeed, what *if* God is capable, as Mr. Berlinski writes, of "inadvertence, anger, mockery, incompetence" and so on? What if he is a surrealist painter? Whatever hypothesis is capable of explaining everything perforce explains nothing, which is why the intelligent-design movement resembles nothing so much as an obsessive hamster entertaining itself on an exercise wheel.

Of course, it is still possible that intelligent-design theory might explain the bare fact of the world's seeming design—which brings us to Mr. Berlinski's discussion of William Dembski's theory of complex specified information. Here he makes the obvious but important observation that specifications are human gestures.

Dembski often argues as if the passage from conventional kinds of specification involving, for example, archers and targets to the physical or biological patterns that chiefly concern him follows an untroubled continuum of inductive inference. Inductions are never untroubled, however, because they always depend on the human judgment that all of the examples to be tested in an experiment are sufficiently similar to warrant being included in the test set. From our observations of sparrows, pigeons, and blue jays we may feel entitled to infer that all feathered bipeds fly; but then we discover ostriches and kiwis. The question then becomes what the *relevant* analogy is; but relevance, unfortunately for Dembski's theory, is a very large concept that probably cannot be formalized.

As for Michael Behe, another intelligent-design prophet, Mr. Berlinski avers in reference to his work that "the origins of irreducibly complex biological systems remain... an utter mystery." The only mystery here is why Mr. Berlinski has failed to read the scientific literature relevant to this problem in the last five years. In a series of articles in *Science* and elsewhere, Albert Barabási and his colleagues have shown that a scale-free, power-law topology is ubiquitous in the genomic regulatory, protein-interaction, and metabolic networks of the cells of dozens of organisms.

These biochemical networks are irreducibly complex in exactly the sense Behe intends: when the most highly connected nodes are removed from them they stop functioning. Because they neither develop nor evolve following Behe's naïve assembly-line model, however, this fact is perfectly irrelevant. Rather, they increase in complexity over time in response to chance symmetry-breaking processes—in particular through gene duplication, which causes the preferential attachment of certain nodes to each other within the network, thereby producing the characteristic topology.

At least it is good to see Mr. Berlinski backpedaling from his 1996 *Commentary* article, "The Deniable Darwin." At this rate, by 2008 he may even backpedal into the truth.

Alexander Eterman
Jerusalem, Israel

THE INTELLIGENT-DESIGN THEORISTS DISCUSSED by David Berlinski have their work cut out for them. They must decide whether the designer monitored his creation for a long time—possibly to this day—intervening in its workings, or whether he detached himself immediately upon the "launch." If the presumed designer has long since detached himself from his creation, it is crucial to determine whether he left it to its own devices, like an animal set free, or provided it with an in-built long-playing program that predetermined its operations, the way we leave the computer on over the weekend, having programmed it to perform certain calculations.

Before a design theoretician pits his strength against Darwin, he should express his own narrative in an intelligible manner. Thus, he should specify whether the evolution of biological species (regardless of its exact mechanism) took place at all, or whether the different species were produced by the designer gradually and independently of one another, or, perhaps, whether they all appeared simultaneously and went on in peaceful coexistence until some of them gradually became extinct.

If he accepts an evolutionary process of any kind, he must decide that blind natural selection is capable of creating things that are new and sensible or, on the contrary, that any biological change of the slightest complexity is the outcome of enlightened design. The proponent of design theory will most likely reject the suggestion that the mammalian eye is the product of natural selection; yet he might concede that the hearing apparatus of mammals evolved in a natural fashion, with no outside help; or at least that the Indian and African elephants, so similar and yet so different, are two naturally divergent branches of the same tree, rather than the product of a sophisticated design.

In short, a hypothesis of biological design must be built not merely as a negation of the theory of natural selection but as a series of diachronically expanded theoretical models of design, consistent in all of their links. It should be noted that building such a hypothesis is a thankless task, in the course of which a good half of "designers" empirically go over to the evolutionist side.

Geoffrey Kent
Chappaqua, New York

DAVID BERLINSKI PURPORTS TO present evidence showing that the natural world is the result of necessity or chance, rather than design, and at the end of his article pronounces the ideas of those espousing intelligent design to be moribund. Though I find it difficult to defend the ideas with which Mr. Berlinski takes issue, I do not really think that he has proved anything, either—nor could he by reference only to the natural world.

Theoretical physicists now believe that our universe is one of an infinite number of universes that are constantly being created—that it exists, in effect, as a bubble formed from some other universe. Accordingly, it is entirely consistent with chance that the charges assigned to the subatomic particles that formed the material foundation of our universe at the "time" of the Big Bang should be the only charges (within a very small variation) that would sustain our universe for a sufficient period of time to permit the development of intelligent life.

Even given the extremely low probability that the particles would have the proper charges following the Big Bang, it is not improbable that, with an infinite number of universes and an infinite period of time, one explosion would eventually take place which would support a universe that would not either collapse back on itself or expand at such a tremendous velocity that no orderly development of stars and galaxies could ever take place.

That is precisely the problem with the argument that chance disproves design. Chance exists only within a limited period of time and space; it cannot exist in infinity. In this case, chance is the design. And the designer, if there is one, may exist outside time and space, outside our universe, outside any universe.

Jonathan Wells

Discovery Institute, Seattle, Washington

DAVID BERLINSKI'S "HAS DARWIN MET HIS MATCH?" is a breath of fresh air. Darwinism has become a sort of anti-religion, and defenders of the faith tend to demonize unbelievers rather than try to understand what they are actually saying. As a result, Darwinists typically distort intelligent-design theory—the latest and most powerful challenge to their orthodoxy—beyond recognition. Though a critic of intelligent-design theory, Mr. Berlinski is no Darwinist, and he is refreshingly fair-minded in his analysis.

I think Mr. Berlinski is mistaken, however, in characterizing the work that I and others have done as an attempt to resurrect William Paley's natural theology. Paley argued that design proves the existence of the Christian deity, but I, for one, have never been convinced by his argument. Design necessarily entails a designer; but the Christian deity is much more than a designer, and additional premises and evidences must be adduced to prove His existence. I happen to believe in the God of Moses and Jesus, but my belief is based on much more than design.

The basic issue for intelligent-design theorists is not whether design gets us to God, but whether design is real. Darwinists contend that it is not. For example, Richard Dawkins, in *The Blind Watchmaker*, argues that though organisms are "complicated things" and "give the appearance of having been designed," "the evidence of evolution reveals a universe without design." Dawkins's claim is false, despite the Darwinists' mantra that they have "overwhelming evidence" for their theory. I pointed this out in my book, *Icons of Evolution* (2000), and Mr. Berlinski agrees, writing that Darwinism is little more than a "fantastic extrapolation" in which the mechanism responsible for some minor changes within species "is read into the global record of life itself."

The evidence of evolution does not reveal a universe without design, and it remains a possibility that some features of living things really are designed. Intelligent-design theorists like Michael Behe and William Dembski have proposed ways to determine which features are designed and which are not. Mr. Berlinski argues that they have not succeeded. In any case, though, their proposals for establishing criteria to detect design are not attempts to prove the existence of the Christian deity.

If intelligent-design theory really were a reincarnation of Paley's natural theology, then Mr. Berlinski might be right that it is in danger of collapsing without a glimpse into the inscrutable mind of God. But it is not. It is an attempt to give a better explanation than chance and necessity for what our senses tell us is evidence for design in living things. In that light, it is poor Darwin who is (as Mr. Berlinski declares of Paley) "dead at last, or at least not very vigorously alive."

Michael J. Behe

Lehigh University, Bethlehem, Pennsylvania

I ALWAYS FIND DAVID Berlinski's writing delightful, and I agree with much that he says in "Has Darwin Met His Match?" Specific claims about how life arose in the murky past should always be examined skeptically, especially if accompanied by grand philosophizing. On the other hand, the fact that much remains mysterious does not mean we cannot conclude anything at all with reasonable certainty.

On the general question of the sufficiency of unintelligent physical processes to produce the astonishing complexity of life, I think a nega-

tive answer is justified, for reasons I gave in my book, *Darwin's Black Box*.
I quite agree with Mr. Berlinski that my argument against Darwinism
does not add up to a logical proof. No argument that rests on empirical
observations can have such force. Yet—despite my sloppy prose in suggesting that, "by definition," irreducibly complex systems cannot be approached gradually—I intended the argument to be a scientific one, not
a purely logical one. In a scientific argument, conclusions are tentative,
based on the preponderance of the physical evidence, and potentially
falsifiable.

Here is my thumbnail sketch of the modern design argument as I
see it: either unintelligent processes can explain all of life or they cannot. Virtually everyone (including Darwinists) agrees that life *appears*
to be intelligently designed. The only physical mechanism ever proposed
that could plausibly mimic design is Darwinian natural selection. Yet,
as I have argued, the irreducible complexity of biochemical systems is a
barrier to direct evolutionary construction by natural selection, leaving
Darwinists to hope for circuitous, indirect routes. No plausible indirect
routes have been proposed, let alone experimentally demonstrated.

That leaves us with biological features that look designed, but only
promissory notes for how they can be explained by unintelligent processes. What's more, we know why the features we see in biological systems look designed: they are at once functional and very unlikely—exactly what William Dembski, whose work Mr. Berlinski also discusses,
means by his phrase "specified complexity." They look designed for the
same reason that nonbiological artifacts like mousetraps look designed,
and non-design explanations have turned out to be so much bluster.

It seems reasonable to me to conclude, while acknowledging our fal-
libility, that at least some features of life were really designed by an intelligent agent.

William A. Dembski
Baylor University, Waco, Texas

DAVID BERLINSKI PROVIDES A clear and popular summary of my work on the theoretical basis for detecting design in nature. He also articulates several criticisms of my work. As it turns out, I specifically formulated my theory to meet the concerns that he raises. I am thus grateful for the opportunity to clarify several key points about my theory.

(1) As Mr. Berlinski explains, central to my theory of design detection are the twin notions of small probability and specification. I argue that highly improbable events that conform to independently given patterns are correctly attributed to intelligent design. Mr. Berlinski correctly points out, however, that some patterns are subjectively imposed upon events (or perceived in events) and do not justify inferring design. He is absolutely correct as far as he goes.

But he misses a critical distinction in my work. In *The Design Inference*, I explain that there are artificially constructed patterns—I call them fabrications—that do not justify design inferences. There are other kinds of patterns that I call specifications, and these, in the presence of small-probability events, *do* justify design inferences. Moreover, I show that there is a clear way to distinguish specifications from fabrications. Specifications are patterns that, in the parlance of probabilists, are conditionally independent of the outcomes that they characterize and that, in the parlance of complexity theorists, exhibit a low minimum-description length (see www.mdlresearch.org). Fabrications fail this test.

The distinction between specifications and fabrications is readily illustrated. Consider an archer who shoots at a target. If the target is fixed and the archer repeatedly hits the bull's-eye, then we rightly draw a design inference by attributing skill to the archer. On the other hand, if the target is movable and always moves to where the arrow lands, then we may not draw a design inference by attributing skill to the archer. In the latter case, the target is a fabrication, in the former a specification.

(2) Mr. Berlinski is right that probabilities sometimes cannot be objectively assigned to various events. But sometimes they can be. And sometimes, when exact probabilities cannot be assigned, credible upper bounds can be. This suffices for a design inference. Sometimes probabilities can be determined on theoretical grounds. Sometimes they can be determined only on empirical grounds, as by running experiments or performing computer simulations. Assigning probabilities to biological systems to determine whether they are designed is an exciting area of research opened up by intelligent design.

(3) Throughout my writings I stress that the absence of specified improbability cannot rule out design, because a designing intelligence can act carelessly, or even deliberately, in ways that do not produce specified events of small probability. For instance, I might deliberately tip an inkwell so that the resulting ink stain is indistinguishable from a random accident. But with that same ink I might also spell a Shakespearean sonnet. In the latter case, the resulting ink "stain" would exhibit specified improbability and could not reasonably be referred to chance.

Thus, while I have always conceded that specified improbability is not a necessary condition for design, I have consistently argued that specified improbability is a sufficient condition for it. Mr. Berlinski argues against this, but his argument hinges on a failure to distinguish specifications from fabrications.

(4) According to Mr. Berlinski, highly improbable events happen "precisely as many times as one might expect, given their probabilities." This claim is easily disproved by flipping a coin a thousand times. The probability of the sequence you get is around one in ten raised to the 300th power. How often should you have expected this sequence to occur? Not at all! With all the elementary particles in the universe furiously flipping coins for trillions of years, the expected waiting time for a given sequence places it well beyond the predicted heat death (or big crunch) of the universe.

I still hope to persuade Mr. Berlinski that his concerns about detecting design can be (or have already been) satisfactorily addressed. In any case, he has identified several key issues raised by my theory, and I look forward to the critical conversation that his piece will engender in the design-theoretic research community.

Paul A. Nelson
Discovery Institute, Seattle, Washington

As an admirer of David Berlinski's intellectual stubbornness and independence, I welcome his critical scrutiny of the theory of intelligent design. No theory was ever improved by being coddled.

Still, when Mr. Berlinski writes that design theorists "underestimate the enduring intellectual force behind [Jacques] Monod's claim that the categories of chance and necessity are mutually exclusive and jointly exhaustive," I must note that Monod himself did not rely solely on these categories. No sane human being does. The argument of Monod's masterpiece, *Chance and Necessity*, leans heavily on the notion of "teleonomy," which he defines as a "characteristic property" of organisms as "objects endowed with a purpose or project." In other words: objects marked by teleology—or, if you will, by design.

But renaming a property to take away its metaphysical sting should fool no one. Monod claimed that he could dissolve teleonomy—the unmistakable designedness of organisms—into chance and necessity. Well, he did not, and if design theorists are right, he could not. *Chance and Necessity* is one long dance around the problem, ending with Monod's leaping into the arms of "a unique occurrence": the causally inexplicable origin of life on earth, an event, Monod observed, whose "probability was virtually zero."

Chance abused in this way can "explain" anything. As a philosophical naturalist, Monod was of course being true to his principles. But let us not give this sort of reasoning the good name of science. It is a philosophy—one contender among many—and not much of a contender at that.

The great theme that unites the intelligent-design community is the falsity of naturalism as a philosophy of explanation. Chance and necessity do not exhaust the causes that we know. The task facing design theorists is to turn this intuition into knowledge, by showing that their theory provides understanding and discoveries not forthcoming within a strictly naturalistic framework.

Mr. Berlinski doubts that this is possible. Time will tell. Paley is dead; so is Darwin; so, too, is Monod. Let the dead bury the dead.

Leonard Levin

Jewish Theological Seminary, New York City

DAVID BERLINSKI'S DECLARATION OF a stalemate between Darwinism and intelligent-design theory strikes me as premature. After so many innovative moves in the past decade, the players on both sides are warming to what may be a long and interesting match. My bets and sympathies are on the pro-design side.

I question Mr. Berlinski's argument that because improbability cannot be quantified, it cannot therefore be qualitatively asserted. Let me counter with an analogy. The improbability of a monkey typing a Shakespearean sonnet is quantifiable because the set of typewriter-strokes is finite. The improbability of the same monkey writing out a sonnet longhand is not precisely quantifiable, yet it is obviously more improbable than typing it. (The monkey can miss a "G" only 25 different ways when typing but an indeterminately large number of ways when writing by hand.) The typing case thus sets a lower limit to the longhand case. Similarly, the improbability of producing a specific gene within the set of DNA nucleotide permutations, though not defining the actual world of possibility, sets a finite lower limit to the improbability of the actual case of producing it within the indeterminate set of all matter configurations. It thus justifies our qualitative judgment that the latter is extremely improbable.

On the other hand, the molecular "Scrabble" theorists have also ignored a crucial factor: determining the percent of syntactically correct formations within permutations of a given alphabet. What are the odds that a monkey will type not a specific sonnet but any syntactically correct string of 200 characters in English (or any language)? Not as small, but extremely small nonetheless. What are the odds (given the machinery of producing DNA) of randomly generating any syntactically correct string of nucleotides that will produce a viable protein? I have not seen this question addressed by either side. It should be amenable to analytical and experimental approaches and may move the debate forward.

Theologically, we must still distinguish between proving God's existence and presuming to describe the divine attributes. We may satisfy ourselves, rationally or subjectively, that life overcomes astronomical odds, yet the next step of inference is much harder. It is not chance—but what is it? Call it the anthropic principle, *élan vital*, quantum reduction, negentropy, or "design"—we have only the token outline of an answer, and the frontier of the eternal mystery. To know what "design" is may be to presume to see God's face.

Michael Sherman

School of Medicine, Boston University, Boston, Massachusetts

AS A PROFESSIONAL BIOLOGIST, I have always wondered why the Darwinian idea of evolution is so accepted among my colleagues. If one were to poll biologists, I would bet that almost all of them would say that there has been evolution and that it has taken place in accord with Darwin's theory. Probably 95 percent of them, however, have never thought seriously about evolution, and the rest are convinced of it because it is a clear, simple, materialistic idea.

On the other hand, reading the papers on evolution published in respected science journals like *Proceedings of the National Academy of Science* or *Nature*, one is surprised at the weakness of the arguments. Indeed, the standards of proof in the field are much lower than in the rest of biology. Such papers would never make it through the peer-review process if they concerned molecular or cellular biology. Of course, there are obvious reasons for such low standards, including the difficulty of testing evolutionary hypotheses through experimentation. But if the theory is based on poor arguments, why have criticisms of it not succeeded in convincing mainstream scientists?

I see a number of reasons. The first is that in arguments against Darwinism, people usually assume that the alternative is creationism—that is, the creation of nature by God without an element of evolution. This idea cannot be accepted by scientists because (a) it is not interesting from a scientific point of view (it does not point to further scientific research) and, more importantly, (b) there is an overwhelming body of arguments from many fields in favor of evolution.

From this follows the second main problem: that despite the large number of flaws in Darwinism, there is no scientifically sound alternative hypothesis. Criticisms of Darwinism will not convince professional biologists until the critics can describe a strong alternative mechanism of evolution that can be tested experimentally.

A third problem is that authors of publications against Darwinism mainly base their arguments on formal logic, e.g., the idea that complex systems cannot evolve in multiple minor steps. Such arguments may convince physicists or mathematicians, but they do not sway experimental biologists. From their experience dealing with enormously complicated biological systems, biologists know that hypotheses based purely on formal logic never work (especially if they involve some mathematics).

Finally, there is the problem of where criticisms of Darwinian theory are made. To convince the scientific community of something, one should not publish books; one should publish in peer-reviewed scientific journals of high profile.

To conclude on a positive note: I do believe that we can draw experimentally testable predictions from the theory of intelligent design. I say this based on three groups of recent findings: (1) paleontological data showing that all major groups of multicellular animals appeared almost simultaneously, indicating a lack of gradual evolution of large groups; (2) the very high homology of regulatory proteins that control development of systems with similar functions that evolved independently (e.g., the mammalian eye and the eye of a fly); (3) the fact that the number of genes in the human genome is not very different from that of worms or flies.

I would suggest that since complex systems cannot, in fact, evolve by random changes, there was a design. When outlining multicellular animals, the designer would have introduced into the genomes of primitive species the information about complex organ systems required for future organisms. These complex organ systems are silent in the primitive organisms but can be activated, giving rise to new, more complicated organisms through the process of evolution.

This idea does not reject the possibility that these newly developed organisms were fine-tuned through random mutation and natural selection, but it does assume that the major complex organ systems were predesigned. With our present understanding of molecular biology, it should not be too difficult or expensive to test this idea by finding information about complex organ systems in the genomes of primitive multicellular organisms and trying to find ways to activate these systems.

David E. Safir

Los Gatos, California

I ENJOYED DAVID BERLINSKI's thoughtful and erudite article about intelligent design. As a scientist, I had found Michael Behe's *Darwin's Black Box* a refreshing critique of Darwin. It always troubled me that much of evolutionary theory seemed to be built on faith alone, with enormous leaps required to accept its view of how life formed and changed.

What strikes me after 56 years as a biologist is how improbable it is that life occurred randomly—improbable but not impossible. We are still left to make a "faith" choice. To me, it seems most likely that some sort of high intelligence designed life as we know it.

Douglas Porter

Albuquerque, New Mexico

DAVID BERLINSKI'S OTHERWISE EXCELLENT article is flawed by his example of a three-legged stool as an attempt to refute Michael Behe's argument that irreducibly complex systems cannot arise by small, random steps. On this point, it is Mr. Berlinski's logic that falls apart.

In the first place, Behe's argument pertains to *systems* requiring complex processes to work, the failure of any one of which dooms the system. A three-legged stool, however, is not a system but a static structure. Moreover, removing one leg from a three-legged stool only causes it to fail because of gravity, a force outside the "system" and therefore irrelevant to the argument. Finally, it may be true, as Mr. Berlinski argues, that a three-legged stool can be constructed by "numerous, successive, slight modifications" of a cylindrical block of wood, but such modifications can hardly be random or Darwinian in nature. They require a designer. Mr. Berlinski merely demonstrates one way in which a three-legged stool can be *designed*—an argument very much in support of Behe's position.

The refusal of biologists to come to terms with the colossal improbabilities of evolution is the reason that Behe's *Darwin's Black Box* has attracted so much attention.

George Jochnowitz

College of Staten Island, Staten Island, New York

DAVID BERLINSKI WRITES THAT "Darwin's theory of evolution and theories of intelligent design are in conceptual conflict. Although they may both be false, they cannot both be true." Not so. To an intelligent designer, evolution would be a brilliant invention, a way of forever expanding and refining creation.

Rev. Edward T. Oakes

University of St. Mary of the Lake, Mundelein, Illinois

ACCORDING TO A STORY that is perhaps *bien trouvé*, Alfred North Whitehead was once asked why he did not write more clearly, to which he is supposed to have replied: "Because I don't think more clearly." Based on the evidence of his various articles in *Commentary*, if David Berlinski were ever to be asked why he writes so clearly, he could well reply: "Because I think clearly."

Clear thinking is especially evident in his latest essay, where he dissects the flaws in the arguments of those who claim that both the universe and biological systems have been intelligently designed (by God presumably, although some authors are annoyingly coy about saying so). True, Mr. Berlinski admits that members of the intelligent-design movement have highlighted genuine dilemmas in Darwinian theory. But more importantly, he has exposed how their own positive proposals cannot really provide an adequate explanation for the inexplicable mystery of life—or the existence of the universe. Complexity, even irreducible complexity, is not the same thing as a consciously intended effect.

I would only add that complexity, whether specified or not, cannot emerge at all except from a prior background of order, and that fact constitutes the real theological point that seems to animate the intelligent-design movement. The implicit (and sometimes explicit) theological agenda of this brigade leads Mr. Berlinski to his final thoughts on theism and natural theology, and here too I think I am largely in agreement with his reflections. The same issue that piqued the curiosity of Albert Einstein—whether the good Lord had any choice in creating the world or not—can be expressed in terms that dominated the debate among theist philosophers and theologians of the Middle Ages.

By the time of Thomas Aquinas it was assumed by all Jewish, Christian, and Muslim thinkers that Aristotle was right in defining God as Pure Act. But if that is the case, then how can it be possible for God to create a *possible* world, since possibility is excluded in God? Either God has a choice in creating the world or He doesn't. But if He *can* create, then possibility is embedded in Pure Act, a contradiction. This dilemma remained unsettled down to the days of Gottfried Leibniz; his way of resolving the tension was to admit a range of possibilities facing the deity, who would then be constrained by his reality as Pure Act to choose the *best* of all possible worlds.

The disaster to theodicy that this hypothesis led to is well known and gave Voltaire his great opening to attack the notion of divine providence. My own point is that Leibniz's dilemma still lives on in precisely the antinomies pointed out by Mr. Berlinski. The only "solution" to that dilemma, really, is a line from Aquinas right after he concludes his famous five proofs for the existence of God: "As Augustine says, since God is the highest good, He would not allow any evil to exist in His works unless His omnipotence and goodness were such as to bring good even out of evil. This is part of the infinite goodness of God, that He should allow evil to exist, and out of it produce good."

Needless to say, such an assertion cannot be grounded in science, since it requires for its verification a view of the final outcome of the universe's career, a view not given to the finite human mind. Only faith avails here. To bring in science as a kind of almost literal *deus ex machina* only gums up the issue, for both theology and science.

Yaffa Ganz
Jerusalem, Israel

DAVID BERLINSKI SUMS UP his long article by writing, "God alone knows what God is thinking.... We are in the position of observers contemplating a vast, cosmic lottery." Not quite. What we contemplate is a vast, mind-boggling, perfectly orchestrated universe. Although God chose which "necessities" would govern this universe and we are not privy to His secrets, He Himself is not governed by the reality He has created. If ever He chose to do so, He is perfectly capable of changing the rules of the game.

But what difference does it make? Why are so many serious scientists determined to prove—refute—God's copyright of creation? As Mr. Berlinski himself writes, "Could a designer whose nature we cannot fathom, using principles we cannot specify, construct a system we cannot characterize? If the question is unyielding, so too is its answer: who knows?"

But this does not mean, as Mr. Berlinski states, that "chance now returns as the default hypothesis, if only because it is the only hypothesis that is completely consistent with our ignorance." After all, if we are so ignorant, how do we know that chance is a more likely source of creation than God? Our ignorance is not the problem; it is our absolute incapacity to confront God. Trying to analyze God's "mind" through human logic and science is doomed to failure: "Where were you when I laid the foundations of the earth? Speak—if you have wisdom!" (Job 38:2–4). We can neither fathom nor confront nor comprehend God. We can only view His world, attempt to describe the mechanics that make it run, and serve Him in humility.

I would suggest breaking our scientific heads over more seemingly mundane, but more profitable, matters that explain and affect the world we live in. We should leave what Mr. Berlinski calls the "ineffable inimitable" (and I would call the Ineffable Inimitable) to theology.

David Berlinski Replies

ON READING "HAS DARWIN Met His Match?," a number of my correspondents seem to have concluded that, like Saul on the road to Damascus, I have seen the light and changed my mind. A conversion to Darwinian orthodoxy is said to be imminent. These impressions I must correct at once. I have *never* expressed support for theories of intelligent design, much less for creationism, and my essay, far from representing a change of mind—no bad thing, in any case—does nothing more than amplify objections I have long held and often voiced.

Six and a half years ago, in responding to critics of "The Deniable Darwin" (*Commentary*, June 1996), I made the point explicitly. "Some readers seem to be persuaded," I wrote in the September 1996 issue, "that in criticizing the Darwinian theory of evolution, I intended to uphold a doctrine of creationism. This is a mistake, supported by nothing that I have written." A few years later (September 2001), responding to critics of "What Brings a World into Being?" (*Commentary*, April 2001), I was even more forthright: "If I thought that intelligent design, or any artful contrivance like it, explained anything in any depth, I would leap to the cannon's mouth and say so. I do not and I did not."

For the record: I do not believe that theories of intelligent design explain those features of living systems that Darwin's theory of evolution fails to explain. And vice-versa. I wrote "The Deniable Darwin" and "Has Darwin Met His Match?" to say why.

IN "HAS DARWIN MET HIS MATCH?," I suggested that theories of intelligent design and Darwin's theory of evolution shared a strong family resemblance—the same guppy eyes, the same small ears, the same potato nose. PAUL R. GROSS will have none of it. Intelligent design is wrong and I am right to affirm the fact, but Darwin's theory is right and I am wrong to deny it.

Mr. Gross's animadversions begin with a reminder. It is only those "who do not know much evolutionary biology" who refer to something called "Darwinism." The professionals know better. I quite understand Mr. Gross's concern. The term "Darwinism" conveys the suggestion of a secular ideology, a global system of belief. So it does and so it surely is. Darwin's theory has been variously used—by Darwinian biologists—to explain the development of a bipedal gait, the tendency to laugh when amused, obesity, *anorexia nervosa*, business negotiations, a preference for tropical landscapes, the evolutionary roots of political rhetoric, maternal love, infanticide, clan formation, marriage, divorce, certain comical sounds, funeral rites, the formation of regular verb forms, altruism, homosexuality, feminism, greed, romantic love, jealousy, warfare, monogamy, polygamy, adultery, the fact that men are pigs, recursion, sexual display, abstract art, and religious beliefs of every description. If Darwinian biologists have not yet appropriated the class struggle, this is only because of their respect for competing ideological prerogatives.

I am also hardly the only one to use the term "Darwinism" and so convey the suggestion of an ideological agenda. Adding his mite to D. S. Bendall's collection, *Evolution from Molecules to Men* (1983), Richard Dawkins entitled his essay "Universal Darwinism." Dawkins liked the word well enough to use it again in "Darwin and Darwinism," the title of his contribution to Microsoft's *Encarta Encyclopedia.* Then there is the series of short books appearing under the title *Darwinism Today* and published by Yale University Press. The first book in the series is by the eminent Darwinian biologist John Maynard Smith.

With regard to his other claims, Mr. Gross rather resembles a standard fixture of the schoolyard brawl: the boy who refuses actually to fight but instead adverts to the remarkable pugilistic powers of his older brother. My criticism of Darwin's theory? "Creationist pablum," Mr. Gross declares with a snort. A mighty host is prepared to enforce the point: tens of thousand of papers, dozens of books, scores of websites, courses in all the better colleges. Specialists are on call. In resting his case on what others are said to have said, Mr. Gross has issued a challenge that it is not possible rationally to meet. Let us by all means have the details—my claims, those refutations—and *then* you and I can fight.

In still other respects, Mr. Gross is concerned to show that his vigor in combating the evils of intelligent design exceeds my own. The men whom I criticize, he complains, have not published their work in peer-reviewed journals. Quite true. They have not. But anyone who understands how science works institutionally will find this unsurprising. "Being right," as one shrewd critic has observed, "isn't enough. What you say, however right, must be said in a currently acceptable language, must not violate too brutally currently acceptable taste, and must somehow signify your membership in a respectable club." That shrewd critic was

Paul Gross himself, writing in a 1998 publication of the Marine Biological Laboratory. Allow me to introduce one Gross to the other.

AS MARK PERAKH OBSERVES, I have indeed endorsed books by both Michael Behe and William Dembski. I would do so again. Behe's *Darwin's Black Box* and Dembski's *The Design Inference* challenge received opinion; they are carefully argued; and they address important issues. I agree with some of the claims made in both books, but not with all of the claims made in either. A man may admire a book without endorsing it completely. Had Darwin's publishers in 1859 asked for a blurb, I would gladly have said that *The Origin of Species* is both quirky and provocative. Sales might well have improved. I would do as much now for Richard Dawkins's *The Blind Watchmaker*, a book whose thesis I reject but whose title I admire.

JASON ROSENHOUSE ASSUMES THAT I harbor an ongoing animus against "mainstream biology." Not so. Molecular biology, one of the glories of modern science, is where the mainstream lies; evolutionary biology remains what it has always been, a distant and rather muddy tributary. It is not molecular biology with which I scruple, needless to say, but Darwin's theory of evolution.

In my most recent essay—the one under discussion, I might remind Mr. Rosenhouse—I introduced Darwin only to suggest that both his theory and theories of intelligent design often lapse at the same points: the fossil record, for example. Mr. Rosenhouse denies this because he denies that Darwin's theory lapses at all. "Blubbering about gaps in the fossil record," he writes, his snort echoing Mr. Gross's, "cannot change the fact that, with millions of fossils collected and classified, not one is out of place from a Darwinian standpoint."

But what is at issue for Darwin's theory is not the fossils that exist but the ones that do not. The Cambrian explosion is mysterious precisely because the phyla that emerge during the Cambrian era have no obvious physical antecedents. By the same token, what is at issue for theories of design is not the fossils that do not exist but the ones that do. The reptile-to-mammal sequence is confounding to intelligent-design theorists precisely because the organisms, slotted head to tail, seem to form an unbroken Darwinian sequence. The fossil record is a puzzle for both views—a point urged on me, I should add, by Phillip Johnson.

Counterexamples are in any case open to challenge. In an essay entitled "Phylogenetic Hypotheses of the Relationship of Arthropods to Precambrian and Cambrian Problematic Taxa" (*Systematic Biology* 45, pp. 190–222), B. M. Waggoner argues that at least some Edicaran fossils fit into known phylogenetic groups and are thus ancestral to lower Cambrian metazoans: frond-like fossils died dreaming of hitting the big time as cnidarians, he suggests, and other members of the Edicara may have had ambitious plans to become annelids and arthropods. L. W. Buss and A. Seilacher, in "The Phylum Vendobionta: A Sister Group of the Eumetzoa?" (*Paleobiology* 20, pp. 1–4), have held the reverse, proposing gloomily that Edicaran fossils represent life forms that went nowhere because they had nowhere to go.

These are paleontological debates, of which the literature is by now considerable. The case against the Cambrian explosion has also been made from a theoretical perspective. Evidence from molecular clocks, Daniel Y.-C. Wang, Sudhir Kumar, and S. Blair Hedges have argued in the *Proceedings of the Royal Society of London* (Series B, January 22, 1999), suggests that a great many organisms from at least three phyla must have been present on earth *before* the Cambrian era. Indeed, Hedges, the paper's principal author, is persuaded that the emergence of so many phyla during the Cambrian era is no longer a mystery—though in reaching this conclusion he has replaced one mystery by another, since the organisms whose existence he champions on theoretical grounds remain undiscovered. "Why don't we see any fossils of these species long before the Cambrian era?," Hedges asks, thus returning the discussion to the point at issue.

On the other hand, the reptile-to-mammal sequence, the jewel in the crown of Darwinian paleontology, is not without critics of its own in the intelligent-design camp. The indefatigable Phillip Johnson has drawn my attention to a paper by John Woodmorappe in *TJ* 15(1), 2001, pp. 44–52. (*TJ* is self-described as a "creation journal," a fact of no relevance to an assessment of Woodmorappe's arguments.) Using cladistic analysis, Woodmorappe investigated a discrete group of morphological characteristics that paleontologists have offered as evidence for the evolutionary nature of the reptile-to-mammal sequence. At issue is the claim that mammal-like reptiles, when arranged in succession from the pelycosaurs on up, "show an essentially unbroken chain of progressively more mammal-like fossils." With respect to 165 of the 181 anatomical characteristics cited in C. A. Sidor and J. A. Hopson's "Ghost Lineages and 'Mammalness': Assessing the Temporal Pattern of Character Acquisition in the Synapsida" (*Paleontology*, 24 (2), 1998, Appendix 2, pp. 269–270), Woodmorappe argues that "the majority... do *not* show a unidirectional progression toward the mammalian condition" (emphasis added).

I have not studied Woodmorappe's paper thoroughly, but I am quite sure that both his conclusions and the methodology upon which they rest will be widely disputed, if they are ever widely noted. I return to my own starting point: neither Darwinians nor design theorists can look to the fossil record with perfect equanimity.

"Darwinism," Mr. Rosenhouse writes, apparently unaware of Paul Gross's terminological strictures, "requires continuity at the level of the genotype, not the phenotype." I am not sure what this might mean, but I am willing to guess: small changes in the genotype may well give rise to large changes in the phenotype. I suspect that Mr. Rosenhouse is correct; I have argued the point myself, including in my response to Arthur Shapiro in the correspondence on "The Deniable Darwin."

A fascinating paper in the January 10, 2003, issue of *Cell* deals with this topic. In "Molecular Rheostats Control Expression of Genes Cell," Richard Freiman and Robert Tijian observe that the machinery of genetic regulation—which gene goes on, which goes off—constitutes a magnificently subtle and exquisitely complicated system, and one by no means completely understood. The small differences between a human being and an earthworm, they suggest, may owe as much to parametric changes in their respective regulatory systems as to the absolute difference in the number of their genes, which in any case is not very great.

What a remarkable shift in view this paper represents! The gene has been demoted: it is still the raw stuff of life—what else is there?—but not the source of life's variety. The picture now emerging suggests that the genome is like a billboard made up of thousands of lights. When the cell's rheostats are in one position, the billboard spells earthworm; when in another position, human being. How these regulatory systems themselves may have evolved, the authors do not say, largely because they do not know; indeed, since many of the structural elements in the systems have been conserved across long periods of evolutionary time, they do not know *whether* the systems have really evolved at all.

In concluding his defense of Darwin's theory, Mr. Rosenhouse appeals to thousands of satisfied researchers, rather as if he were framing advertising copy for arch supports. "Numerous complex systems have been studied," he writes, "and the major steps of their evolution revealed." I would ask Mr. Rosenhouse to supply a single example of a complex biochemical system whose major evolutionary steps have been revealed and then explained by Darwin's theory. A good place to start would be with the systems discussed by Michael Behe in *Darwin's Black Box*.

What Behe and I both require in this regard is quite simple: a detailed, step-by-step biochemical account demonstrating a plausible Darwinian pathway for the emergence of any irreducibly complex system. Plausible—meaning, no appeals to unlikely events; and Darwinian—meaning, incremental improvement at each step. Before Mr. Rosenhouse proposes to pester me during office hours, a copy of Kenneth Miller's *Finding Darwin's God* in hand, I would suggest he consult Behe's own expert demolition of Miller's proposals. References are available online at the Discovery Institute's website: www.discovery.org.

CLAY SHIRKY IS QUITE right to observe that physicists do not practice Einsteinism; but as I have already noted, Darwinists do refer constantly to Darwinism. "Like Freud and Marx," A. S. Byatt remarked recently, "Darwin has suffered from becoming a belief system, when he was simply a very original thinker." Mr. Shirky might ask himself why this is so.

I do not believe that Darwinism is a fixed philosophical system. Like some primordial jelly, the thing is both squishy and constantly in motion. Terms, claims, and stories multiply unceasingly.

An example: some organisms lose certain functional properties over time. There are wingless insects, flightless birds, blind moles. Evolutionary biologists have long maintained that what is lost is destined to stay lost: it has been an article of their faith. And one that has apparently been misplaced. In a recent research report entitled "Loss and Recovery of Wings in Stick Insects," Michael E. Whiting, Sven Bradley, and Taylor Maxwell demonstrate that stick insects of the order Phasmatodea, having lost their wings over evolutionary time, nonetheless retain the capacity to reacquire them later (*Nature* 421, January 16, 2003, pp. 264–267). "These results," they write, "suggest that wing-development pathways are conserved in wingless phasmids, and that 'reevolution' of wings has had an unrecognized role in insect diversification." Will evolutionary biologists be forced as a result to reassess evolutionary theory, concluding perhaps that a theory that has failed to account for the facts is in need of revision or, perhaps, rejection? What a thought. Within a few years, "Loss and Recovery of Wings in Stick Insects" will be counted a Darwinian triumph, as the theory successfully manages once again to adapt to alien circumstances.

At the time I wrote "The Deniable Darwin" in 1996, biologists were still concerned to keep their disputes from the public eye. Their reticence has given way. They are now inclined to exhibit their battle scars like stigmata; the late Stephen Jay Gould was a virtuoso of the form, his mournful recriminations filling entire issues of the *New York Review of Books*. Although Mr. Shirky is persuaded that the factional debates within Darwinian biology are a sign of glowing good health, another view is possible. These debates represent the conceptual confusions of a discipline that has simply not achieved sufficient clarity to count as a serious science.

I failed to discuss the work of R. A. Fisher or William Hamilton in my essay for the same reason that I did not discuss late Sung poetry: neither their work nor that of the late Sung poets has anything pertinent to say about theories of intelligent design.

The concluding paragraphs of Mr. Shirky's letter prompt me to wonder whose essay he has been reading. Why on earth should he imagine that I desire a world in which science stops trying to explain things? I am in favor of *scientific* explanations wherever they may lead.

No less perplexing is that S. L. Baccus should believe that I—of all people—have written a brief for the existence of the Christian God, and that my ultimate aim "appears to be getting Christian beliefs taught in our schools." I urge Mr. Baccus to consider my name, my devotion to Zion, and my dark Semitic good looks. Phillip Johnson's opinions—the ones that Mr. Baccus quotes—are interesting, but they have nothing directly to do with what I have written.

Under the mistaken impression that my essay was a critique of Darwin's theory of evolution—perhaps he has confused it with "The Deniable Darwin"—Morton Rosoff is eager to establish my ignorance. Under the no less mistaken impression that I have endorsed intelligent design, he next offers his own criticism of the movement: "no empirical evidence, no models, no verifiable predictions, no possibility of correction or elaboration..., not intended to improve our knowledge or extend scientific horizons." When he catches his breath, Mr. Rosoff might wish to consult the essay I actually wrote for a few additional suggestions. I did not discuss the anthropic principle, but as for the other methodological tools he lists, I am all for them.

The issues that Matt Young raises with respect to Nilsson and Pelger's evolutionary model of the development of the mammalian eye are of greater moment than they may appear to the casual reader. They require a detailed response, one that cannot be fitted into even a lengthy exchange of letters. Here I might just note in passing that while Mr. Young has obviously looked at the footnote in which I accuse him of reading carelessly, he seems to have read carelessly the essay to which it is attached: how else to explain his reference to me and my "neo-creationist colleagues"?

"NATURALISTS," TONY DOYLE WRITES, "contend that for any event or phenomenon that needs explaining, we should seek only physical causes." This is indeed the claim often made, and I assume that Mr. Doyle endorses it. But naturalism, broadly conceived, is neither a premise to any of the great physical theories nor one of their conclusions. Its role in contemporary discourse is, as Phillip Johnson argued, wholly extrascientific.

Mr. Doyle's defense of naturalistic doctrine is thus philosophical and historical. That is fine by me; I am in favor of what the *New York Times* calls the living arts. Still, I believe his philosophical views are incorrect and his historical analysis wrong. Mr. Doyle may have "a robust idea of physical causation," but David Hume, and the great Arabic theologian Abu Hamid al-Ghazzali before him, argued forcefully that, when analyzed, causation dwindles into temporal succession and constant conjunction: one thing following another. I urge Mr. Doyle to reread al-Ghazzali's *The Incoherence of the Philosophers* or Hume's *Inquiry Concerning Human Understanding*. The inner connection that attaches a cause to its effect remains beyond our grasp, part of the mystery of nature. When Mr. Doyle writes that "we do not have a clue about how the theist's nonphysical causation is supposed to work" (emphasis added), he is simply specifying a general lack of understanding. We are pretty much clueless across the board.

By the same token, it seems to me profoundly incorrect to say that in mathematical physics it is the "search for physical causes and only physical causes [that] has paid off." When Isaac Newton introduced the universal force of gravitation in the *Principia*, he was appealing to a power in nature that acted both instantaneously and at a distance.

The connection between any "robust idea of physical causation" and the assumptions needed to explain gravity was thus broken from the very first. Mathematical physics has proceeded inexorably in the direction that Newton indicated. String theorists now explain the charges carried by D-branes in terms of algebraic K-theory. Every *direct* connection to a world of physical experience or causation has been lost. So too, for the moment, has every *indirect* connection, since string theory is not yet amenable to experimental tests of any sort.

I agree that Shakespeare was "physical from head to toe." So is a cat, and as every cat owner knows, there is absolutely no saying what a cat is going to do next. The laws of physics are both unavailing and irrelevant.

As CHRIS BEALL NOTES, the design inference, inasmuch as it goes beyond the existence of a designer to conclusions about his nature, very often topples over into incoherence. I agree. Very often it does.

IT IS GOOD TO be reminded by the distinguished biologist GEORGE C. WILLIAMS that complex structures that are optimal in one dimension may well be sub-optimal in another. As well as we may be able to see, we cannot see what is going on behind our backs, not to mention what is often beneath our noses. I am not sure what follows.

May I also observe that no one with two ex-wives could possibly be unaware of the deity's less attractive features?

CONTRARY TO WHAT KARL WESSEL writes, I did not provide a "critique" of philosophical naturalism in my essay. I mentioned it in passing, and paused only to agree with Phillip Johnson that as a principle it is wholly extra-scientific.

In commenting on my discussion of Michael Behe's work, Mr. Wessel suggests that I have failed to read the scientific literature, and in particular the work of A.-L. Barabási. Indeed? In a paper entitled "Hierarchical Organization of Modularity in Metabolic Networks" (*Science*, Vol. 297, August 30, 2002), Barabási and his colleagues argue that "spatially or chemically isolated functional modules composed of several cellular components and carrying discrete functions" should be considered "fundamental building blocks of cellular organization."

The result is a model of the cell resembling one of those horrible Swedish factories in which workers sit glumly at red and white Ikea desks and are isolated by glass partitions. But Barabási *et al.* also recognize that the "thousands of components of a living cell are dynamically interconnected," the Swedish factory now giving way to a second model, one resembling an equally horrible discotheque, rather like the old Studio 54, in which everyone is simultaneously involved with everyone else, frequently in ways no one wishes to know. Reconciling the Swedish factory to the discotheque is the burden of their paper.

This is modestly interesting stuff, if hardly calculated to induce an ardent desire to see a sequel. But its relevance to Mr. Wessel's concern is somewhat limited. In the first place, Barabási *et al.* do not often tie their work to biochemistry, and when they do, the results are inconclusive. "[I]t is… apparent," they write, "that putative module-boundaries do not always overlap with intuitive 'biochemistry-based boundaries.'" This prompts Barabási to the conclusion favored by every researcher facing a gap between his theories and the facts: "further experimental and theoretical analysis will be needed." No doubt.

In the second place, Barabási *et al.* are perfectly aware of a point that has escaped Mr. Wessel, namely, that their work does nothing to settle any issue that Michael Behe may have raised. "Understanding the evolutionary mechanism that explains the simultaneous emergence of the observed hierarchical and scale-free topology of the metabolism, as well as its generality [*sic*] to cellular organization, is now a prime challenge." A *prime challenge*—meaning a challenge that has not been met.

I THANK ALEXANDER ETERMAN for his thoughtful letter, with which I agree. Intelligent design is doomed to disappear as a movement if it can do no more than criticize Darwin's theory of evolution. Still, the purely negative criticisms made by members of the intelligent-design community have often been considerable in their effect—certainly more so than any criticisms I may have made. As a result, Darwin's theory has lost some of its intellectual respectability even as it has continued to extend its popular reach.

A great many scholars are now willing to say in public what they have long believed in private: that random variation and natural selection do not suffice to explain the observed facts in biology. An advertisement to this effect, placed by the Discovery Institute in the *New York Review of Books*,[53] was signed by one hundred academics: under the letter "S" alone, the list stretched from Henry E. Schaefer, the director of the center for computational quantum chemistry at the University of Georgia, to Richard Sternberg of the department of invertebrate zoology at the Smithsonian Institution. *Geshmak*, as we right-wing Christian fundamentalists say.

SOME PHYSICISTS MAY WELL entertain the idea that "our universe is one of an infinite number of universes," as GEOFFREY KENT writes—striking evidence that during hard times, some physicists will entertain anything and welcome anyone. But surely scientists who are happy to urge Occam's razor against design theory should hesitate before heedlessly multiplying universes themselves, the more so since the existence of the damn things is often invoked precisely to avoid the inference to design in the first place.

I am not sure what Mr. Kent means by asserting that chance "cannot exist in infinity." A random variable can certainly have a continuous distribution—but that may not be what he has in mind.

AS JONATHAN WELLS WRITES in his admirably clear letter, the gravamen of the intelligent-design movement is the inference to design, an inference that carries us from the observed properties of biological systems to the existence of a designer. The identification of the designer with the Christian deity requires a separate inference, one that design theorists

---

53. The text reads: "We are skeptical of claims for the ability of random mutation and natural selection to account for the complexity of life. Careful examination of the evidence for Darwinian theory should be encouraged."

need not make. I accept the point. In my essay, I tried to suggest as much, at least implicitly, by incorporating Fred Hoyle's supercalculating intellect into the class of possible designers. Hoyle was a life-long atheist, and in reading him I find no reason to believe that in fundamental matters he ever changed his mind.

Still, I admit to a tickle of discontent at Mr. Wells's attempt to dissociate the designer's existence from the designer's identity. It may be helpful to put the matter in deductive form. If there is design in nature, then there is a designer: that is Mr. Wells's leading premise. A second premise follows, embodying a factual claim: there is design in nature. The conclusion that there is a designer follows as a matter of logic.

But, articulated in this way, the conclusion seems suspiciously trivial. If the designer's nature is not known, then in what sense does his existence tell us anything that is not expressed in the second premise? To be sure, the twin propositions that there is design in nature and that there is a designer do not say quite the same thing; but, like certain houses in Paris, they have a tendency to collapse toward one another. Which designer? The one handling the design. Which design? The one handled by the designer. In the end, we are left with the fact that certain properties of living systems have not been explained. On this point Mr. Wells and I agree completely.

IN CHARACTERIZING HIS OWN work, MICHAEL J. BEHE is entirely too unassuming. By demonstrating that irreducibly complex systems cannot arise along one Darwinian path—the assembly-line model of construction—he has indeed provided a logical argument against Darwin's theory. This is an important achievement. My point was only that his argument has not been generalized to include *all* Darwinian paths. This is a modest criticism.

As long as I find myself explaining Mr. Behe's achievements to Mr. Behe, let me go a bit further. It seems to me that *Darwin's Black Box* is destined to play the same role in evolutionary thought that Karl Lashley's "The Serial Order of Behavior" played in behavioral psychology over fifty years ago (cf. L. A. Jefress, editor, *Cerebral Mechanisms in Behavior*, 1951). Although a behaviorist by training—and a student of John Watson, the founder of the field—Lashley came to understand that a certain class of familiar psychological acts involve a complicated serial order. In formulating a sentence, we typically see to the end of the sentence before venturing on its beginning, adjusting our stream of speech accordingly to account for grammatical relations of subordination and deferred placement. Serial order, Lashley realized, could not be explained by any system of associative "chaining" in which each act in a behavioral repertoire is explained by an act that has already taken place.

Psychologists were slow to appreciate the force of Lashley's critique. But when Noam Chomsky, writing ten years later, pointed to certain features of natural language that were inaccessible to finite-state mechanisms or even Markov processes, he was sharpening and extending Lashley's insight. Thereafter, a shrewd insight became a movement in thought: the so-called cognitive revolution.

It is curious that so few biologists appreciate the formal similarities between Darwin's theory of evolution and behavioral theories in psychology. But the same basic idea is at work, with "reinforcement" in psychology called "natural selection" in biology. And here is an odd point. Although behaviorism is widely thought to have been stabbed through the heart, most especially by Chomsky's criticism of B. F Skinner, the fact of the matter is that in the end, Skinner has proved more durable than anyone might have guessed, and with bat wings flapping has burst buoyantly from his crypt.

We know that behavioral psychology provides an explanation for a limited class of experimental results; beyond those results, it fails. What an individual does in acquiring a natural language cannot be explained in terms of any schedule of reinforcement. Indeed, it cannot be explained in terms of the environment at all; reference must be made to the individual's innate endowment. But linguists who came early to scoff at Skinner—Chomsky now included—are involved in making Skinner's argument on the level of the species. For what are random variation and natural selection but a form of behavioral *biology?*

The intellectual history of the past half-century is not without its ironical aspects.

IN HIS GRACEFUL LETTER, WILLIAM A. DEMBSKI argues that in some respects I have misunderstood his views, and in other respects I have insufficiently appreciated their force. It is certainly possible; the issues that Mr. Dembski raises are subtle, and his letter and my response should both be regarded as efforts to get things a little clearer.

Under what conditions may we rationally eliminate chance as the explanation for certain events? This is Mr. Dembski's question. Two things, he argues, are necessary: small probabilities and reasonably tight specifications. When these criteria are met, we are entitled to strike off chance; if the event in question is also not explicable by natural laws, design emerges as the only plausible alternative.

In developing his argument, Mr. Dembski has a certain model in mind. The design comes first, expressed perhaps as a blueprint, agenda, schedule, or even a system of thought. Next comes the designed event or object. "How a designer," he writes in *No Free Lunch*, "gets from a thought to a thing is, at least in broad strokes, straightforward: (1) A designer conceives a purpose. (2) To accomplish that purpose, the designer forms a plan. (3) To execute that plan, the designer specifies the building materials and assembly instructions. (4) Finally, the designer or some surrogate applies the assembly instructions to the building materials. What emerges is a designed object, and the designer is successful to the degree that the object fulfills the designer's purposes."

This is *executive* design, to coin a phrase and mark a distinction. In my essay, I suggested that there is a range of states, acts, or processes that are clearly intentional—they are brought about by intelligent agency— and yet share none of the features of executive design. The design of a painting is very often revealed *in* its execution and not before. Design in this sense might well be called *immanent*. The painter Francis Bacon often stressed just this point in commenting on his own work (see *Francis Bacon*, 1975), and the distinction between executive and immanent design appears as well in Nelson Goodman's *Languages of Art*, a book that design theorists might study with profit. With respect to immanent design, there are no prior purposes, no plans, and no application of assembly instructions to building materials. For this class of artifacts, probabilities are not relevant and specifications are inapplicable.

I am well aware of Mr. Dembski's distinction between a specification and a fabrication, the more so since I discussed the distinction at length in my book *Black Mischief: Language, Life, Logic & Luck* (1986), where Mr. Dembski's "fabrications" appear as "retroactive specifications." But this distinction has nothing to do with the distinction between executive and immanent design. Both specifications and fabrications apply, and apply only, to events capable of possessing an executive design. Velasquez's painting of the royal court, which I cited in my *Commentary* essay, lacks both a specification *and* a fabrication. A painting is not like a point in archery. It cannot be specified before it exists, and specifying it after it exists is useless, voiding the intended contrast between a specification and a fabrication.

Although both Mr. Dembski and I agree that specified improbabilities are not necessary to trigger a design inference, we do so for different reasons. On the question of whether they are sufficient, we part company altogether. In "Has Darwin Met His Match?" I argued that specifications are not necessary to trigger a sense that an unlikely event has occurred, and that if specified improbabilities are not necessary, they are also not sufficient to trigger an inference to design, since plainly improbability *by itself* tells us nothing about design.

The issue, then, turns on a tight circle. My argument for the irrelevance of Mr. Dembski's specifications begins with the obvious. A specification is a human gesture, a bit of descriptive apparatus. It does not change a system of probabilities, and so it has nothing to do with expectations based on those probabilities. Highly improbable specified events happen precisely as many times as one would expect, given their probabilities, but so, for that matter, do highly improbable *un*specified events.

It is this last claim that Mr. Dembski would deny. His counterexample is a fair coin flipped 1,000 times. The particular sequence of heads and tails that results has the probability of roughly one in ten to the 300th power. "How often should [I] have expected this sequence to occur?" Mr. Dembski asks. Not at all, he responds cheerfully. But the sequence has occurred, and the relevant record of heads and tails is there in plain sight. This persuades Mr. Dembski that, if left unspecified, unlikely events tend to occur more often than any assignment of probabilities would suggest.

I understand Mr. Dembski's reasoning, but I reject it as unwholesome. He is stuck, after all, with an expectation—a particular sequence should not occur at all—that is in plain contradiction to the facts—a particular sequence has indeed occurred. This is not necessarily a fatal flaw, but it does suggest that something has to give. What has to give is the nice coincidence between what the theory of probability affirms and what the facts reveal. Here Mr. Dembski's intuitions seem to be making certain clanging noises while emitting a good deal of smoke.

My own intuitions, by contrast, are smoke-free and purr like a charm. Is the sequence that has just been revealed highly improbable? It is. How often should it have occurred? Precisely as many times as one might expect, given its probability. In a sequence of identical and independent experiments, *this* particular outcome would not be realized again until the heat death of the universe or the rehabilitation of Senator Trent Lott, whichever comes first. The fact that the sequence has already occurred is of no significance. Nothing in the theory of probability prevents an extremely unlikely event from being realized in the very first experiment, just as nothing in the same theory prevents a man from winning the lottery after purchasing his very first ticket. I see no reason to be swayed from my conclusion that unlikely events happen as many times as one might expect, given their probability.

With the distinction between specified and unspecified improbabilities wiped out, the inference to design sputters. Improbabilities are not by themselves sufficient to trigger that inference, and, *a fortiori*, neither are specified improbabilities.

PAUL A. NELSON REMINDS me that, in *Chance and Necessity*, Jacques Monod *began* by acknowledging the fact that living creatures are driven by a sense of their purpose in life, and *then* persuaded himself that this obvious property was not so obvious after all. What seems to be their design, Monod concluded, is an illusion, or an artifact; there is necessity, and there is chance, and there is nothing more. Richard Dawkins has made almost the same argument in *The Blind Watchmaker*, and Mr. Nelson is right to remark that there is in all this something nutty. Monod's efforts to explain away the obvious did not succeed.

Of course, there is a distinction between saying that Monod did not succeed and saying that he *could* not have succeeded. Drawing that distinction, Mr. Nelson places his hopes for the latter possibility on the prospects for intelligent design. But to the extent that design theorists have overestimated their own arguments, to that extent—precisely—have they *under*estimated the enduring force of Monod's claim that necessity and chance exhaust our powers of explanation. In my essay, I went no further than this, and I am not prepared to go further now.

Whatever the issue between Mr. Nelson and Jacques Monod, it is (as he says) naturalism that is the real target of the intelligent-design movement, and, I presume, his target as well. If I cheer him on only weakly, that is because it is not entirely clear to me what the target is or how to fight it. From what he writes, I gather that naturalism is embodied in Monod's disjunction itself. But an inference to design is consistent with Monod's disjunction. A designer may have no choice in his design, a point I make in the concluding paragraphs of my essay. But, equally, a denial of Monod's disjunction is consistent with the failure of an inference to design. There may be facts in nature that are the result of neither necessity, chance, nor design, and we must accept the possibility, at least, that the most obvious facts about living systems—their existence and their nature—may have no deeper explanation than that this is the way things are.

Something more is at issue in the assessment of naturalism because something more must be at issue. I encourage Mr. Nelson to make that something clearer.

LEONARD LEVIN'S REMARKS ARE subtle and compelling. What he says about quantitative and qualitative aspects of probability is correct. We *do* make qualitative judgments of likelihood even when we are unable to promote those judgments quantitatively. I may not be able to determine the probability that Basque will be declared the official language of the People's Republic of China, but I can say that it is unlikely. Judgments of this sort are in constant use; they are the currency of trade.

But if commonly made, they are also commonly flawed. Our qualitative judgments are often insecure and frequently unstable, a point that emerges more clearly when artificial examples are put aside. Is it more or less likely that North Korea will prove a greater danger than Iraq? Some guesses are possible, but answers may change by the minute as information is presented or withdrawn, one reason that policy analysts often try frantically to place numbers on their judgments.

In pursuing this argument, Mr. Levin draws a suggestive distinction between two cases. In the first, the proverbial monkeys are typing; in the second, they are writing (or scrawling) by hand. It is only the first case, he argues, that permits a quantitative judgment; the second case trails off because, given the monkey's scrawl, we cannot specify any set of discrete objects on which to peg a probability. "The improbability of the... monkey writing out a sonnet longhand is not precisely quantifiable, yet it is obviously more improbable than typing it."

Is it? Two events are being compared. The first may be quantified by means of the calculations of probability. The second, Mr. Levin argues, is "not precisely quantifiable." In fact, it is not quantifiable at all. "The monkey may miss a 'G,'" Mr. Levin writes, "in only 25 different ways when typing but an indeterminately large number of ways when writing." But if the monkey may miss that "G" in any number of ways, he may also find that "G" in any number of ways. It depends on who is looking and who is counting and what criteria of success are in force. If we cannot assign precise probabilities to one of two events, then we cannot draw comparative judgments between them, either. To say of these events—one quantifiable, the other not—that the second is more improbable than the first is very much like saying of two men that one is taller than the other, given only that one is six feet tall and the other trim.

Mr. Levin complains that biologists have not addressed his questions about molecular syntax. His questions are precisely the questions I discussed in "The Deniable Darwin," and before that in *Black Mischief.*

MICHAEL SHERMAN'S LETTER RAISES a great many interesting questions, most notably why evolutionary biologists are so dopey. If *he* is "surprised at the weakness of their arguments," imagine how I must feel. But having asked this question, he then asks a more difficult one: why has intellectual change not been forthcoming? His own answers are canny but unsatisfying.

Biologists stick to Darwin, he suggests, "because the alternatives do not lead to further research." Now, this is not quite true, as Mr. Sherman himself indicates. The alternatives *do* lead to further research, and he has suggested the agenda. In fact, what is really at stake for most biologists is their chances of being funded, their place at the common trough. But here Mr. Sherman breaks the flow of his own argument. Biologists are not inclined to explore the options he has outlined, he writes, because there is "an overwhelming body of arguments from many fields in favor of evolution." If that is so, then Mr. Sherman has answered his own question in the most straightforward way possible: biologists stick with Darwin's theory because it is *true.* In fact, the "overwhelming body of arguments" to which Mr. Sherman alludes are more overpowering than overwhelming, and like Mexican food they keep coming up without ever quite going down. In this regard, Mr. Sherman has seen the light, but he has not used the light to see.

I am more than prepared to believe that, as Mr. Sherman asserts, formal logic does not often sway evolutionary biologists; as far as I can tell, it does not *ever* sway them. I am also prepared to believe that evolutionary biologists tend to dismiss theories of intelligent design because the authors of those theories publish books rather than peer-reviewed essays. I have already addressed this issue in my reply to Paul Gross, but I would point out that when Richard Dawkins and Daniel Dennett make *their* views known through trade books, which are, of course, not peer-reviewed, other evolutionary biologists tend to regard those efforts with quiet pride.

In his concluding paragraphs, Mr. Sherman takes an immensely daring position—something like old-fashioned preformationism. Now that he has ventured so far out on a limb, I might observe that something like this is just what recent work on genetic regulatory systems also suggests. Who knows what dreams of glory the earthworm harbors? But for a few parametric changes in its regulatory apparatus, and a missing gene or two, it might rule the world.

I AGREE WITH DAVID E. SAFIR that is it improbable but not impossible that life arose randomly. I would add only that the origins of mind and of matter are equally mysterious. I am not sure, however, what it would mean to make a "choice" of faith on these matters.

DOUGLAS PORTER IS MAKING too much of my stool. I invoked the thing only to make a modest point: small incremental changes may lead to an irreducibly complex system by paths that Michael Behe did not consider. I agree that biologists have not come to terms with the "colossal improbabilities" of evolution.

I SUPPOSE THAT, AS GEORGE JOCHNOWITZ suggests, an intelligent designer might have favored evolution as a "brilliant invention," but this claim, which is often made by evolutionary biologists concerned to cover all of their bets, places the conclusion to certain arguments before their premises. How the designer works is one question. What he did is another, and this question is logically prior. If certain structures cannot arise by Darwinian means, then, no matter the designer's intentions, he could not have realized them by means of a Darwinian mechanism. He, too, has his limits.

I THANK FATHER EDWARD T. OAKES for his very generous comments, and would not dream of disputing theology with him. Still, in writing that by the time of Aquinas, Jewish, Christian, and Muslim thinkers had come to agree with Aristotle that God is Pure Act, it seems to me that he has underestimated the complexity of Muslim thought. Aquinas, after all, criticized Avicenna's distinction between essence and existence on the grounds that it came perilously close to implying that existence is an accident. Indeed, the more I study the extraordinary record of medieval Arabic thought, the less I am inclined to emphasize its unity and the more to stress its diversity.

The course that Father Oakes traces between the Aristotelian doctrine of Pure Act and its apparent collapse some four centuries later is fascinating, but beyond my competence to assess.

YAFFA GANZ WRITES TO remind me that the concerns expressed in my essay may not matter very much in the end. We do what we can; we ask the questions that we can answer; and as for the rest, we face the illimitable, the unanswerable, and the incomprehensible.

So we do.

Borges on Intelligent Design

THE CAFÉ HAS GROWN QUIET. THE DISCUSSION THAT NIGHT HAD turned to language and its nature.

"It is odd," I remarked, "how extraordinary the coincidence between biological evolution and the evolution of languages. It often seems as if the various languages were living organisms. One can almost see the ancestor of all the Romance languages."

"A common thesis," Borges said, signaling the waiter with his upraised cane. "And one championed for reasons of their own by certain members of the Papal Curia. But it is nonetheless certainly false."

The waiter had come gliding over to our table and placed another *demitasse* of bitter espresso before Borges's sightless eyes.

Borges sensed my astonishment. "The chief European languages," he said, "were in fact designed and then specifically created. They did not evolve at all, and most certainly not from a common ancestor. Research establishing this conclusion has been undertaken by linguists in Montevideo. Documents are available both in the Bodleian Library at Oxford University and in the *Institut des Hautes Études Scientifiques* just outside of Paris."

"You cannot be serious," I sputtered. "Just consider the striking similarities among the European languages. How else to interpret this if not be means of a common ancestor?"

"Various features seem similar only if one supposes that the various European languages are descended from a common ancestor," Borges said. "Once this superstition is given up, those features no longer seem similar."

Borges sipped delicately at his coffee, and then with elaborate attention to detail, lit one of the thin cheroots he had been smoking all evening.

"In fact," Borges said, "recent evidence indicates that all European languages arose quite suddenly in a single linguistic explosion taking place in the fifteenth century."

"But that is absurd," I said. "Surely there is evidence for the existence of medieval versions of all the European languages?"

"To be sure," Borges replies, "but such primitive devices as are used in the construction of *El Cid* bear no relationship to the complex and sophisticated language of the *Quixote*. In the same fashion one can say quite confidently that there is no relationship whatsoever between the language used by Chrétien de Troyes and Ronsard."

"By the same token," he added, "certain features of Spanish indicate quite conclusively that it could not have arisen by any gradualistic process whatsoever."

"An example?" I said.

"Spanish possesses the future subjunctive, a sign of its sophistication. No process of gradual change could possibly change a language lacking a future subjunctive into one possessing it. There is, after all, no such thing as a functional *partial* future subjunctive. We know of no way in which the requisite information could be acquired in small stages. It is quite clear."

"There is an additional point," Borges added. "Spanish is obviously designed in the sense that its words fit naturally into the human mouth. Considering the virtually infinite number of ways in which a language could have been organized otherwise, this cannot be the result of chance. Of course, in the case of other languages, the design may not have been quite so happy. It may well be that the Deity, having designed Spanish, lost some of his enthusiasm when it came to Rumanian."

The café grew even more quiet, the waiters standing solemnly at their stations.

"Why," I asked, "haven't these facts become more widely known?"

"The research *is* widely known," Borges answered, "but owing to the interference of the Knights Templar, it has been prevented from appearing in print."

A Scientific Scandal

I N SCIENCE, AS IN LIFE, IT IS ALWAYS AN EXCELLENT IDEA TO CUT THE cards after the deck has been shuffled. One may admire the dealer, but trust is another matter.

In an essay in *Commentary*, "Has Darwin Met His Match?" (December 2002), I discussed, evaluated, and criticized theories of intelligent design, which have presented the latest challenge to Darwin's theory of evolution. In the course of the discussion I observed that the evolution of the mammalian eye has always seemed difficult to imagine. It is an issue that Darwin himself raised, and although he settled the matter to his own satisfaction, biologists have long wished for a *direct* demonstration that something like a functional eye could be formed in reasonable periods of time by means of the Darwinian principles of random variation and natural selection.

Just such a demonstration, I noted in my essay, is what the biologists Dan-Erik Nilsson and Susanne Pelger seemed to provide in a 1994 paper.[54] Given nothing more than time and chance, a "light-sensitive patch," they affirmed, *can* "gradually turn into a focused-lens eye," and in the space of only a few hundred thousand years—a mere moment, as such things go.

Nilsson and Pelger's paper has, for understandable reasons, been widely circulated and widely praised, and in the literature of evolutionary biology it is now regularly cited as definitive. Not the least of its remark-

---

54. "A Pessimistic Estimate of the Time Required for an Eye to Evolve," *Proceedings of the Royal Society*, London B (1994) 256, 53–58. In my essay I twice misspelled Susanne Pelger's name, for which I apologize.

able authority is derived from the belief that it contains, in the words of one of its defenders, a "computer simulation of the eye's evolution."

If this were true, it would provide an extremely important defense of Darwin's theory. Although a computer simulation is not by itself conclusive—a simulation is one thing, reality another—it is often an important link in an inferential chain. In the case of Darwin's theory, the matter is especially pressing since in the nature of things the theory cannot be confirmed over geological time by any experimental procedure, and it has proved very difficult to confirm under laboratory conditions. The claim that the eye's evolution has been successfully simulated by means of Darwinian principles, with results falling well within time scales required by the theory, is thus a matter of exceptional scientific importance.

And not just *scientific* importance, I might add; so dramatic a confirmation of Darwinian theory carries large implications for our understanding of the human species and its origins. This is no doubt why the story of Nilsson and Pelger's computer simulation has spread throughout the world. Their study has been cited in essays, textbooks, and popular treatments of Darwinism like *River Out of Eden* by the famous Oxford evolutionist Richard Dawkins; accounts of it have made their way onto the Internet in several languages; it has been promoted to the status of a certainty and reported as fact in the press, where it is inevitably used to champion and vindicate Darwin's theory of evolution.

In my essay, I suggested that Nilsson and Pelger's arguments are trivial and their conclusions unsubstantiated. I also claimed that representations of their paper by the scientific community have involved a serious, indeed a flagrant, distortion of their work. But in a letter published in the March issue of *Commentary*, the physicist Matt Young, whom I singled out for criticism (and whose words I have quoted here), repeated and defended his characterization of Nilsson and Pelger's work as a "computer simulation of the eye's evolution." It is therefore necessary to set the matter straight in some detail.

I hope this exercise will help to reveal, with a certain uncomfortable clarity, just how scientific orthodoxy works, and how it imposes its opinions on the faithful.

HERE IN THEIR OWN words is the main argument of Nilsson and Pelger's paper:

> Theoretical considerations of eye design allow us to find routes along which the optical structure of the eye may have evolved. If selection constantly favors an increase in the amount of detectable spatial information, a light-sensitive patch will gradually turn into a focused-lens eye through continuous small improvements in design. An upper limit for the number of generations required for the complete transformation can be calculated with a minimum number of assumptions. Even with a consistently pessimistic approach, the time required becomes amazingly short: only a few hundred thousand years.

And here is how they arrived at their conclusions. The setting is "a single circular patch of light-sensitive cells"—a retina, in effect—"which is bracketed and surrounded by dark pigment." A "protective layer" lies above these light-sensitive cells, so that the pigment, the light-sensitive cells, and the protective layer form a kind of sandwich. Concerning the light-sensitive patch itself, Nilsson and Pelger provide no further details, indicating neither its size nor the number of cells it might contain.

What they do assume, if only implicitly, is that changes to the initial patch involve either a deformation of its shape or a thickening of its cells. The patch can be stretched, dimpled, and pulled or pushed around, and cells may move closer to one another, like bond salesmen converging on a customer.

So much for what changes. What is the change worth? Assuming (reasonably enough) that an eye is an organ used in order to see, Nilsson and Pelger represent its value to an organism by a single quantitative character or function, which they designate as "spatial resolution" or "visual acuity"—sharp sight, in short. Visual acuity confers an advantage on an organism, and so, in any generation, natural selection "constantly favors an increase in the amount of detectable spatial information."

There are two ways in which visual acuity may be increased in an initial light-sensitive patch: a) by the "invagination" of the patch, so that it becomes progressively more concave and eventually forms the enclosed interior of a sphere; and b) by the constriction of the sphere's aperture (the two rounded boundaries formed as the flat patch undergoes invagination). These changes may be represented on sheets of high-school graph paper on which two straight lines—the x and y axes of the system—have been crossed. On the first sheet, representing invagination, visual acuity moves upward on one axis as invagination moves to the right on the other; on the second sheet, visual acuity moves upward as constriction moves to the right. The curves that result, Nilsson and Pelger assert, are continuous and increasing. They do not hurdle over any gaps, and they go steadily upward until they reach a theoretical maximum.

The similar shape of the two graphs notwithstanding, invagination and aperture constriction exercise different effects on visual acuity. "Initially, deepening of the pit"—i.e., invagination—"is by far the most efficient strategy," Nilsson and Pelger write; "but when the pit depth equals the width, aperture constriction becomes more efficient than continued deepening of the pit." From this, they conclude that natural selection would act "first to favor depression and invagination of the light-sensitive patch, and then gradually change to favor constriction of the aperture."

THE RESULT IS A pin-hole eye, which is surely an improvement on no eye at all. But there exists an aperture size beyond which visual acuity cannot be improved without the introduction of a lens. Having done all that it can do, the pin-hole eye lapses. Cells within the light-sensitive sphere now obligingly begin to thicken themselves, bringing about a "local increase" in the eye's refractive index and so forming a lens. When the focal length of the lens is 2.55 times its radius—the so-called Mattiessen ratio—the eye will have achieved, Nilsson and Pelger write, the "ideal solution for a graded-index lens with a central refractive index of 1.52."[55]

55. A graded-index lens is a lens that is not optically homogeneous; the figure of 1.52 is "the value close to the upper limit for biological material."

Thereafter, the lens "changes its shape from ellipsoid to spherical and moves to the center of curvature of the retina." A flat iris "gradually forms by stretching of the original aperture," while the "focal length of the lens... gradually shortens, [until] it equals the distance to the retina... producing a sharply focused image." The appearance of this spherical, graded-index lens, when placed in the center of curvature of the retina, produces "virtually aberration-free imaging over the full 180 degrees of the visual field."

The same assumptions that governed invagination and aperture constriction hold sway here as well. Plotted against increasing lens formation, visual acuity moves smoothly and steadily upward as a graded-index lens makes its appearance, changes its shape, and moves to center stage. When these transformations have been completed, the result is a "focused camera-type eye with the geometry typical for aquatic animals."

One step remains. Nilsson and Pelger now amalgamate invagination, constriction, and lens formation into a single "transformation," which they represent by juxtaposing, against changes in visual acuity, changes to the original patch in increments of 1 percent. The resulting curve, specifying quantitatively how much visual acuity may be purchased for each 1-percent unit of change, is ascending, increasing, and straight, rising like an arrow at an angle of roughly 45 degrees from its point of origin. Transformations are "optimal" in the sense that they bring about as much visual acuity as theoretically possible, with the "geometry of each stage [setting] an upper limit to the spatial resolution of the eye."

It is the existence and shape of this fourth curve that justify their claim that "a light-sensitive patch will gradually turn into a focused-lens eye through continuous small improvements in *design*" (emphasis added). This is not the happiest formulation they could have chosen.

How much does the initial light-sensitive patch have to change in order to realize a focused camera-type eye? And how long will it take to do so? These are the questions now before us.

As I have mentioned, Nilsson and Pelger assume that their initial light-sensitive patch changes in 1-percent steps. They illustrate the procedure with the example of a flat one-foot ruler that also changes in 1-percent steps. After the first step, the ruler will be one foot plus 1 percent of one foot long; after the second step, it will be 1-percent longer than the length just achieved; and so forth. It requires roughly 70 steps to double a one-foot ruler in length. Putting the matter into symbols, $1.01^{70} \cong 2$.

Nilsson and Pelger undertake a very similar calculation with respect to their initial light-sensitive patch. But since the patch is a three-dimensional object, they are obliged to deal with three dimensions of change. Growing in steps of 1 percent, their blob increases its length, its curvature, and its volume. When all of these changes are shoehorned together, the patch will have increased in magnitude along some overall (but unspecified) dimension.

The chief claim of their paper now follows: to achieve the visual acuity that is characteristic of a "focused camera-type eye with the geometry typical for aquatic animals," it is necessary that an initial patch be made 80,129,540 times larger (or greater or grander) than it originally was. This number represents the *magnitude* of the blob's increase in size. How many *steps* does that figure represent? Since $80{,}129{,}540 = 1.01^{1{,}829}$, Nilsson and Pelger conclude that "altogether 1,829 steps of 1 percent are required" to bring about the requisite transformation.

These steps, it is important to remember, do not represent *temporal* intervals. We still need to assess how rapidly the advantages represented by such a transformation would spread in a population of organisms, and so answer the question of how long the process takes. In order to do this, Nilsson and Pelger turn to population genetics. The equation that follows involves the multiplication of four numbers:

$$R = h^2 \times i \times V \times m$$

Here, R is the response (i.e. visual acuity in each generation), h is the coefficient of heredity, i designates the intensity of selection, V is the coefficient of variation (the ratio of the standard deviation to the mean), and *m*, the mean value for visual acuity. These four numbers designate the extent to which heredity is responsible for visual acuity, the intensity with which selection acts to prize it, the way its mean or average value is spread over a population, and the mean or average value itself. Values are assigned as estimates to the first three numbers; the mean is left undetermined, rising through each generation.

As for the estimates themselves, Nilsson and Pelger assume that h^2 = .50; that $i = 0.01$; and that $V = 0.01$. On this basis, they conclude that $R = 0.00005m$. The response in each new generation of light-sensitive patches is 0.00005 times the mean value of visual acuity in the previous generation of light-sensitive patches.

Their overall estimate—the conclusion of their paper—now follows in two stages. Assume that *n* represents the number of generations required to transform a light-sensitive patch into a "focused camera-type eye with the geometry typical for aquatic animals." (In small aquatic animals, a generation is roughly a year.) If, as we have seen, the mean value of visual acuity of such an eye is $1.01^{1,829} = 80,129,540$, where 1,829 represents the number of steps required and 80,129,540 describes the extent of the change those steps bring about; and if $1.00005^n = 1.01^{1,829}$ = 80,129,540, then it follows that $n = 363,992$.

It is this figure—363,992—that allows Nilsson and Pelger to conclude at last that "the time required [is] amazingly short: only a few hundred thousand years." And this also completes my exposition of Nilsson and Pelger's paper. Business before pleasure.

NILSSON AND PELGER'S WORK is a critic's smorgasbord. Questions are free and there are second helpings.

Every scientific paper must begin somewhere. Nilsson and Pelger begin with their assumption that, with respect to the eye, morphological change comes about by invagination, aperture constriction, and lens formation. Specialists may wish to know where those light-sensitive cells came from and why there are no other biological structures coordinated with or contained within the interior of the initial patch—for example, blood vessels, nerves, or bones. But these issues may be sensibly deferred.

Not so the issues that remain. Nilsson and Pelger treat a biological organ as a physical system, one that is subject to the laws of theoretical optics. There is nothing amiss in that. But while theoretical optics justifies a *qualitative* relationship between visual acuity on the one hand and invagination, aperture constriction, and lens formation on the other, the relationships that Nilsson and Pelger specify are tightly *quantitative*. Numbers make an appearance in each of their graphs: the result, it is claimed, of certain elaborate calculations. But no details are given either in their paper or in its bibliography. The calculations to which they allude remain out of sight, if not out of mind.

The 1-percent steps: in what units are they expressed? And how much biological change is represented by each step? Nilsson and Pelger do not say. Nor do they coordinate morphological change, which they treat as simple, with biochemical change, which in the case of light sensitivity is known to be monstrously complex.

Does invagination represent a process in which the patch changes as a whole, like a balloon being dimpled, or is it the result of various local processes going off independently as light-sensitive cells jostle with one another and change their position? Are the original light-sensitive cells the complete package, or are new light-sensitive cells added to the ensemble as time proceeds? Do some cells lose their sensitivity and get out of the light-sensing business altogether? We do not know, because Nilsson and Pelger do not say.

Biologists commenting on Darwin's theory have almost always assumed that evolution reflects what the French biologist François Jacob called *bricolage*—a process of tinkering. Biological structures are put together out of pieces; they adapt their function to changes in their circumstances; they get by. This suggests that in the case of eye formation, morphological change might well purchase *less* visual acuity than Nilsson and Pelger assume, the eye being tinkered into existence instead of flogged up an adaptive peak. But if, say, only half as much visual acuity is purchased for each of Nilsson and Pelger's 1-percent steps, twice as many steps will be needed to achieve the effect they claim. What is their justification for the remarkably strong assertion that morphological transformations purchase an optimal amount of visual acuity at each step?

Again we do not know, because they do not say.

More questions—and we have not even finished the *hors d'oeuvres*. The plausibility of Nilsson and Pelger's paper rests on a single number: 1,829. But without knowing precisely how the number 1,829 has been derived, the reader has no way of determining whether it is reasonable or even meaningful.

If nothing else, the number 1,829 represents the maximum point of a curve juxtaposing visual acuity against morphological transformation. Now, a respect for the ordinary mathematical decencies would suggest that the curve is derived from the number, and the number from various calculations. But all such calculations are missing from Nilsson and Pelger's paper. And if the calculations are not given, neither are any data. Have Nilsson and Pelger, for example, *verified* their estimate, either by showing that 1,829 1-percent steps do suffice to transform a patch into an eye, or by showing that such an eye may, in 1,829 1-percent steps, be resolved backward into an initial light-sensitive patch? Once again, we do not know because they do not say.

Still other questions suggest themselves. Although natural selection is mentioned by Nilsson and Pelger, it is a force that plays no role in their reasoning. Beyond saying that it "constantly favors an increase in the amount of detectable spatial information," they say nothing at all. This is an ignominious omission in a paper defending Darwinian principles. An improvement in visual acuity is no doubt a fine thing for an organism; but no form of biological change is without cost.

Let us agree that in the development of an eye, an initial light-sensitive patch in a given organism becomes invaginated over time. Such a change requires a corresponding structural change to the organism's anatomy. If nothing else, the development of an eye requires the formation of an eye socket—hardly a minor matter in biological terms. Is it really the case that an organism otherwise adapted to its environment would discover that the costs involved in the reconstruction of its skull are nicely balanced by what would initially be a very modest improvement in sensitivity to light? I can imagine the argument going either way, but surely an argument is needed.

Then there is Nilsson and Pelger's data-free way with statistics. What is the basis of the mathematical values chosen for the numbers they use in assessing how rapidly transformation spreads in a population of eye patches? The coefficient of variation is the ratio of the standard deviation to the mean. The standard deviation, one might ask, of *what*? No population figures are given; there are no quantitative estimates of any relevant numerical parameter. Why is selection pressure held constant over the course of 300,000 years or so, when plainly the advantages to an organism of increasing light sensitivity will change at every step up the adaptive slope? Why do they call their estimates pessimistic (that is, conservative) rather than wildly optimistic?

Finally, Nilsson and Pelger offer an estimate of the number of *steps*, computed in 1-percent (actually, .005-percent)[56] intervals, that are required to transform their initial patch. At one point, they convert the steps into generations. But a step is not a temporal unit, and, for all anyone knows, each step could well require half again or twice the number of generations they suggest. Why do Nilsson and Pelger match steps to generations in the way they do? I have no idea, and they do not say.

WE ARE AT LAST at the main course. Curiously enough, it is the intellectual demands imposed by Darwin's theory of evolution that serve to empty Nilsson and Pelger's claims of their remaining plausibility.

Nilsson and Pelger assert that only 363,992 generations are required to generate an eye from an initial light-sensitive patch. As I have already

---

56. Editor's note: In the original, this was given as 1.0005 percent. Berlinski corrected that to the number given here (.005 percent) in his response to critics, attributing the error to an editorial slip.

observed, the number 363,992 is derived from the number 80,129,540, which is derived from the number 1,829—which in turn is derived from nothing at all. Never mind. Let us accept 1,829 *pour le sport*. If Nilsson and Pelger intend their model to be a vindication of Darwin's theory, then changes from one step to another must be governed by random changes in the model's geometry, followed by some mechanism standing in for natural selection. These are, after all, the crucial features of *any* Darwinian theory. But in their paper there is no mention *whatsoever* of randomly occurring changes, and natural selection plays only a ceremonial role in their deliberations.

At the beginning of their paper, Nilsson and Pelger write of their initial light-sensitive patch that "we *expose* this structure to selection pressure favoring spatial resolution" (emphasis added), and later that "[a]s the lens approaches focused conditions, *selection pressure* gradually appears to... adjust its size to agree with Mattiesen's ratio" (emphasis added). But whatever Nilsson and Pelger may have been doing to their patch, they have not been exposing it to "selection pressure." The patch does only what they have told it to do. By the same token, selection pressures play no role in adjusting the size of their lenses to agree with Mattiesen's ratio. That agreement is guaranteed, since it is Nilsson and Pelger who bring it about, drawing the curve and establishing the relevant results. What Nilsson and Pelger *assume* is that natural selection would track their results; but this assumption is never defended in their paper, nor does it play the slightest role in their theory.

And for an obvious reason: if there are no random variations occurring in their initial light-sensitive patch, then natural selection has nothing to do. And there are no random variations in that patch, their model succeeding as a defense of Darwin's theory only by first emptying the theory of its content.

An example may make clearer both the point and its importance. Only two steps are required to change the English word "at" to the English word "do": "at" to "ao" and "ao" to "do." The changes are obvious: they have been *designed* to achieve the specified effect. But such design is forbidden in Darwinian theory. So let us say instead, as Darwin must, that letters are chosen randomly, for instance by being fished from an urn. In that case, it will take, on average, 676 changes (26 letters times 26) to bring about the same two steps.

Similarly, depending on assessments of probability, the number of changes required to bring about a single step in Nilsson and Pelger's theory may range widely. It may, in fact, be anything at all. How long would it take to transform a light-sensitive patch into a fully functioning eye? It all depends. It all depends on how *likely* each morphological change happens to be. If cells in their initial light-sensitive patch must discover their appointed role by chance, all estimates of the time required to bring about just the transformations their theory demands—invagination, aperture construction, and lens formation—will increase by orders of magnitude.

If Darwin were restored to pride of place in Nilsson and Pelger's work, the brief moment involved in their story would stretch on and on and on.

FINALLY, THERE IS THE matter of Nilsson and Pelger's computer simulation, in many ways the gravamen of my complaints and the dessert of this discussion.

A computer simulation of an evolutionary process is not a mysterious matter. A theory is given, most often in ordinary mathematical language. The theory's elements are then mapped to elements that a computer can recognize, and its dynamical laws, or laws of change, are replicated at a distance by a program. When the computer has run the program, it has simulated the theory.

Although easy to grasp as a concept, a computer simulation must meet certain nontrivial requirements. The computer is a harsh taskmaster, and programming demands a degree of specificity not ordinarily required of a mathematical theory. The great virtue of a computer simulation is that if the set of objects is large, and the probability distribution and fitness function complicated, the computer is capable of illustrating the implications of the theory in a way that would be impossible using ordinary methods of calculation. "Hand calculations may be sufficient for very simple models," as Robert E. Keen and James Spain write in their standard text, *Computer Simulation in Biology* (1992), "but computer simulation is almost essential for understanding multi-component models and their complex interrelationships."

Whatever the merits of computer simulation, however, they are beside the point in assessing Nilsson and Pelger's work. In its six pages, their paper contains no mention of the words "computer" or "simulation." There are no footnotes indicating that a computer simulation of their work exists, and their bibliography makes no reference to any work containing such a simulation.

Curious about this point, I wrote to Dan-Erik Nilsson in the late summer of 2001. "Dear David," he wrote back courteously and at once,

> You are right that my article with Pelger is not based on computer simulation of eye evolution. I do not know of anyone else who [has] successfully tried to make such a simulation either. But we are currently working on it. To make it behave like real evolution is not a simple task. At present our model does produce eyes gradually on the screen, but it does not look pretty, and the genetic algorithms need a fair amount of work before the model will be useful. But we are working on it, and it looks both promising and exciting.

These are explicit words, and they are the words of the paper's senior author. I urge readers to keep them in mind as we return to the luckless physicist Matt Young. In my *Commentary* essay of December 2002, I quoted these remarks by Mr. Young: "Creationists used to argue that... there was not enough time for an eye to develop. A computer simulation by Dan-Erik Nilsson and Susanne Pelger gave the lie to that claim."

These, too, are forthright words, but as I have just shown, they are false: Nilsson and Pelger's paper contains no computer simulation, and no computer simulation has been forthcoming from them in all the years since its initial publication. Sheer carelessness, perhaps? But now, in responding to my *Commentary* article, Matt Young has redoubled his misreading and proportionately augmented his indignation. The full text of his remarks appears in last month's *Commentary*; here are the relevant passages:

> In describing the paper by Nilsson and Pelger..., I wrote that they had performed a computer simulation of the development of the eye. I did not write, as Mr. Berlinski suggests, that they used nothing more than random variation and natural selection, and I know of no reference that says they did.
>
> ... The paper by Nilsson and Pelger is a sophisticated simulation that even includes quantum noise; it is not, contrary to Mr. Berlinski's assertion, a back-of-the-envelope calculation. It begins with a flat, light-sensitive patch, which they allow to become concave in increments of 1 percent, calculating the visual acuity along the way. When some other mechanism will improve acuity faster, they allow, at various stages, the formation of a graded-index lens and an iris, and then optimize the focus. Unless Nilsson and Pelger performed the calculations in closed form or by hand, theirs was, as I wrote, a "computer simulation." Computer-*aided* simulation might have been a slightly better description, but not enough to justify Mr. Berlinski's sarcasm at my expense....

And here is my familiar refrain: there is *no* simulation, "sophisticated" or otherwise, in Nilsson and Pelger's paper, and their work rests on no such simulation; on this point, Nilsson and I are in complete agreement. Moreover, Nilsson and Pelger do *not* calculate the visual acuity of any structure, and certainly not over the full 1,829 steps of their sequence. They suggest that various calculations have been made, but they do not show how they were made or tell us where they might be found. At the very best, they have made such calculations for a handful of data points, and then joined those points by a continuous curve.

There are two equations in Nilsson and Pelger's paper, and neither requires a computer for its solution; and *there are no others*. Using procedures very much like Nilsson and Pelger's own, Mr. Young has nevertheless deduced the existence of a missing computer simulation on theoretical grounds: "Unless Nilsson and Pelger performed the calculations in closed form or by hand, theirs was, as I wrote, a computer simulation."

But another possibility at once suggests itself: that Nilsson and Pelger did not require a computer simulation to undertake their calculations because they made no such calculations, their figure of 1,829 steps representing an overall guess based on the known optical characteristics of existing aquatic eyes.

Whatever the truth—and I do not know it—Mr. Young's inference is pointless. One judges a paper by what it contains and one trusts an author by what he says. No doubt Matt Young is correct to observe that "computer-*aided* simulation might have been a better description" of Nilsson and Pelger's work. I suppose one could say that had Dan-Erik Nilsson and Susanne Pelger rested their heads on a computer console while trying to guess at the number of steps involved in transforming a light-sensitive patch into a fully functioning eyeball, their work could also be represented as computer-aided.

Matt Young is hardly alone in his lavish misreadings. The mathematician Ian Stewart, who should certainly know better, has made virtually the same patently false claims in *Nature's Numbers* (1995). So have many other prominent figures.[57] But misreadings are one thing, misrepresentations another. More than anyone else, it has been Richard Dawkins who has been responsible for actively misrepresenting Nilsson and Pelger's work, and for disseminating worldwide the notion that it offers a triumphant vindication of Darwinian principles.

In a chapter of his 1995 book, *River Out of Eden*, Dawkins writes warmly and at length about Nilsson and Pelger's research.[58] Here is what he says (emphasis added throughout):

> [Their] task was to set up *computer models* of evolving eyes to answer two questions...[:] is there a smooth gradient of change, from flat skin to full camera eye, such that every intermediate is an improvement?... [and] how long would the necessary quantity of evolutionary change take?

---

57. Among those who, by contrast, have raised points similar to my own, I would single out especially Brian Harper, a professor of mechanical engineering at Ohio State University.

58. A version of the same material by Dawkins, "Where D'you Get Those Peepers," was published in the *New Statesman* (July 16, 1995).

In their *computer models*, Nilsson and Pelger made no attempt to simulate the internal workings of cells.

... Nilsson and Pelger began with a flat retina atop a flat pigment layer and surmounted by a flat, protective transparent layer. The transparent layer was allowed to *undergo localized random mutations of its refractive index*. They then let *the model transform itself at random*, constrained only by the requirement that any change must be small and must be an improvement on what went before.

The results were swift and decisive. A trajectory of steadily mounting acuity led unhesitatingly from the flat beginning through a shallow indentation to a steadily deepening cup, *as the shape of the model eye deformed itself on the computer screen....* And then, *almost like a conjuring trick*, a portion of this transparent filling condensed into a local, spherical region of higher refractive index.

... This ratio is called Mattiessen's ratio. Nilsson and Pelger's *computer-simulation model homed in* unerringly on Mattiessen's ratio.

How very remarkable all this is—inasmuch as there are no computer models mentioned, cited, or contained in Nilsson and Pelger's paper; inasmuch as Dan-Erik Nilsson denies having based his work on any computer simulations; inasmuch as Nilsson and Pelger never state that their task was to "set up computer models of evolving eyes" for any reason whatsoever; inasmuch as Nilsson and Pelger assume but do not prove the existence of "a smooth gradient of change, from flat skin to full camera eye, such that every intermediate is an improvement"; and inasmuch as the original light-sensitive patch in Nilsson and Pelger's paper was never allowed to undergo "localized random mutations of its refractive index."

And how very remarkable again—inasmuch as there are no computer "screens" mentioned or cited by Nilsson and Pelger, no indication that their illustrations were computer-generated, and no evidence that they ever provided anyone with a real-time simulation of their paper where one could observe, "almost like a conjuring trick," the "swift and decisive" results of a process that they also happen to have designed.

And yet again how very remarkable—inasmuch as Nilsson and Pelger's "computer-simulation model" did not home in unerringly on Mattiessen's ratio, Nilsson and Pelger having done all the homing themselves and thus sparing their model the trouble.

Each and every one of these very remarkable asseverations can be explained as the result of carelessness only if one first indicts their author for gross incompetence.

FINAL QUESTIONS. WHY, IN the nine years since their work appeared, have Nilsson and Pelger never dissociated themselves from claims about their work that they know are unfounded? This may not exactly be dishonest, but it hardly elicits admiration. More seriously, what of the various masters of indignation, those who are usually so quick to denounce critics of Darwin's theory as carrying out the devil's work? Eugenie Scott, Barbara Forrest, Lawrence Krauss, Robert T. Pennock, Philip Kitcher, Kelly Smith, Daniel Dennett, Paul Gross, Ken Miller, Steven Pinker—they are all warm from combat. Why have they never found reason to bring up the matter of the mammalian eye and the computer simulation that does not exist?

And what should we call such a state of affairs? I suggest that scientific fraud will do as well as any other term.

A Scientific Scandal?
David Berlinski & Critics

In April 2003, Commentary *magazine published David Berlinski's essay "A Scientific Scandal"; page 367 in this volume. The article, like its recent predecessors in the same journal, provoked a passionate response from scientists and philosophers. Below are the letters, and at the bottom is Dr. Berlinski's reply. Originally published in July–August 2003 in* Commentary.

—Editor

Dan-E. Nilsson
Lund University, Lund, Sweden

I appreciate the opportunity to respond to David Berlinski's essay on the 1994 paper I authored with Susanne Pelger called "A Pessimistic Estimate of the Time Required for an Eye to Evolve." Because it gives them credibility, I generally do not debate pseudo-scientists, but I have decided to make an exception here.

Apart from a mix-up in chronology and some other minor peculiarities, the only major flaw in Mr. Berlinski's description of our paper is his misunderstanding of the response variable R, which he calls a measure of "visual acuity." It is not, and the original paper does not say so. This is his first serious mistake—and it gets worse.

Mr. Berlinski's next move is to list the important information he claims is missing in our paper. (At regular intervals he repeats the phrase: "they do not say.") But all the necessary information is there. I cannot reply individually to every point here, but two examples will do. Mr. Berlinski claims that there is no unit for morphological change and that we do not explain how we arrive at a sum of 1,829 steps of 1 percent, but explanations for both are given on page 56 of our paper. He further claims that we fail to explain how morphological change relates to improvements in visual acuity, though pages 54 through 56 (together with the graphs and legends in figures 1 and 3) deal with exactly that, and in great detail.

For the rest of his essay Mr. Berlinski focuses on issues where he believes he has detected logical flaws. He is not right in a single case, and instead reveals an insufficient background in visual optics, sampling theory, basic evolutionary theory, and more. Nor does he seem to have read key references such as Warrant and McIntyre (1993), Falconer (1989), or Futuyma (1986). Without such knowledge it would be hard to grasp the details of our paper, but it is standard scientific practice not to repeat lengthy reasoning when a short reference can be given.

But there is more. Mr. Berlinski has a problem with definitions. "Morphological change" becomes "biological change." "Spatial resolution" (visual acuity) becomes "sensitivity of vision." He does not distinguish between selection and intensity of selection. He is obviously confused by the difference between the 1-percent steps that we use as a unit of measure for morphological change and the 0.005-percent change per generation that is our conservative estimate of evolutionary rate.

Mr. Berlinski attempts a peculiar probability argument involving the random substitution into the word "at" of letters "fished from an urn," but he does not realize that his example implies a single individual in the population, in which case there can of course be no selection at all. Again, he badly needs to read Falconer's standard work.

Contrary to Mr. Berlinski's claim, we calculate the spatial resolution (visual acuity) for all parts of our eye-evolution sequence, and the results are displayed in figure 1 of our paper. The underlying theory is explained in the main text, including the important equation 1 and a reference to Warrant and McIntyre (1993), where this theory is derived. Yet Mr. Berlinski insists that "Nilsson and Pelger do *not* calculate the visual acuity of any structure." It would be much simpler for Mr. Berlinski if he went just a tiny step farther and denied the existence of our paper altogether.

Had these and other points been unfortunate misunderstandings, I would have been only too happy to help, but I have the distinct impression that they are deliberate attempts to eliminate uncomfortable scientific results. Why does Mr. Berlinski not read up on the necessary scientific background? Why does he so blatantly misquote our paper? Why has he never asked me for the details of the calculation he claims to want so badly? It is simply impossible to take Mr. Berlinski seriously.

Mr. Berlinski is right on one point only: the paper I wrote with Pelger has been incorrectly cited as containing a computer simulation of eye evolution. I have not considered this to be a very serious problem, because a simulation would be a mere automation of the logic in our paper. A complete simulation is thus of moderate scientific interest, although it would be useful from an educational point of view.

Our paper remains scientifically sound, and has not been challenged in any peer-reviewed scientific journal. I do not intend to take any further part in a meaningless debate with David Berlinski.

Paul R. Gross
Jamaica Plain, Massachusetts

"A SCIENTIFIC SCANDAL" is itself a scientific scandal: the continued publication, in a political-cultural opinion journal, of David Berlinski's uninformed bellyaching about evolutionary biology. *Commentary* is not the place for quasi-technical arguments against Darwinism, or for reprinting the scientific papers or textbook chapters that disprove them.

Mr. Berlinski has several times found fault with me. The method is characteristic, and it is salient in this latest article. I had written earlier that his disparagements of Darwinism are old and naïve, refuted in the literature. Responding in the March issue ("Darwinism versus Intelligent Design"), he dismissed this airily as an unanswerable gripe. But it is not a gripe. Nor was it meant to be answered in *Commentary*. It is just a fact about the scientific literature. Any reader can check for himself. Examples include: Mark Ridley, *Evolution*, 2nd Edition (1996); John Gerhart and Marc Kirschner, *Cells, Embryos, and Evolution* (1997); Rudolf A. Raff, *The Shape of Life* (1996).

Only once, in the eleven years since the start of their anti-evolution PR-blitz, have any arguments of Mr. Berlinski's colleagues at the Discovery Institute's Center for Science and Culture appeared in the primary literature. That was an early philosophical monograph by the Christian apologist William Dembski. Mr. Berlinski (in "Has Darwin Met His Match?," December 2002) now rejects that argument as applied to biology, although he gave Dembski's book a glowing blurb. The rest of their anti-evolution kvetching has been in trade books mainly from religious publishers, in nonscientific journals, testimony to legislators, interviews, speeches, and rallies for the faithful. For this, Mr. Berlinski offered the crank excuse: scientific prejudice. And as *coup de main*, he quoted lines from a 1986 essay of mine. But the burden of that essay is precisely the opposite of Mr. Berlinski's reason for quoting it. It was about a distinguished regular contributor *to the scientific literature*.

The obvious purpose of "A Scientific Scandal," like Mr. Berlinski's other adventures in evolutionary thought, is to belittle Darwinism. He cites Darwin himself, who worried a little that his theory might not be able to account for the eye. Mr. Berlinski's real case is that Darwin's fears were justified: evolutionary theory cannot explain the eye, and there has been a cover-up. But Darwin's fears are ancient history: Darwin was still haunted by Paley's 1802 version of the argument from design. A century and a half have passed.

In the twenty-first century there is no question that eyes, endlessly varied in structure and quality, have evolved. Most of the intermediates between a primitive patch of photosensitive cells and the camera eye of a fish or a mammal exist. Many more have existed in the past, during the 540 million years since there have been eyes.

So what is the fuss about? In their 1994 theoretical paper, Nilsson and Pelger *modeled* one possible evolutionary pathway to the geometry of a fish-like eye from a patch of photoresponsive cells. There were already such cells on Earth a billion years before there were eyes. Nilsson and Pelger used pessimistic estimates of such relevant parameters as the intensity of selection for their number-crunching. The point was to determine how many plausible, populational micro-steps of variation would be needed for very weak selection to yield a fish-like eye—and then under reasonable assumptions to convert micro-steps into generations and years. The answer was about 350,000—a geological blink of the eye. This answer is just one of many to the failed nineteenth-century complaint of insufficient time for evolution to have taken place.

Mr. Berlinski misunderstands or misinterprets critical elements of the paper. Then he quibbles ponderously about terms and assumptions—and about a popular gloss of the paper by Richard Dawkins. He accuses some of his critics of fraud for having failed to denounce Dawkins's use in a trade book of certain of those terms. Mr. Berlinski's arguments *are* quibbles.

But these quibbles are beside the real point, which is that we lack grounds for believing that eyes evolved. That is false. Eyes, like anything else, could have been invented at a stroke by a supernatural designer. But there is no evidence of it. Neither can it ever be disproved. The only *explanation*, however, that we have for the structure of eyes—as solid as any explanation in science—is Darwinian evolution.

Like the intelligent-design group as a whole, Mr. Berlinski seems unable or unwilling to understand the newest branch of biology: evolutionary *developmental* biology. There, with the discovery of the developmental regulatory genes, we have learned how subtle, how versatile, and yet how simple the mechanisms can be for transforming one biological structure to another. (A professional but accessible account can be found in *From DNA to Diversity: Molecular Genetics and the Evolution of Animal Design* [2001] by Sean B. Carroll, Jennifer K. Grenier, and Scott D. Weatherbee. A popular but sound insight is available in: "Which Came

First, the Feather or the Bird?" by Richard O. Prum and Alan H. Brush, *Scientific American*, March 2003.) A reader whose view of science comes only from Mr. Berlinski will never know of such things.

Matt Young
Colorado School of Mines, Golden, Colorado

CREATIONISTS OFTEN CLAIM, WITHOUT presenting evidence, that there has not been enough time for a complex organ such as the eye to have evolved. To examine that claim empirically, Nilsson and Pelger devised a scenario in which an eye could have evolved through stages that are known to exist in the animal kingdom. I described their scenario in my last letter to *Commentary* (March), and David Berlinski has it almost right in "A Scientific Scandal."

Briefly, Nilsson and Pelger formed an eye by changing various parameters, such as aperture diameter, in 1-percent increments, until no improvement could be made. One percent is an arbitrary number (any small increment will suffice) and does not represent the change in a single generation. Using what they and Richard Dawkins describe as conservative numbers, Nilsson and Pelger calculated an average change of 0.005 percent per generation. The relative change in n generations is therefore $(1.00005)^n$, which they set equal to the overall change of morphology in their simulation ($1.01^{1.829}$, where 1,829 is the number of 1-percent steps required to form an eye). The number 1.00005 is not, contrary to Mr. Berlinski, a percentage; it is the relative change of a given parameter in a single generation. Nilsson and Pelger concluded that an eye could have evolved in approximately 350,000 years.

Does anyone claim that an eye evolved precisely as Nilsson and Pelger's simulation suggests? No. But I stand by my statement that they have given the lie to the creationists' claim and firmly made the case that an eye could have evolved within a geologically short time.

Mr. Berlinski argues, for example, that morphological changes of the skull might slow the process. Never mind that only vertebrates have skulls, and Nilsson and Pelger's eye is, again contrary to Mr. Berlinski, an invertebrate eye. The development of an eye will require not only morphological changes but also advancements to the nervous system and the brain. Will these requirements bring evolution to a halt? Georges Cuvier asked the same question in 1812, and the answer is, "no." We now know that evolution progresses in a modular way, with different systems evolving in parallel and nearly independently. If Mr. Berlinski thinks that various modules could not have co-evolved, he needs to support his argument quantitatively, not just proclaim it. Nilsson and Pelger have shown precisely what they set out to show: that an eye could have evolved in a geologically short time and that the eye itself is not a limiting factor. Mr. Berlinski holds against them that they did not perform the full-fledged simulation he wants them to have done and seems to think that their calculation is therefore somehow faulty.

I will not respond to Mr. Berlinski's disdainful tone, nor to the cheap shots directed at me personally. Nor will I continue the pointless distraction of whether Nilsson and Pelger performed a simulation or a calculation. I am, however, concerned with Mr. Berlinski's contention that reputable scientists have conspired to support a technical paper that he finds "unfounded"; charging specific individuals with "fraud" is not to be taken lightly. The paper has survived peer review, and has not been shown to be unfounded in any peer-reviewed journal. If Mr. Berlinski thinks the paper is unfounded, let him submit a paper of his own to a peer-reviewed journal and find out what the scientific community thinks of his ideas. It is unlikely that scientific journals, which have occasionally published papers on homeopathic medicine and the Bible codes, would reject Mr. Berlinski's paper out of sheer prejudice.

Mark Perakh
Bonsall, California

It is funny that *Commentary*—by no measure a scientific publication—has allocated so much space to recent articles by David Berlinski.

With no record of scientific research, either in biology or in computer science, he sets out to pronounce judgment on topics within these two fields.

But contrary to Mr. Berlinski's rhetoric, any scandal related to Nilsson and Pelger's paper occurred only in Mr. Berlinski's imagination. Nilsson and Pelger estimate the time necessary for the development of an eye, a calculation that entails certain assumptions but which is viewed by many scientists as sufficiently sound. (According to the *Science Citation Index*, Nilsson and Pelger's article has been positively referenced in at least 25 peer-reviewed scientific publications.)

But Mr. Berlinski, unlike all these scientists, does not like Nilsson and Pelger's conclusion, and obfuscates the issue by discussing the distinctions among computer simulations, models, and calculations. These semantic exercises are inconsequential to the real question: whether an eye could have developed in a geologically short time via a Darwinian mechanism, as Nilsson and Pelger and scores of biologists familiar with their work think.

A reader cannot fail to notice an especially appalling feature of Mr. Berlinski's escapade: he accuses ten respected scientists of "scientific fraud." The reason for that preposterous accusation is that they did not repudiate Nilsson and Pelger's work. Mr. Berlinski apparently cannot imagine that these scientists, among them professional biologists and physicists with records of substantial achievement, can have an opinion of Nilsson and Pelger's work different from his own. His accusation sounds even odder coming from a man who provided rave blurbs for books by William Dembski and Michael Behe even though, as is clear from his article in the December 2002 *Commentary*, he is actually in disagreement with them regarding essential parts of their assertions. Maybe by his standards this is a manifestation of integrity, but to me it looks more like an expediency whose roots are not exactly in the search for scientific truth.

Jason Rosenhouse

Kansas State University, Manhattan, Kansas

CONNOISSEURS OF PSEUDOSCIENCE WILL recognize in David Berlinski's latest essay the standard tropes of the crank's playbook: the smug sarcastic tone, the barrage of bullet-point criticisms to create the illusion that something truly rotten is being exposed (criticisms he knows will be answered by nothing more formidable than a few indignant letters), the crude baiting of scholars of vastly greater accomplishment than he, and the presentation of minor errors as tantamount to fraud.

Mr. Berlinski has no interest in bringing clarity to difficult scientific issues. If he did, he would not have made so many misrepresentations in describing Nilsson and Pelger's work. Two examples: Mr. Berlinski's claim that their model eyes were simply "flogged up an adaptive peak" ignores the fact that establishing the existence of such a peak was one of the primary accomplishments of the paper. That there is a smooth gradient of increasing visual acuity linking a light-sensitive spot to a lens-bearing eye is a discovery that they made, not a foregone conclusion. And his claim that "in their paper there is no mention whatsoever of randomly occurring changes" falls flat, since the need for such changes is explicitly mentioned in the discussion section of the paper, and is plainly implied throughout.

In addition, Mr. Berlinski would not have unloaded so many spurious criticisms. For example, his query—"why is selection pressure held constant over the course of 300,000 years"—is easily answered by noting that it was held constant at a value that was ludicrously low for almost any environment.

Once we have swept the field of Mr. Berlinski's distortions we are left with a few simple facts. (1) Several decades of research on the evolution of eyes has not only made it plain that eyes have evolved, but has also revealed the major steps through which they did so. (2) Nilsson and Pelger's paper provides an elegant capstone for this research, by providing a convincing calculation for an upper limit on the time required for an eye to evolve. (3) Minor errors in popular treatments of Nilsson and Pelger's paper do nothing to change facts (1) and (2). (4) Finally, David Berlinski is not a reliable source for scientific information.

Nick Matzke
Goleta, California

DAVID BERLINSKI SHOULD BE congratulated for pointing out Richard Dawkins's inaccurate description of Nilsson and Pelger's paper as a stochastic computer simulation of the evolution of the eye (it was actually a mathematical model). But Mr. Berlinski should remove the plank from his own (discussion of the) eye. He asserts that one of the problems that Nilsson and Pelger did not consider was how the skull would be "reconstructed" to include eye sockets. But as any decent student of even high-school biology would know, eyes evolved before bones. Cephalochordates, the closest invertebrate relatives of vertebrates, have primitive eyes but no bones. In fact, based on genetic evidence, many biologists now think that vertebrate eyes share a common ancestral eyespot with insect eyes.

To envision the evolution of the eye as occurring on some kind of mythical eyeless fish with a fully formed skull and brain is a typical creationist straw man. Biologists know that all manner of gradations of eye complexity exist in extant organisms, from creatures with a single photoreceptor cell, through the various stages that Nilsson and Pelger depict, to the advanced camera-eyes of mammals and cephalopods. Sometimes the whole sequence from eyespot to advanced eye with lens can be seen in a single group (e.g., snails), yet another thing Mr. Berlinski would have known had he followed Nilsson and Pelger's reference to the classic work on eye evolution, a 56-page article by Salvini-Plawen and Mayr in *Evolutionary Biology* (vol. 10, 1977) called "On the Evolution of Photoreceptors and Eyes." That paper answers many of the questions that Mr. Berlinski asserts are unanswered or unanswerable.

If Mr. Berlinski is going to declare as bunk the central organizing theory of biology, he should take the matter up with biologists in the professional literature rather than in forums like *Commentary*, wherein elementary questions like "which came first, skulls or eyes?" can be botched and yet still be published.

David Safir
Los Gatos, California

Once again, David Berlinski has shown how a truly scientific inquiry can expose academic and intellectual fraud by evolutionists. As a physician, I have always been made uneasy by the assertions offered by proponents of evolution to explain complex biological life. Mr. Berlinski shows exactly how the process works: start with the belief that no other possible explanation for the diversity of life on earth could exist other than what we think we know about evolution; demonstrate utter contempt for other ideas (*ad-hominem* attacks are often employed here); then simply invent a pathway describing how it might have been possible to get from point A to point B—from a light-sensitive spot, say, to a complex eye. Where I come from this is called nonsense.

I would feel better about a theorist like Richard Dawkins if he did not pontificate about how gloriously perfect his explanations are. I cast my fate instead with scientists like Mr. Berlinski who keep an open mind. The jury is still out, after all, and will be for a very long time.

Norman P. Gentieu
Philadelphia, Pennsylvania

As a retired science writer, I appreciated David Berlinski's superb analysis refuting Nilsson and Pelger's simplistic scenario of the evolution of the mammalian eye. To account for the perfection of that incredibly complex organ by means of formulaic fumblings is nothing less than preposterous. I wonder if Nilsson and Pelger might some day use this iffy method to explain the development of stereoscopic color vision.

"A Scientific Scandal" is an apt name for the docile acceptance of a dubious theory. What has happened to vetting? Back in the 1950s, the science establishment did not hesitate to zap Immanuel Velikovsky and his *Worlds in Collision*.

David Berlinski Replies

IN "A SCIENTIFIC SCANDAL," I observed that Dan-E. Nilsson and Susanne Pelger's paper, "A Pessimistic Estimate of the Time Required for an Eye to Evolve," was a critic's smorgasbord. There are so many things wrong with it that even the finickiest of eaters could leave the table well-satisfied and ready for a round of Alka-Seltzer. But, in itself, there is nothing here that suggests a scandal. Dan-E. Nilsson is a distinguished scientist. Witness his discovery that the mysid shrimp, *Dioptromysis pauciponisa*, is an organism whose eyes are at once simple and compound (D. Nilsson, R.F. Modlin, "A Mysid Shrimp Carrying a Pair of Binoculars," *Journal of Experimental Biology*, Vol. 189, pp. 213–236, 1994), or his precise work on the optical system of the butterfly (D. Nilsson, M.F. Land, J. Howard, "Optics of the Butterfly Eye," *Journal of Comparative Physiology*, A 162, 341–366, 1988). Together with Susanne Pelger, he has simply written a silly paper. It happens. And in the literature of evolutionary biology, it happens very often.

No, the scientific scandal lies elsewhere. Nilsson and Pelger's paper has gained currency in both the popular and the scientific press because it has been misrepresented as a computer simulation, most notably by Richard Dawkins. Word spread from Dawkins's mouth to any number of eagerly cupped but woefully gullible ears. Subsequent references to Nilsson and Pelger's work have ignored what they actually wrote in favor of that missing computer simulation, in a nice example of a virtual form of virtual reality finally displacing the real thing altogether. This misrepresentation of scientific work is a species of fraud, no different in kind from plagiarism in journalism or the fabrication of data in experimental physics. It is the *indifference* to this fraud that I denounced as scandalous.

Recognizing so many fond familiar faces among my critics—Paul Gross, Jason Rosenhouse, Matt Young, and Mark Perakh have replied to previous essays of mine in *Commentary*—I hoped that self-interest, if nothing else, might have prompted a moment of critical self-reflection. No very delicate moral sense is involved in determining that fraud is fraud. If Richard Dawkins is one of their own, all the more reason to apply to him the moral standards that Messrs. Gross, Rosenhouse, Young, and Perakh are accustomed to applying to their intellectual enemies.

Reading their letters, I realize that they had no intention of saying boo. What could I have been thinking?

DAN-E. NILSSON IS PERSUADED that I wrote my essay because I am moved to reject "uncomfortable scientific results." He is mistaken. The length of time required to form an eye is a matter of perfect indifference to me; had he and Susanne Pelger been able to demonstrate that the eye was in fact formed over the course of a long weekend in the Hamptons, I would have warmly congratulated them. As I have many times remarked, I have no creationist agenda whatsoever and, beyond respecting the injunction to have a good time all the time, no religious principles, either. Evolution long, evolution short—it is all the same to me. I criticized their work not because its conclusions are unwelcome but because they are absurd.

The vertebrate eye, Nilsson and Pelger claim, emerged from a patch of light-sensitive cells. Climbing up evolution's greasy pole, or adaptive peak, those cells got to where they were going by invagination, aperture constriction, and lens formation. In explaining the evolution of the eye in terms of such global geometrical processes, Nilsson and Pelger rather resemble an art historian prepared to explain the emergence of the *Mona Lisa* in terms of preparing the wood, mixing the paint, and filling in the details. The conclusion—that Leonardo completed his masterpiece in more than a minute and less than a lifetime—while based squarely on the facts, seems rather less than a contribution to understanding.

It is hardly surprising, then, that while theoretical optics serves *qualitatively* to justify the overall connection Nilsson and Pelger draw between morphology and visual acuity, nothing in their paper and nothing in their references justifies the *quantitative* relationships they employ to reach their quantitative conclusion. To be sure, Mr. Nilsson denies that this is so. "Contrary to Mr. Berlinski's claim," he writes,

> we calculate the spatial resolution (visual acuity) for all parts of our eye-evolution sequence, and the results are displayed in figure 1 of our paper. The underlying theory is explained in the main text, including the important equation 1 and a reference to Warrant and McIntyre (1993), where this theory is derived.

In fact, no underlying theory whatsoever is explained in Nilsson and Pelger's main text, or in the legend to figure 1; and while they do assert that calculations were made, they do not say where they were made or how they were carried out. The burden of Mr. Nilsson's denials is conveyed entirely by equation 1 and by his references.

Let us start with equation 1, and with figure 1b that this equation is said to control. It is in figure 1b that aperture constriction takes over from invagination in getting an imaginary eye to see better. The graph juxtaposes aperture size against detectable spatial resolution. Having dimpled itself in figure 1a, Nilsson and Pelger's blob is now busy puckering its topmost surface to form a pinhole in figure 1b.[59] In a general way, the curve they present is unremarkable. No one doubts that spatial resolution is improved in an eye when its aperture is constricted. But why is it improved in just the way that Nilsson and Pelger's graph indicates?

Equation 1 is of scant help in this regard, despite Nilsson's insistence that it is important. Drawing a connection among visual acuity, focal length, light intensity, and noise, the equation specifies the local maximum of a curve, the place where it stops rising. In other words, it

---

59. Three curves are given in figure 1b, representing three different levels of light intensity, but this plays no role in what follows.

specifies a point; and it does nothing more. "We can now use this relationship," Nilsson and Pelger nevertheless declare, "to plot resolution against aperture diameter." They can do nothing of the sort, at least not in my calculus class. Knowing that a man has reached the summit of Mt. Everest, we still know nothing about the route he has taken to get there. What is needed if Nilsson and Pelger are to justify their graph is the equation from which equation 1 has been derived by differentiation. It is not there, just where I said it would not be.

Similarly with Nilsson and Pelger's references, which do nothing to support their argument. Quite the contrary. Three papers are at issue: (1) A. W. Snyder, S. Laughlin, and D. Stavenga, "Information Capacity of the Eyes" (*Vision Research*, vol. 17, 1163–1175, 1977); (2) A. W. Snyder, "Physics of Vision in Compound Eyes" (in *Vision in Invertebrates*, Handbook of Sensory Physiology, edited by H. Autrum, vol. VII/6A, pp. 225–313, 1979); and (3) E. J. Warrant & P. D. McIntyre, "Arthropod Eye Design and the Physical Limits to Spatial Resolving Power" (*Progress in Neurobiology*, vol. 40, pp. 413–461, 1993). Of these papers, the first is recapitulated (and corrected) in the second, and the second is summarized in the third. In what follows, references to Snyder are always to the Snyder of his second paper.

As their titles might suggest, both "Physics of Vision in Compound Eyes" and "Arthropod Eye Design and the Physical Limits to Spatial Resolving Power" deal with *compound invertebrate eyes*. Nilsson and Pelger's work is devoted to the evolution of the *camera* eye characteristic of fish and cephalopods. Theoretical considerations that apply to bugs do not necessarily apply to fish or octopuses, the more so since their eyes are structurally different, as are their evolutionary histories. Writing about the compound eye, Nilsson himself has remarked that "it is only a small exaggeration to say that evolution seems to be fighting a desperate battle to improve a basically disastrous design" (Dan-E. Nilsson, "Optics and Evolution of the Compound Eye," in *Facets of Vision*, edited by D. G. Stavenga & R. C. Hardie, p. 3075, 1989). Whatever the desperate battle going on among the arthropods, there is no battle at all taking place among the vertebrates or the cephalopods. Nilsson and Pelger's eye moves from triumph to triumph with serene and remarkable celerity.

If the papers by Snyder and by Warrant and McIntyre say nothing about fish or octopuses, neither do they say anything about evolution. No mention there of Darwin's theory, no discussion of morphology, not a word about invagination, aperture constriction, or lens formation, and *nothing* about the time required to form an eye, whether simple, compound, or camera-like.

The purpose of these three papers is otherwise. No less than any other system of communication, the eye represents a balance struck between signal and noise. There is the object out there in the real world—whether a point source like a star, or an extended source like a grating of light and dark lines—and there is its image trembling on the tips of the retina's budded nerve cells. Slippage arises between what the object is and how it is seen. Noise occurs in the visual system as the result of the random nature of photon emission, and it also occurs as the result of inherent imperfections in the eye's optical system. The theoretical optician abbreviates these limitations in one mathematical instrument.

Imagine one of Nilsson and Pelger's plucky light-sensitive cells, and then extend two flanking lines from the cell up past the constricted aperture and out into space, so that the cell and those two flanking lines form a cone with a flat top. In the center of the cone, where a cherry would sit atop the ice cream, there is a light source. The cherry moves to the sides of the cone in angular steps; the cell dutifully responds. The correlation between moving cherry and twitching cell constitutes the optician's "angular-sensitivity function."

Equation B15 (p. 238) in Snyder's "Physics of Vision in Compound Eyes" defines the signal-to-noise ratio of a hypothetical eye in terms of noise, modulation contrast (the difference in intensity between black and white stripes in a grating), and the modulation-transfer function, which is simply a mathematical transformation of the eye's angular-sensitivity function (its Fourier transform). Lumbering in Snyder's footsteps, Warrant and McIntyre split his equation into two of their own (equations 10 and 11 in Warrant and McIntyre, p. 430), the one describing the signal, the other the noise in a hypothetical visual system. They observe what is in any case obvious: whatever the parameters affecting visual acuity, signal and noise will always reach a point where the first is drowned out by the second and the system fails, a point evident enough to anyone trying to see in the dark.

These equations lead by primogeniture to Nilsson and Pelger's equation 1, which, as it happens, does not appear anywhere in their sources in the form in which they express it. But neither Snyder's original equation nor Warrant and McIntyre's bright bursting clones in any way suggest that the tipping point between signal and noise is unique. The ratio of signal to noise in an optical system depends on a host of factors, including head size and eye movement, most of which Nilsson and Pelger ignore. Nor, for that matter, do these equations taken in isolation justify any particular quantitative conclusions. Until the angular-sensitivity function is specified, whether theoretically or experimentally, its role is ceremonial.

Such specification is no easy business. Determining the shape of the angular-sensitivity function is a little like trying to guess an astronaut's weight in space. Scales are not likely to be of use. In an early paper dealing with this subject and devoted experimentally to flies, K. G. Götz noted that the angular-sensitivity function in *Drosophila* seemed to follow what is known mathematically as a Gaussian probability distribution (K. G. Götz, "Die optischen Übertragungseigenschaften der Komplexaugen von *Drosophila*," *Kybernetik* 2, pp. 215–221, 1965). It was an interesting idea, but one that led to very considerable computational difficulties.

Looking Götz-ward, and understandably recoiling, Snyder adopted a different strategy. In assessing the weight of an astronaut in space, it is simpler to count the calories he consumes and the exercise he undergoes than to try to measure his weight directly. His weight, although unmeasured, follows inferentially. In just the same way, Snyder thought to consider the angular-sensitivity function indirectly by considering the structures that determined its shape. These, he assumed, were the eye's retinal receptive field—the area of the retina responding to signals—and its optical "blur spot"—the smeared image represented on the retina corresponding to the sharp object being seen. Let them both, he declared, be identically Gaussian. Why not? Both parameters had simple mathematical natures. The retinal receptive field is given as the ratio of the rhabdom's diameter to its posterior nodal distance, the optical blur as the ratio of the wavelength of stimulating light to the eye's aperture. From this the shape of the angular-sensitivity function followed.

The result is known as the Snyder model. "*The great beauty of this model,*" Warrant and McIntyre remark (in words that they have italicized), "*is that if one knows some very simple anatomical information about the eye*"—i.e., the nature of its optical blur spot and retinal receptive field—"*one has the ability to predict… the approximate shape of the angular-sensitivity function*" (p. 434). In referring to Warrant and McIntyre, Nilsson and Pelger are, in fact, appealing to Snyder, the *maître* behind their masters—for, like Snyder, they, too, assume that retinal receptive fields and optical blur spots are identically Gaussian (p. 54).

But theory is one thing, and living flesh another. Staking their all on Snyder's model, Nilsson and Pelger must live with its consequences. "Having considered the physical limitations to resolving power," Snyder wrote, "in addition to the absolute sensitivity of eyes, we now apply our concepts to real compound eyes." This is something that Nilsson and Pelger never do. And no wonder. For Snyder then added the rather important caveat that bringing theory to bear on life "requires *precise* knowledge [of various optical parameters] in the various regions of the eye" (Snyder, p. 276, emphasis in the original).

If precise knowledge is needed in applying Snyder's model, precise detail is what is lacking in Nilsson and Pelger's paper. Precise detail? *Any* detail whatsoever.

And for obvious reasons. When tested, Snyder's model turns out to be *false* across a wide range of arthropods. As Warrant and McIntyre note glumly, "The model, on the whole, works best for those eyes for which it was originally formulated—apposition compound eyes functioning according to geometrical optics—but recent careful and sensitive measurements of angular sensitivity reveal that even in these types of eye, the model often performs poorly." Readers may consult figure 34 (p. 441) of Warrant and McIntyre's paper to see how poorly the Snyder model does. In studies of the locust *Locustia*, real and predicted angular-sensitivity functions do not even share the same qualitative shape.

Responding to my observation that no quantitative argument supports their quantitative conclusions—no argument at all, in fact—Mr. Nilsson has thus (1) offered a mathematically incoherent appeal to his only equation; (2) cited references that make no mention of any morphological or evolutionary process; (3) defended a theory intended to describe the evolution of vertebrate camera eyes by referring to a theory describing the theoretical optics of compound invertebrate eyes; (4) failed to explain why his own work has neglected to specify any relevant biological parameter precisely; and (5) championed his results by means of assumptions that his own sources indicate are false across a wide range of organisms.

In acknowledgments to *their* paper, Nilsson & Pelger thank E. J. Warrant for help with their computations; in the acknowledgments to their paper, Warrant and McIntyre thank Mr. Nilsson for critically reading what they have written.

Schnapps all around, I am sure.

I TURN NEXT TO the morphological units that are missing from Nilsson and Pelger's paper. It makes no sense to say of a ruler that it is one long. One what? When the "what" has been specified, a physical unit has been indicated: one inch, say, in the case of length, one pound in the case of weight. If one inch and one pound are units, length and weight are their dimensions. Only an origin in zero remains to be specified to complete the picture.

In my essay, I observed that Nilsson and Pelger had not specified their unit of morphological change. Nilsson now asks me to consider again their remarks on p. 56 of their paper. There, he is certain, I will find the missing unit carefully explained. Here is what they write, and it is *all* that they write: "Our principles have been to use whole-length measurements of straight structures, arc lengths of curved structures, and height and width of voluminous structures."

Very well. These are the fundamental units. They are none too clearly explained—try estimating the volume of a donut by looking at its height and width—but I know roughly what Nilsson and Pelger are getting at. What they do not say is *how* these three separate fundamental units are combined in a single overall derived unit of change.

A homely example may make this more vivid. Except for the fact that it cannot see, a Swedish meatball is rather like an eye. And plainly it makes no sense to ask of two Swedish meatballs, one of them twice as greasy but half as wide as the other, which of them is bigger—at least not until units of grease and length have been *combined*. But this is, in general, no easy task, not even when shape alone is under consideration. "It is important to keep in mind," C. P. Klingenberg and L. J. Leamy write ("Quantitative Genetics of Geometric Shape in the Mouse Mandible," *Evolution*, 55(11), pp. 2342–2352, 2001), "that shape is a multivariate feature and cannot be easily divided into scalar traits without imposing arbitrary constraints on the results of the analysis." To see how difficult a conceptual problem Nilsson and Pelger have set themselves, readers may follow the trail of Klingenberg and Leamy's references to the badlands of current work on geometric morphometrics.

Operating perhaps on the principle that a difficulty disclosed is a difficulty denied, Nilsson and Pelger do mention this very point, citing an example of their own on p. 56 to show just how arbitrary can be the business of calculating combined or derived units. In then justifying their own procedure, which is never explained, they remark: "As we are going to relate our measure of morphological change only to general estimates of phenotypic variation" in visual acuity, "we will be safe as long we avoid unorthodox and strange ways of comparing origin and product."

Origin and product? I am sure they meant origin and unit. No matter. The remark speaks for itself.

THERE IS NEXT THE matter of random variation: the heart of the matter so far as I am concerned. Nilsson and Pelger's paper is not an exercise in theoretical optics. It is intended to serve polemical purposes. Thus, they write: "In this context it is obvious that the eye was never a real threat to Darwin's theory of evolution" (p. 58). By "this context," they mean one in which only "eye geometry" and "optical structures" are up for grabs. But whether in this context or any other, it is as a defense of Darwin's theory that Nilsson and Pelger's theory fails most obviously.

Let me review the chief steps in their argument. There is morphological change on the one hand, visual acuity on the other. As their population of light-sensitive cells alters its geometry—by means never specified—visual acuity perks up. In all, they assert, 1,829 steps are involved in tracing a path from their first patch to their final "product."

Just how do Nilsson and Pelger's light-sensitive cells move from one step on that path to the next? I am not asking for the details, but for the odds. There are two possibilities. Having reached the first step on the path, the probability that they will reach the second (and so on to the last) is either one or less than one. If one, their theory cannot be Darwinian—there are no random changes. If less than one, it cannot be right—there is no way to cover 1,829 steps in roughly 300,000 generations if each step must be discounted by the probability of its occurrence.

Demonstrating the existence of a path between two points in the history of life is in general not hard. What is hard is determining how the path was *discovered*. (This was the point of the linguistic example I offered in my essay.) If one assumes, as Nilsson and Pelger do, that probabilities need not be taken into account because all transitions occur with a probability of one, there is no problem to be discussed—but nothing of any conceivable interest, either. In responding to this obvious point by generously suggesting that I need to spend more time by the lamp with D. S. Falconer's *Principles of Quantitative Genetics*, Mr. Nilsson has covered an embarrassment by addressing an irrelevance. Neither population size nor natural selection is at issue.

A few minor matters. Falconer's response variable R is a measure, all right: a measure of the extent to which the mean of some quantitative phenotypic character—snout length, crop yield, scab color, or scrotum size (examples from the literature, I am afraid)—rises or falls as the result of natural selection. Just what I said, just as I explained. Although I offered no definitions in my essay, the paraphrases I employed were harmless. Why not say "sensitivity to vision" instead of "visual acuity," just to vary pace and prose? But in one respect, Mr. Nilsson is right: I did not distinguish between selection and intensity of selection. Neither does he. Neither does Falconer's response statistic, which contains only one selectional parameter, and that one measuring the intensity of selection. Neither does anyone else in this context.

His paper with Susanne Pelger, Mr. Nilsson writes, has never been criticized in the peer-reviewed literature. I am certain that this is so.

PAUL R. GROSS TAKES the occasion of his current letter to assure readers that what he meant in his last letter he did not say and what he said he did not mean. Like golf, Mr. Gross suggested in the 1988 essay from which I uncharitably quoted in the March *Commentary*, science is rather a clubby affair, and just as a great many men prefer to cover the links sedately in the company of men like themselves—tassels on their shoes, alligators on their polo shirts—so scientists prefer to keep company with their own, men and women who share their tastes, point of view, outlook on life.

These are sentiments so candid that I was surprised to find Mr. Gross expressing them. But he is now prepared to disown what he said. The club is just fine, and just look at those splendid greens! The admissions board is to be faulted only when, by accident or inadvertence, it excludes one of its own, a scientist who like L. V. Heilbrunn has *published in the literature*. Such men are entitled to wear the gold cufflinks with the crossed golf clubs; keeping *them* out would be irresponsible. But keeping out the others is not only good science but good sense. *Ipse dixit.*

A few other points deserve comment. In offering Nilsson and Pelger the oil of his approval, Mr. Gross affirms that I have misunderstood or misinterpreted critical elements of their paper. In keeping with his long-standing policy of never documenting his discontent, he does not say which elements. As I keep reminding him, this is not sporting. Still, it is inconceivably droll to see Mr. Gross excusing Richard Dawkins's misrepresentation of Nilsson and Pelger's work by appealing to the fact that Dawkins expressed his views in a *trade* book. Mr. Gross apparently believes that outside the country club, a man can say anything he wants, a policy that he would not dream of applying to critics of Darwin's theory.

A few of Mr. Gross's remarks suggest a need for remedial reading. I have never argued that "evolutionary theory cannot explain the eye." How on earth would I know *that*? And explain what in particular? Its emergence, its structure, its physiology, its biochemistry? What I contended specifically is that Nilsson and Pelger's paper is just nuts. Conspiracies and cover-ups are, in any case, not in my line, and I never suggested or supposed that evolutionary biologists who failed to criticize Richard Dawkins for misrepresenting Nilsson and Pelger did so as part of a conspiracy. Like *droshky* horses, they were only doing what comes naturally: turning a blind eye.

If the burden of Nilsson and Pelger's paper was to demonstrate the existence of "one possible evolutionary pathway to the geometry of a fish-like eye from a patch of photoresponsive cells," as Mr. Gross writes, they have surely wasted their time. The existence of such a path is hardly in doubt. Every normal human being creates an eye from a patch of photo-responsive cells in nine months.

I certainly agree that the "only explanation we have for the structure of the eye... is Darwinian evolution." But neither an orchestra nor an explanation becomes good by being the only game in town.

On the other hand, I disagree that Darwin's theory is as "solid as any explanation in science." Disagree? I regard the claim as preposterous.

Quantum electrodynamics is accurate to thirteen or so decimal places; so, too, general relativity. A leaf trembling in the wrong way would suffice to shatter either theory. What can Darwinian theory offer in comparison?

Finally, I would hardly dispute Mr. Gross's claim that "with the discovery of the developmental regulatory genes, we have learned how subtle, how versatile, and yet how simple the mechanism can be for transforming one biological structure to another." If he were to re-read the correspondence (*Commentary*, September 1996) following the publication of my "The Deniable Darwin" (June 1996), he could not fail to be struck by my reply to his own letter, in which I specifically called attention to work on regulatory genes and eye formation—the very work that he now suggests I am keeping from my readers. Subtle and versatile, those genes? Yes, indeed. Absolutely astonishing? That, too. But hardly a triumph of Darwin's theory. For one thing, no Darwinian theorist had predicted the existence of these genes; for another, no Darwinian theorist has explained their emergence. The facts are simply far more fascinating than anything that poor drab Darwin, endlessly sifting time and chance, could possibly have imagined.

CITING THOSE EVER USEFUL but eternally anonymous "creationists," MATT YOUNG argues yet again, as he did in our earlier exchange, that Nilsson and Pelger have given the lie to creationist claims. If it was their computer simulation that originally lent ardor to his asseverations, now it is their paper itself. Mr. Young is a man plainly prepared to rely on an endless series of fallback positions. In the end, he may have to argue that his refutation is its own best friend, and that Nilsson and Pelger's paper is itself superfluous.

No one doubts that the eye has evolved. Not me, in any event. Fish have eyes; rocks do not. Those eyes came from somewhere—right?—and if coming from somewhere counts as evolution, count me among its champions. No one doubts, furthermore, that the "eye could have evolved in 350,000 generations." As I remarked earlier, the eye could have evolved in a weekend. The issue is whether it could have evolved in 350,000 generations *given the constraints of random-variation and natural selection.*

I have absolutely no idea. Neither do Nilsson and Pelger. And neither does Matt Young.

Arguing now from the last trench before the bunker, Mr. Young writes that Nilsson and Pelger's paper deals with the development of *invertebrate* eyes, and triumphantly chides me for overlooking this point. On p. 56 of their paper, Nilsson and Pelger write: "After constriction of the aperture and the gradual formation of a lens, the final product becomes a focused camera-type eye with the geometry typical for aquatic animals (e.g. fish and cephalopods)." Fish are, of course, vertebrates, as anyone who has picked the flesh from a flounder knows. Perhaps I will be forgiven if I refer to this exchange as shooting fish in a barrel.

Making the point that the emergence of even the most modest eye will require simultaneous and parallel evolutionary development, Mr. Young asks that I defend my claim that this process could not have taken place by quantitative steps. In the first place, I made no such claim, if only because its truth struck me as obvious. But were I to make such a claim I would observe, as Richard Dawkins does, that to the extent that simultaneous and parallel changes are required to form a complex organ, to that extent does the hypothesis of random variation and natural selection become implausible. It is one thing to find a single needle in a haystack, quite another to find a dozen needles in a dozen haystacks at precisely the same time. Surely the burden of proof in such matters is not mine. I am not obliged to defend such mathematical trivialities as the proposition that as independent events are multiplied in number, their joint probability of occurrence plummets.

I have no idea what Mr. Young means when he writes that the number 1.00005 is not a percentage. Every number can be expressed as a percent, and every percent is a pure number. But he gets half credit for spotting a slip: the figure of 1.00005 between parentheses on p. 33 in my text should have been .005. Mr. Nilsson, who also spotted the slip, gets the other half. Me? I blame my editors.

Finally, I did not fault the scientific community for failing to criticize Nilsson and Pelger's work. I did the job of criticism myself. I faulted the Darwinian community—Mr. Young included—for failing to denounce scientific fraud, specifically the misrepresentation of Nilsson and Pelger's work by Richard Dawkins. Now I see that Mr. Young feels I have manhandled him in these exchanges. Too bad. *Commentary* is not some academic mouse hole.

MARK PERAKH, A *SENSEI* of the "noted scientists say" school of self-defense, is right in one respect: the computer simulation missing from Nilsson and Pelger's paper has no bearing on what they actually said and claimed. And right in a second respect: "The real question [is] whether an eye could have developed in a geologically short time *via a Darwinian mechanism*" (emphasis added). But then, although quite confident that I am wrong in my criticisms, he offers nothing by way of rebuttal. Like so many of these martial-arts types, he is too busy preparing himself to run from the field with honor to bother doing battle.

Contrary to what Mr. Perakh asserts, not only can I imagine, I do not doubt, that "distinguished scientists," many with a record of "substantial achievement," can have an opinion different from my own. It happens all the time. I would not dream of accusing ten respected scientists of fraud simply because they passed on the opportunity to have a go at Nilsson and Pelger. The men and women I criticized earned my contempt the hard and dirty way, by saying nothing about scientific misconduct when it was right under their noses.

LIKE MR. PERAKH AND Paul R. Gross, JASON ROSENHOUSE regards Richard Dawkins's misrepresentation of Nilsson and Pelger's work as a "minor error." Some minor, some error. What, may I ask, is the difference between inventing data out of whole cloth and inventing a computer simulation out of whole cloth? Should not evolutionary biologists be held to the same standards as physicists? Or even journalists? What part of the declaration that fraud is fraud does he fail to endorse? These are not semantic issues. If I claimed in print that Mr. Rosenhouse has four eyes, his denials would not turn on what I *meant*. Two eyes, I am sure he would say, are not there. Two eyes, and one computer simulation.

Mr. Rosenhouse believes that Nilsson and Pelger made an important discovery: namely, "that there is a smooth gradient of increasing visual acuity linking a light-sensitive spot to a lens-bearing eye." This is not their discovery, it is a restatement of their chief assumption. "The model sequence is made," they write, "such that every part of it, no matter how small, results in an increase of the spatial information the eye can detect" (p. 53). Note: *made*, not discovered.

To repeat, the flaw in Nilsson and Pelger's work to which I attach the greatest importance is that, as a defense of Darwinian theory, it makes no mention of Darwinian principles. Those principles demand that biological change be driven first by random variation and then by natural selection. There are no random variations in Nilsson and Pelger's theory. Whatever else their light-sensitive cells may be doing, they are not throwing down dice or flipping coins to figure out where they are going next.

Mr. Rosenhouse's conviction that the randomly occurring changes required by Darwin's theory are nevertheless "plainly implied" throughout Nilsson and Pelger's paper owes nothing to the facts and little to common sense. If changes in their model were really random, their temporal estimates would be apt to change by orders of magnitude, a point I made in my essay and again in my reply to Dan-E. Nilsson above. In my essay I also questioned Nilsson and Pelger's decision to hold selection pressure constant over time. In this, I found myself echoing John Gillespie (*The Causes of Molecular Evolution*, 1991, p. 294). "[W]e must be concerned," Gillespie writes, "with models of selection in variable environments. How could it be otherwise? Natural selection is a force adapting species to their environments. Environments are in a constant state of flux; selection coefficients must be in a constant state of flux as well." What is good enough for Gillespie is good enough for me.

In approving of the value chosen by Nilsson and Pelger for selection pressure, Mr. Rosenhouse writes that it is "ludicrously low for almost any environment." Is it indeed? The figure that Mr. Rosenhouse calls ludicrous, Nilsson and Pelger term pessimistic, and Mr. Gross reasonable. The correct term is arbitrary—as in, it is anyone's guess what the variance among a bunch of fish might have been a couple of million years ago. Studies of variance and heredity typically deal with tiny populations and small periods of time. Studying the collard flycatcher, *Ficedula albicollis*, Merilla, Kruuk, and Sheldon collected eighteen years of data for 17,171 nestlings in order to reach some quite modest quantitative conclusions (J. Merilla, L. E. B. Kruuk, and B. C. Sheldon, "Natural Selection on the Genetic Component of Variance in Body Condition in a Wild Bird Population," *Journal of Evolutionary Biology* 14, pp. 918–921, 2001). Nilsson and Pelger's imaginary population ranges over space and time in a way that could not possibly be disciplined by the data.

NICK MATZKE BELIEVES THAT Nilsson and Pelger provide a mathematical model for the development of the eye. Let us be honest: beyond a few finger-counting exercises, there is no mathematics in their model, and while their references do contain some legitimate mathematics (nothing beyond second-semester calculus, but also nothing to sneeze at), their references, as I have shown in patient detail, do not support their theory. The task of modeling the eye's complicated geometry from light-sensitive cell to fully functioning eye is utterly and completely beyond our powers, as a glance at any textbook dealing with embryology would show.

Mr. Matzke devotes the greater part of his otherwise interesting letter to doing battle with various "creationist straw men." It is useful work, I am sure, the more so since the creationists are never named. But whoever they are, I am not among them. Quite the contrary, I am as eager to do right by the snails as he is: why should he think otherwise? It is only when he passes to matters of fact that we part company.

Nilsson and Pelger's theory is intended to encompass the evolution of the eye in fish and cephalapods. Fish indisputably have bones, an attractive skull, and for the most part two staring eyes. The cephalochordate *Branchiostoma* (*Amphioxus* in a now out-of-date system of nomenclature) is widely taken by paleontologists to be a very plausible ancestral model to the vertebrates. It has certain vertebrate features while lacking others. These others include bones, a skull, a brain, and *paired sensory organs*: in other words, it has no eyes. Mr. Matzke's very confident assertion that cephalochordates have "primitive eyes" is simply untrue.

Now that I have swept away a few straw men of my own, let us see what is left to clean up. In my essay I wrote that Nilsson and Pelger made no attempt to discuss the cost-benefit payoffs associated with an improvement in visual acuity. My aim in discussing the reconstruction of the fish skull was not to argue that eyes came first *or* that bones did. Paired sensory organs *and* bones are characteristics of the vertebrates. Plainly they evolved together. Plainly, too, one function of the bony skull in vertebrates is to provide protection for the paired sensory organs located on their heads. The protection racket, as every Mafia boss is aware, does not come cheap; but Nilsson and Pelger, in adding up the benefits of visual acuity, did not ever bother to consider the vigorish. This is such an unobjectionable point that I cannot imagine why Mr. Matzke found it fishy.

I very much appreciate the letters from David Safir and Norman Gentieu.

THE VAMPIRE'S HEART

Having on a number of occasions driven a silver stake through the heart of Nilsson and Pelger's well-known essay[60] about the formation of the eye, I have been alarmed by twitches in the resulting Darwinian corpse, including at the Internet forum Talk Reason.

This note will thus serve as a follow-up stake.

Earlier, I drew attention to the fact that Nilsson and Pelger's study, which was widely considered a computer simulation, contained *no* computer simulation whatsoever, and that in arguing for its existence, Richard Dawkins was engaged in a form of fraud. Commentators at the time responded both that he had done no such thing and that if he had done so it was nothing more than a trivial oversight.

I also drew attention to the fact that Nilsson and Pelger's essay—the thing itself—contained *no* defense whatsoever of its chief assertion, namely that 1,829 steps are required to transform a light-sensitive patch into a functioning eye. In my own words:

> Moreover, Nilsson and Pelger do *not* calculate the visual acuity of any structure, and certainly not over the full 1,829 steps of their sequence. They suggest that various calculations have been made, but they do not show how they were made or tell us where they might be found. At the very best, they have made such calculations for a handful of data points, and then joined those points by a continuous curve.

The calculations to which Nilsson and Pelger appeal are neither in their paper, nor in their footnotes, nor in a technical appendix, nor are

---

60. "A Pessimistic Estimate of the Time Required for an Eye to Evolve," *Proceedings of the Royal Society*, London B (1994) 256, 53–58.

they available on their website. In the twelve years since their paper was published, they have never appeared in any public forum.

IN RESPONDING RECENTLY TO my second observation—no data, no evidence, and thus no reason to assent—Talk Reason's James Downard managed inadvertently to confirm the first: "When I wrote Nilsson to check up on these matters, I did ask about his data set, and he readily supplied a neat summary of the ten variables involved in the *simulation* and the stages of their acquisition" (emphasis added).

It is a great merit of Nilsson and Pelger's study that based as it is on a non-existent simulation, it can be defended on that basis as well.

James Downard, whom I respect as by far the most capable of commentators at Talk Reason, is now prepared to accommodate the obvious: "True, a 'computer' wasn't involved in these calculations, so let's all slap Richie Dawkins for being a bad student." This is progress, and I am grateful for it, even if it is remarkably difficult to account for sheer scientific fraud by an appeal to pedagogical ineptitude.

Downard is nonetheless still persuaded that had I pursued the matter more diligently, I might have discovered the data missing from Nilsson and Pelger's original paper, writing now, and after all these years, that "I confirmed with Nilsson that Berlinski had never even bothered to request the original data summary, let alone establish that there was anything biologically unjustifiable about it." This is perfectly correct. I never bothered. And it merits a perfectly obvious rejoinder. Why on earth should I have done anything of the sort? It is the responsibility of serious scientists making an historically important claim to *publish* their evidence, or in the age of the Internet, to make it publicly available online. This Nilsson and Pelger did not do, and this they have never done.

THOSE PRIVATE COMMUNICATIONS TO which Downard claims access are in any case hardly invigorating. Writing in much the same spirit and, indeed, to the same end, Tom Curtis indicated that he, too, had gained access to Nilsson and Pelger's research, and presented Nilson's account on-line. I reproduce the first of his entries, which covers 176 steps, in precisely the form in which it appeared:

Stage 1 to 2 176
corneal width (curve) 46.5 46.5 0 corneal thickness 3.35 10.1 110 upper retinal surface width 46.6 47.2 1 lower retinal surface width 46.6 51.2 9 upper pigment surface width 46.6 53.9 14 lower pigment surface width 46.6 58.0 21.

On the same forum (Talk Reason), I responded that:

1. The list is incomplete, and that:
2. It is largely incomprehensible. What, for example, does "corneal width (curve) 46.5 46.50" mean? And that:
3. It is irrelevant inasmuch as it was not included in Nilsson and Pelger's original paper nor in their notes nor in any technical appendix. And, finally that:
4. It is absurd, inasmuch as a list is not a calculation. In my essay, I asked *how* Nilsson and Pelger's numbers were derived? They do not say and neither does Mr. Curtis. Neither does James Downard.

I see no reason now to change my opinion. A scientific paper making a historically important claim *must* include its central argument, together with the data that supports it. These requirements are a matter of principle at both *Science* and *Nature*. The point, I should think, is obvious.

I MUST NOW CONSIDER and correct certain *technical* confusions among my critics, altogether a more agreeable undertaking.

Writing on this point, and judiciously changing their simulation to a *study*, Downard claims that Nilsson and Pelger's paper "contained an entirely valid *mathematical* analysis of eye evolution basing each stage of the process on *biologically known* intermediaries" (first emphasis added).

Here, then, is a demonstration of Nilsson and Pelger's mathematical analysis. I have myself filled in the details that are, in fact, missing from Nilsson and Pelger's paper.

Nilsson and Pelger's dynamical model consists of two parts. The first tracks changes in a population of imaginary cells, the second changes in the mean value of a single quantitative characteristic, namely visual acuity.

Their first dynamical model is based on the differential equation

$$1.\ dx/dt = ax,$$

where $x(t)$ is a function measuring the size of a population of imaginary cells (and in units that are never specified), and a is a coefficient representing .01 changes to $x(t)$.

Solutions are exponential

$$2.\ x(t) = Ke^{at},$$

where K is an initial value. But since K =1, and a = .01,

$$3.\ x(t) = 1.01^t$$

since $e^a = e^{.01} \cong 1.01$.

Subject to the condition that

$$4.\ 1.01^t = 80,129,540$$

it follows that $t = 1,829$.

It is equation 3 that actually appears in Nilsson and Pelger's paper, but it is 1, from which it is derived (by entirely trivial means), that functions as their dynamical model.

1 is a first-order, deterministic differential equation and remains a deterministic differential equation for every value of the parameter a.

NILSSON AND PELGER'S SECOND dynamical model is simply equation 1 rewritten with a different value of a.

This new value of a is derived from Falconer's *short-term* response statistic **R**, which is a percentage measure of changes to the mean value M of a quantitative characteristic in a single generation. Thus

$$5.\ \mathbf{R} = h^2 \times i \times V \times M.$$

Values for these parameters given by Nilsson and Pelger are:

a) $h^2 = .50$

b) $i = 0.01$

c) $V = 0.01$

d) $M = 1$.

Substituting 0.00005 for *a* and 1 for K in equation 2, yields

6. $1.00005^t = 80,129,540$,

whence Nilsson and Pelger's conclusion that $t = 363,992$.

IT REMAINS FOR ME to observe that far from being "an entirely valid mathematical analysis," what Nilsson and Pelger have offered involves *no* mathematical analysis whatsoever and since they have not even solved their own differential equation, no *appearance* of mathematical analysis either. What there is instead is the simple, and entirely simple-minded, assumption that biological change, whether measuring size or visual acuity, may best be represented by the same formula used in calculating compound interest.

WHAT IS MORE, THE resulting analysis is strictly non-Darwinian.

If equation 1 is deterministic, so, too, equation 6, since *it represents a solution of the same equation.*

For some reason, this last has proven a difficult point to grasp. Let me stress it again.

Both James Downard and Jason Rosenhouse, writing on the Internet and in the pages of *Commentary*, are persuaded that it is the presence of *V* in equation 5 that suggests the missing stochastic element, since *V* is a measure of variance. But *V* is a fixed parameter, and thus so, too, **R**. *It does not change over the entire course of roughly 330,000 generations.* The underlying dynamical model has not changed. It remains deterministic.

While V is a measure of fixed variance, nothing in Falconer's short-term response statistic or in Nilsson and Pelger's paper indicates *why* it should remain constant over 330,000 generations.

The answer provided by a properly Darwinian theory, of course, would be that variance is a proxy for random changes in the organism's underlying genome. How many such changes might be required to maintain V = constant over 330,000 consecutive generations?

I have no idea, of course, if only because no one does; but let us suppose that each of Nilsson and Pelger's steps involves a single positive mutation. Treating those steps as an example of a pure Bernoulli process in which each positive increase in either size or visual acuity is counted a success, and each negative decrease a failure, it follows on the biologically unrealistic assumption that each such step has the probability .5, that the odds in favor of seeing 1,829 consecutive successes are one in $2^{1,829}$.

It thus follows that Nilsson and Pelger's pessimistic estimate for the formation of an eye is true only if a remarkably rare event is realized.

It remains for me to observe that no population geneticist would use equation 5 as the basis for substitution in equation 1, not least of all because it is well-known that changes in gene frequencies affect variance, often by linkage disequilibrium. It is for this reason that population geneticists take pains to observe that Falconer's response statistic cannot be extrapolated beyond one or two generations. The point is well known to breeders using the breeder's equation. It is generally thought, moreover, that when $4N_e \times i$ is less than 1, where N_e is the effective population size, that stochastic effects will predominate over selection, so that the long-term relevance of equation 5 is compromised from the first.

The use of equations 1 and 6 as the basis of a dynamical theory governing changes to the mean value of a quantitative characteristic is obviously absurd inasmuch as it suggests that any quantitative characteristic whatsoever exhibiting a positive response statistic will grow exponentially forever.

2004

On the Origins of the Mind

It's all scientific stuff; it's been proved.
—Tom Buchanan in *The Great Gatsby*

AT SOME TIME IN THE HISTORY OF THE UNIVERSE, THERE WERE NO human minds, and at some time later, there were. Within the blink of a cosmic eye, a universe in which all was chaos and void came to include hunches, beliefs, sentiments, raw sensations, pains, emotions, wishes, ideas, images, inferences, the feel of rubber, *Schadenfreude*, and the taste of banana ice cream.

A sense of surprise is surely in order. How did *that* get *here*?

If the origin of the human mind is mysterious, so too is its nature. There are, Descartes argued, two substances in the universe, one physical and the other mental.

To many contemporary philosophers, this has seemed rather an embarrassment of riches. But no sooner have they ejected mental substances from their analyses than mental properties pop up to take their place, and if not mental properties then mental functions. As a conceptual category, the mental is apparently unwilling to remain expunged.

And no wonder. Although I may be struck by a thought, or moved by a memory, or distracted by a craving, these familiar descriptions suggest an effect with no obvious physical cause. Thoughts, memories, cravings—they are what? Crossing space and time effortlessly, the human mind deliberates, reckons, assesses, and totes things up; it reacts, registers, reflects, and responds. In some cases, like inattention or carelessness, it invites censure by doing nothing at all or doing something in the wrong way; in other cases, like vision, it acts unhesitatingly and without reflection; and in still other cases, the human mind manages both to slip itself into and stay aloof from the great causal stream that makes the real world boom, so that when *it* gives the go-ahead, what *I* do is, as Thomas Aquinas observed, "inclined but not compelled."

These are not properties commonly found in the physical world. They are, in fact, not found at all.

And yet, the impression remains widespread that whoever is responsible for figuring out the world's deep things seems to have figured out the human mind as well. Commenting on negative advertising in political campaigns, Kathleen Hall Jamieson, the director of the Annenberg Public Policy Center at the University of Pennsylvania, remarked that "there appears to be something hard-wired into humans that gives special attention to negative information." There followed what is by now a characteristic note: "I think it's evolutionary biology."

Negative campaign advertisements are the least of it. There is, in addition, war and male aggression, the human sensitivity to beauty, gossip, a preference for suburban landscapes, love, altruism, marriage, jealousy, adultery, road rage, religious belief, fear of snakes, disgust, night sweats, infanticide, and the fact that parents are often fond of their children. The idea that human behavior is "the product of evolution," as the *Washington Post* puts the matter, is now more than a theory: it is a popular conviction.

It is a conviction that reflects a modest consensus of opinion among otherwise disputatious philosophers and psychologists: Steven Pinker, Daniel Dennett, David Buss, Henry Plotkin, Leda Cosmides, John Tooby, Peter Gärdenfors, Gary Marcus. The consensus is constructed, as such things often are, on the basis of a great hope and a handful of similes. The great hope is that the human mind will in the end find an unobtrusive place in the larger world in which purely material causes chase purely material effects throughout the endless night. The similes are, in turn, designed to promote the hope.

THREE SIMILES ARE AT work, each more encompassing than the one before. They give a natural division of labor to what is now called evolutionary psychology.

First, the human mind is *like* a computer in the way that it works. And it is just because the mind *is* like a computer that the computer comprises a model of the mind. "My central thesis," the cognitive psychologist H. A. Simon has written, is that "conventional computers can be, and have been, programmed to represent symbol structures and carry out processes on those structures that parallel, step by step, the way the human brain does it."

Second, the individual human mind is *like* the individual human kidney, or any other organ of the body, in the way that it is created anew in every human being. "Information," Gary Marcus writes, "copied into the nucleus of every newly formed cell, guides the gradual but powerful process of successive approximation that shapes each of the body's organs." This is no less true of the "organ of thought and language" than of the organs of excretion and elimination.

Third, the universal human mind—the expression in matter of human nature—is *like* any other complicated biological artifact in the way that it arose in the human species by means of random variation and natural selection. These forces, as Steven Pinker argues, comprise "the only explanation we have of how complex life *can* evolve. . . ."

Taken together, these similes do succeed wonderfully in suggesting a coherent narrative. The ultimate origins of the human mind may be found in the property of irritability that is an aspect of living tissue itself. There is a primordial twitch, one that has been lost in time but not in memory; various descendant twitches then enlarged themselves, becoming, among the primates at least, sophisticated organs of perception, cognition, and computation. The great Era of Evolutionary Adaptation arrived in the late Paleolithic, a veritable genetic Renaissance in which the contingencies of life created, in the words of the evolutionary psychologist Leda Cosmides, "programs that [were] well-engineered for solving problems such as hunting, foraging for plant foods, courting mates, cooperating with kin, forming coalitions for mutual defense, avoiding predators, and the like." There followed the long Era in Which Nothing Happened, the modern human mind retaining in its structure and programs the mark of the time that human beings spent in the savannah or on the forest floor, hunting, gathering, and reproducing with Darwinian gusto.

Three quite separate scientific theories do much to justify this grand narrative and the three similes that support it. In the first instance, computer science; in the second, theories of biological development; in the third, Darwin's theory of evolution. At times, indeed, it must seem that only the width of a cigarette paper separates evolutionary psychology from the power and the glory of the physical sciences themselves.

IF THE CLAIMS OF evolutionary psychology are ambitious, the standard against which they should be assessed must be mature, reasonable, and persuasive. If nothing else, that standard must reflect principles that have worked to brilliant success in the physical sciences themselves. This is more than a gesture of respect; it is simple common sense.

In stressing the importance of their subject, the mathematicians J. H. Hubbard and B. H. West begin their textbook on differential equations by observing that "historically, Newton's spectacular success in describing mechanics by differential equations was a *model for what science should be*" (emphasis added). Hubbard and West then add what is in any case obvious: that "all basic physical laws are stated as differential equations, whether it be Maxwell's equations for electrodynamics, Schrödinger's equation for quantum mechanics, or Einstein's equations for general relativity."

Equations do lie close to the mathematician's heart, and differential equations closer than most. On one side of such an equation, there is a variable denoting an unknown mathematical function; on the other, a description of the rate at which that unknown function is changing at every last moment down to the infinitesimal. Within the physical sciences, such changes express the forces of nature, the moon perpetually falling because perpetually accelerated by the universal force of gravitation. The mathematician's task is to determine the overall, or global, identity of the unknown function from its local rate of change.

In describing the world by means of a differential equation, the mind thus moves from what is local to what is global. It follows that the "model for what science should be" involves an interdiction against action at a distance. "One object," the Russian mathematician Mikhael Gromov observes, "cannot influence another one removed from it without involving local agents located one next to another and making a continuous chain joining the two objects." As for what happens when the interdiction lapses, Gromov, following the French mathematician René Thom, refers to the result as *magic*. This contrast between a disciplined, differential description of a natural process and an essentially magical description is a useful way of describing a fundamental disjunction in thought.

A differential equation, it is important to stress, offers only a general prescription for change. The distance covered by a falling object is a matter of how fast it has been going and how long it has been going fast; this, an equation describes. But how *far* an object has gone depends on how high it was when it began falling, and this the underlying equation does not specify and so cannot describe. The solutions to a differential equation answer the question, how is the process changing? The data themselves answer a quite different question: how or where does the process *start?* Such specifications comprise the initial conditions of a differential equation, and represent the intrusion into the mathematical world of circumstances beyond the mathematical.

It is this that in 1902 suggested to the French mathematician Jacques Hadamard the idea of a "well-posed problem" in analysis. For a differential equation to be physically useful, Hadamard argued, it must meet three requirements. Solutions must in the first place exist. They must be unique. And they must in some reasonable sense be stable, the solutions varying continuously as the initial conditions themselves change.

With these requirements met, a well-posed differential equation achieves a coordination among continuous quantities that is determined for every last crack and crevice in the manifold of time. And is this the standard I am urging on evolutionary psychology? Yes, absolutely.

Nothing but the best.

Although evolutionary psychologists have embraced the computational theory of mind, it is not entirely a new theory; it has been entertained, if not embraced, in other places and at other times. Gottfried Leibniz wrote of universal computing machines in the seventeenth century, and only the limitations of seventeenth-century technology prevented him from toppling into the twenty-first. As it was, he did manage to construct a multipurpose calculator, which, he claimed, could perform the four elementary operations of addition, subtraction, division, and multiplication. But when he demonstrated the device to members of the Royal Society in London, someone in the wings noticed that he was carrying numbers by hand.

I do not know whether this story is true, but it has a very queer power, and in a discussion dominated by any number of similes it constitutes a rhetorical figure—shaped as a warning—all its own.

In 1936, the British logician Alan Turing published the first of his papers on computability. Using nothing more than ink, paper, and the resources of mathematical logic, Turing managed to create an imaginary machine capable of incarnating a very smooth, very suave imitation of the human mind.

Known now as a Turing machine, the device has at its disposal a tape divided into squares and a reading head mounted over the tape. It has, as well, a finite number of physical symbols, most commonly 0's and 1's. The reading head may occupy one of a finite number of distinct physical states. And thereafter the repertoire of its action is extremely limited. A Turing machine can, in the first place, recognize symbols, one square at a time. It can, in the second place, print symbols or erase them from the square it is scanning. And it can, in the third place, change its internal state, and move to the left or to the right of the square it is scanning, one square at a time.

There is no fourth place. A Turing machine can do nothing else. In fact, considered simply as a mechanism, a Turing machine can do nothing whatsoever, the thing existing in that peculiar world—my own, and I suspect others' as well—in which everything is possible but nothing gets done.

A Turing machine gains its powers of imitation only when, by means of a program, or an algorithm, it is told what to do. The requisite instructions consist of a finite series of commands, written in a stylized vocabulary precisely calibrated to take advantage of those operations that a Turing machine can perform. What gives to the program its air of cool command is the fact that its symbols function in a double sense. They are symbols by virtue of their *meaning*, and so reflect the intentions of the human mind that has created them; but they are *causes* by virtue of their structure, and so enter into the rhythms of the real world. Like the word "bark," which both expresses a human command and sets a dog to barking, the symbols do double duty.

Although imaginary at its inception, a Turing machine brilliantly anticipated its own realization in matter. Through a process of intellectual parthenogenesis, Turing's ideas gave rise to the modern digital computer. And once the sheer physical palpability of the computer was recognized—there it is, as real as the rocks, the rifts, and the rills of the physical sciences—there was nothing to stand in the way of the first controlling simile of evolutionary psychology: that the human mind is itself a computer, one embodied in the human brain.

The promotion of the computer from an imaginary to a physical object serves the additional purpose of restoring it to the world that can be understood in terms of the "model for what science should be." As a physical device, nothing more than a collection of electronic circuits, the digital computer can be represented entirely by Clerk Maxwell's theory of the electromagnetic field, with the distinction between a Turing machine and its program duplicated in the distinction between a differential equation and its initial conditions. We are returned to the continuous and infinite world studied by mathematical physics, the world in which differential equations track the evolution of material objects moving through time in response to the eternal forces of nature itself.

THE INTELLECTUAL MANEUVERS THAT I have recounted serve to make the computer an irresistibly compelling object. But they serve, as well, to displace attention from the human mind. The effect is to endow the simile that the human mind is like a computer with a plausibility it might not otherwise enjoy.

A certain "power to alter things," Albertus Magnus observed, "indwells in the human soul." The *existence* of this power is hardly in doubt. It is evident in every human act in which the mind imposes itself on nature by taking material objects from their accustomed place and rearranging them; and it is evident again whenever a human being interacts with a machine. Writing with characteristic concision in the *Principia*, Isaac Newton observed that "the power and use of machines consist only in this, that by diminishing the velocity *we* may augment the force, and the contrary" (emphasis added). Although Newton's analysis was restricted to mechanical forces (he knew no others), his point is nonetheless general. A machine is a material object, a *thing*, and as such, its capacity to do work is determined by the forces governing its behavior and by its initial conditions.

Those initial conditions must themselves be explained, and in the nature of things they cannot be explained by the very device that they serve to explain. This is precisely the problem that Newton faced in the *Principia*. The magnificent "system of the world" that he devised explained why the orbits of the planets around the sun must be represented by a conic section; but Newton was unable to account for the initial conditions that he had himself imposed on his system. Facing an imponderable, he appealed to divine intervention. It was not until Pierre Simon Laplace introduced his nebular hypothesis in 1796 that some form of agency was removed from Newtonian mechanics.

This same pattern, along with the problem it suggests, recurs whenever machines are at issue, and it returns with a vengeance whenever computers are invoked as explanations for the human mind. A computer is simply an electromechanical device, and this is precisely why it is useful as a model of the human brain. By setting its initial conditions, a computer's program allows the machine to do work in the real world. But the normal physical processes by which a computer works are often obscured by their unfamiliarity—who among us *really* understands what a computer is and how it works? No doubt, this is why the thesis that the mind is like a computer resonates with a certain intellectual grandeur.

An abacus conveys no comparable air of mystery. It is a trifle. Made of wood, it consists of a number of wires suspended in a frame and a finite number of beads strung along the wires. Nevertheless, an idealized abacus has precisely the power of a Turing machine, and so both the abacus and the Turing machine serve as models for a working digital computer. By parity of reasoning, they also both serve as models for the human mind.

Yet the thesis that the human mind is like an abacus seems distinctly less plausible than the thesis that the human mind is like a computer, and for obvious reasons. It is precisely when things have been reduced to their essentials that the interaction between a human being and a simple machine emerges clearly. That interaction is naked, a human agent handling an abacus with the same directness of touch that he might employ in handling a lever, a pulley, or an inclined plane. The force that human beings bring to bear on simple machines is muscular and so derived from the chemistry of the human body, the causes ultimately emptying out into the great ocean of physical interactions whose energy binds and loosens the world's large molecules. But what we need to know in the example of the abacus is not the nature of the forces controlling its behavior but the circumstances by which those forces come into play.

No chain of causes known to date accommodates the inconvenient fact that, by setting the initial conditions of a simple machine, a human agent brings about a novel, an unexpected, an entirely idiosyncratic distribution of matter. Every mechanical artifact represents what the anthropologist Mary Douglas calls "matter out of place." The problem that Newton faced but could not solve in the *Principia* returns when an attempt is made to provide a description of the simplest of human acts, the trivial tap or touch that sets a polished wooden bead spinning down a wire. Tracing the causal chain backward leads only to a wilderness of causes, each of them displacing material objects from their proper settings, so that in the end the mystery is simply shoveled back until the point is reached when it can be safely ignored.

A chain of physical causes is thus not obviously useful in explaining how a human agent exhibits the capacity to "alter things." But neither does it help to invoke, as some have done, the hypothesis that another abacus is needed to fix the initial conditions of the first. If each abacus requires yet another abacus in turn, the road lies open to the madness of an infinite regress, a point observed more than 70 years ago by the logicians Kurt Gödel and Alfred Tarski in their epochal papers on incompleteness.

If we are able to explain how the human mind works neither in terms of a series of physical causes nor in terms of a series of infinitely receding mechanical devices, what then is left? There is the ordinary, very rich, infinitely moving account of mental life that without hesitation we apply to ourselves. It is an account frankly magical in its nature. The human mind registers, reacts, and responds; it forms intentions, conceives problems, and then, as Aristotle dryly noted, it *acts*. In analyzing action, we are able to say only, as Albertus Magnus said, that a certain power to alter things inheres in the human soul.

A simile that for its persuasiveness depends on the very process it is intended to explain cannot be counted a great success.

IF THE COMPUTATIONAL ACCOUNT of the human mind cannot be brought under the control of the "model for what science should be," what of the thesis that the human mind can be comprehended by reference to the laws of biological development? Here we come to the second simile of evolutionary psychology.

"As the ruler of the soul," Ptolemy wrote in the *Tetrabiblos*, "Saturn has the power to make men sordid, petty, mean-spirited, indifferent, mean-minded, malignant, cowardly, diffident, evil-speaking, solitary, tearful, shameless, superstitious, fond of toil, unfeeling, devisors of plots against their friends, gloomy, taking no care of their body." We know the type; there is no need to drown the point in words. Some men are just rotten.

The analysis that Ptolemy offers in defense of his thesis is anything but crude. "The chronological starting point of human nativities," he writes, "is naturally the very time of conception, for to the seed is given once and for all the very qualities that will mark the adult and that are expressed in growth." It is Saturn's position that affects the seed, and the seed thereafter that affects the man.

Ptolemy's sophistication notwithstanding, no one today is much minded to study the *Tetrabiblos* as a guide to human psychology. Even if a convincing correlation could be established between the position of the planets and the onset of human rottenness, persuading us that we have identified some remote cause in nature for some human effect, that cause would quite obviously violate the interdiction against action at a distance. Ptolemy himself was sensitive to the distinction between astrological knowledge and real knowledge. In trying to construct a continuous chain between the position of the planets and the advent of human rottenness, he was at as great a loss as we are. It is for this reason that the word he employs to describe the way in which heavenly objects evoke their effects is *influence*; it is a word that does not appear, and is not needed, in the *Almagest*, Ptolemy's great treatise on astronomy.

More than 2,000 years have gone by since Ptolemy composed the *Tetrabiblos*. The stars have withdrawn themselves; their role in human affairs has been assigned to other objects. Under views accepted by every evolutionary psychologist, the source of human rottenness may be found either in the environment or within the human genome.

The first of these, the environment, has been the perpetual Plaintiff of Record in *Nurture v. Nature et al.* But for our purposes it may now be dismissed from further consideration. If some men are made bad, then they are not born that way; and if they are not born that way, an explanation of human rottenness cannot be expressed in evolutionary terms.

The question at hand is thus whether the path initiated by the human genome in development can be understood in terms of "the model for what science should be." A dynamical system is plainly at work, one that transforms what Ptolemy called "the seed" into a fully formed human being in nine months, and then into an accomplished car thief in less than twenty years. What evolutionary psychology requires is a demonstration that this process may itself be brought under control of a description meeting the standard that "one object cannot influence another one removed from it without involving local agents located one next to another and making a continuous chain joining the two objects."

"OUR BASIC PARADIGM," BENJAMIN Levin writes in his textbook on genetics, "is that genes encode proteins, which in turn are responsible for the synthesis of other structures." Levin is a careful and a conscientious writer. By "other structures" he means only the nucleic acids. But his "basic paradigm" is now a part of a great cultural myth, and by "other structures" most evolutionary psychologists mean *all* of the structures that are made from the proteins, most notably the human brain.

The myth begins solidly enough—with the large bio-molecules that make up the human genome. The analysis of the precise, unfathomably delicate steps that take place as the genome undertakes various biochemical activities has been among the glories of modern science. Unfortunately, however, the chain of causes that begins in the human genome gutters out inconclusively long before the chain can be continued to the human brain, let alone the human mind. Consider in this regard the following sequence of representative quotations in which tight causal connections are progressively displaced in favor of an ever more extravagant series of metaphors:

(1) *Quantum chemistry*: "For a molecule, it is reasonable to split the kinetic energy into two summations—one over the electrons, and one over the nuclei."

(2) *Biochemistry*: "Initiation of prokaryotic translation requires a tRNA bearing N-formyl methionne, as well as three initiation factors (IF1,2,3), a 30S ribosomal subunit GTP," etc.

(3) *Molecular biology*: "Once the protein binds one site, it *reaches* the other by *crawling* along the DNA, thus *preserving* its *knowledge* of the orientation of the first site" (emphasis added).

(4) Embryology: "In the embryo, cells divide, *migrate, die, stick to each other, send out* processes, and *form* synapses" (emphasis added).

(5) and (6)   *Developmental genetics*: "But genes are simply regulatory elements, molecules that *arrange* their surrounding environments into an *organism*" (emphasis added).

"Genes *prescribe* epigenetic *rules*, which are the neural *pathways* and *regularities* in *cognitive development* by which the individual *mind assembles* itself" (emphasis added).

(7) *Developmental biology*: "The *pattern* of neural connections (synapses) *enables* the human cortex to *function* as the *center* for *learning, reasoning,* and *memory*, to *develop* the *capacity* for *symbolic expression,* and to *produce voluntary responses* to interpreted stimuli" (emphasis added).

(8) and (9) *Evolutionary psychology*: "Genes, of course, do *influence* human development" (emphasis added).

"[Genes] *created* us, body and mind" (emphasis added).

Now the very sober (1) and (2) are clearly a part of "the model for what science should be." By the time we come to (3), however, very large molecular chains have acquired powers of agency: they are busy reaching, crawling, and knowing; it is by no means clear that these metaphors may be eliminated in favor of a biochemical description. Much the same is true of (4). In (5) and (6), a connection is suggested between genes, on the one hand, and organisms, on the other, but the chain of causes and their effects has become very long, the crucial connections now entirely expressed in language that simply disguises profound gaps in our understanding.

In (7) the physical connection between morphology and the mind is reduced to wind, while (8) defiantly resurrects "influence," Ptolemy's original term of choice. It is the altogether exuberant (9)—the quotation is from Richard Dawkins—that finally drowns out any last faint signal from the facts.

These literary exercises suggest that the longer the chain of causes, the weaker the links between and among them. Whether this represents nothing more than the fact that our knowledge is incomplete, or whether it points to a conceptual deficiency that we have no way of expressing, let alone addressing—these are matters that we cannot now judge.

Curiously enough, it has been evolutionary psychologists themselves who are most willing to give up in practice what they do not have in theory. For were that missing theory to exist, it would cancel—it would *annihilate*—any last lingering claim we might make on behalf of human freedom. The physical sciences, after all, do not simply trifle with determinism: it is the heart and soul of their method. Were Boron salts at liberty to discard their identity, the claims of inorganic chemistry would seem considerably less pertinent than they do.

Thus, when Steven Pinker writes that "nature does not dictate what we should accept or how we should live our lives," he is expressing a hope entirely at odds with his professional commitments. If ordinary men and women are, like the professor himself, perfectly free to tell their genes "to go jump in the lake," why then pay the slightest attention to evolutionary psychology—why pay the slightest attention to Pinker?

Irony aside, a pattern is at work here. Where (in the first simile) computational accounts of the mind are clear enough to be encompassed by the model for what science should be, they are incomplete—radically so. They embody what they should explain. Where (in the second simile) biochemical and quantum chemical accounts of development are similarly clear and compelling, they extend no farther than a few large molecules. They defer what they cannot explain. In both cases, something remains unexplained.

This is a disappointing but perhaps not unexpected conclusion. We are talking, after all, about the human mind.

EVOLUTIONARY PSYCHOLOGISTS BELIEVE THAT the only force in nature adequate to the generation of biological complexity is natural selection. It is an axiom of their faith. But although natural selection is often described as a force, it is certainly not a force of nature. There are four such forces in all: gravitational, electromagnetic, and the strong and weak forces. Natural selection is not one of them. It appears, for the most part, as a free-floating form of agency, one whose identity can only be determined by field studies among living creatures—the ant, the field mouse, and the vole.

But field studies have proved notoriously inconclusive when it comes to natural selection. After three decades spent observing Darwin's finches in the Galapagos, P. R. and B. R. Grant were in the end able to state only that "further continuous long-term studies are needed." It is the conclusion invariably established by evolutionary field studies, and it is the only conclusion established with a high degree of reliability.

The largest story told by evolutionary psychology is therefore anecdotal. Like other such stories, it subordinates itself to the principle that we are what we are because we were what we were. Who could argue otherwise? All too often, however, this principle is itself supported by the counter-principle that we were what we were because we are what we are, a circle not calculated to engender confidence.

Thus, in tests of preference, Victor Johnson, a bio-psychologist at New Mexico State University, has reported that men throughout the world designate as attractive women with the most feminine faces. Their lips are large and lustrous, their jaws narrow, their eyes wide. On display in every magazine and on every billboard, such faces convey "accented hormonal markers." These are a guide to fertility, and it is the promise of fertility that prompts the enthusiastic male response.

There is no reason to doubt Johnson's claim that on the whole men prefer pretty young women to all the others—the result, I am sure, of research extending over a score of years. It is the connection to fertility that remains puzzling. If male standards of beauty are rooted in the late Paleolithic era, men worldwide should now be looking for stout muscular women with broad backs, sturdy legs, a high threshold to pain, and a welcome eagerness to resume foraging directly after parturition. It has not been widely documented that they do.

In any case, an analysis of human sexual preferences that goes no farther than preferences is an exercise in tiptoeing to the threshold of something important and never peering over. The promise of evolutionary psychology is nothing less than an explanation of the human *mind*. No psychological theory could possibly be considered complete or even interesting that did not ask *why* men exhibit the tastes or undertake the choices they do. When it comes to sexual "preferences," what is involved is the full apparatus of the passions—beliefs, desires, sentiments, wishes, hopes, longings, aching tenderness. To study preferences without invoking the passions is like studying lightning without ever mentioning electricity.

This is one of those instances where evolutionary psychology betrays a queer family resemblance to certain theories in philosophy and psychology that (as we have seen in the case of determinism) evolutionary psychologists are themselves eager to disown. Behaviorism in psychology, as in the work of John Watson and B. F. Skinner, came to grief because human behavior is itself a contested category, and one that lapses into irrelevance once it is enlarged to accommodate the sources of behavior in the mind itself. It may be possible to analyze the mating strategies of the vole, the subject of much current research, by means of a simple assessment of what the vole does: a single genetic switch seems sufficient to persuade an otherwise uxorious male vole to become flamboyantly promiscuous. But human beings, it goes without saying, are not voles, and what they do becomes intelligible to them only when it is coordinated with what they are.

DESPITE THE PALPABLY UNRELIABLE stories that evolutionary psychologists tell about the past, is there, nevertheless, a scientifically reasonable structure that may be invoked to support those stories (as fine bones may support an otherwise frivolous face)?

The underlying tissue that connects the late Paleolithic and the modern era is the gene pool. Changes to that pool reflect a dynamic process in which genes undergo change, duplicate themselves, surge into the future or shuffle off, and by means of all the contingencies of life serve in each generation the purpose of creating yet another generation. This is the province of population genetics, a discipline given a remarkably sophisticated formulation in the 1930s and 40s by Ronald Fisher, J. B. S. Haldane, and Sewall Wright. Excellent mathematicians, these men were interested in treating evolution as a process expressed by some underlying system of equations. In the 1970s and 80s, the Japanese population geneticist Motoo Kimura revived and then extended their theories.

Kimura's treatise, *The Neutral Theory of Molecular Evolution* (1983), opens with words that should prove sobering to any evolutionary psychologist: "The neutral theory asserts that the great majority of evolutionary changes at the molecular level, as revealed by comparative studies of protein and DNA sequences, are caused not by Darwinian selection but by random drift of selectively neutral or nearly neutral mutants."

If Darwin's theory is a matter of random variation *and* natural selection, it is natural selection that is demoted on Kimura's view. Random variation is paramount; chance is the driving force. This is carefully qualified: Kimura is writing about "the great majority of evolutionary changes," not all. In addition, he is willing to accept the Darwinian disjunction: either complex adaptations are the result of natural selection or they are the result of nothing at all. But the effect of his work is clear: insofar as evolution is neutral, it is not adaptive, and insofar as it is not adaptive, natural selection plays no role in life.

Like his predecessors, Kimura writes within a particular tradition, one whose confines are fixed by the "model for what science should be." Thus, in trying to describe the fate of a mutant gene, Kimura is led to a differential equation—the Fokker-Planck equation, previously used to model diffusion processes. Although complicated, the equation has a straightforward interpretation. It describes the evolution of a probability distribution, tracking the likelihood over every instant of time that a specific gene will change its representation in a population of genes. Kimura is able to provide an explicit solution for the equation, and thus to treat molecular evolution as a well-posed problem in analysis.

BUT IF THE "MODEL for what science should be" is powerful, it is also limited. Stretching it beyond its natural limits often turns out to be an exercise in misapplied force, like a blow delivered to the empty air.

As I have noted several times, the power of a differential equation to govern the flow of time is contingent on some specification of its initial conditions. It is precisely these initial conditions that anecdotal accounts of human evolution cannot supply. We can say of those hunters and gatherers only that they hunted and they gathered, and we can say this only because it seems obvious that there was nothing else for them to do. The gene pool that they embodied cannot be directly recovered.

The question very naturally arises: might that gene pool be recovered from the differential equations of mathematical genetics, much as the original position and momentum of a system of particles moving under the influence of gravitational forces might be recovered from their present position and momentum? This is the question posed by Richard Lewontin.[61] Writing in a recent issue of the *Annual Review of Genetics*, Lewontin observes that if Kimura's equations carry "a population forward in time from some initial conditions," then what is needed is a second theory, one "that can reverse the deductions of the first theory and infer backward from a particular observed state at present."

Lewontin is correct: this is precisely what is needed. Given the trajectory described by the solution of the Fokker-Planck equation, it is certainly possible to track the equation backward, past the middle ages, well

---

61. I am grateful to Robert Berwick of MIT for calling my attention to this article, and for insisting on its importance.

past the Roman and then the Sumerian empires, and then into the era of the hunter-gatherers. There is nothing troubling about this. Kimura's equation has an explicit solution, and seeing where it led from is like running a film backward.

But whether, in running this particular film backward, we inevitably channel the temporal stream into a *unique* set of initial conditions is not altogether clear. With questions of this sort, we are in the domain of inverse problems, in which the past is contingent on the present. The solution to an inverse problem, the Russian mathematician Oleg Alifanov remarked, "entails determining unknown causes based on observation of their effects." It is this problem that evolutionary psychology must solve if its engaging stories about the Paleolithic era are to command credibility at the molecular level.

And it is this problem that Lewontin argues cannot be solved in the context of mathematical genetics. "A dynamical theory that predicts the present state generally requires that we know not only the nature and magnitude of the forces that have operated, but also the initial conditions and how long the process has been in operation." This double requirement—*know the forces, specify the initial conditions*—cannot simultaneously be met in going backward from the present. One item of knowledge is needed for the other.

This specific argument may now be enlarged to accommodate the general case. Inverse problems arise in mathematics when the attempt is made to run various mathematical films backward, and they are by now sufficiently well understood so that something may be said about them in a rough-and-ready way. Inverse problems are *not* in general well posed. Observing a pot of boiling liquid, we cannot use the heat equations to determine its identity. Many liquids reach the same boiling point in roughly the same time.

With inverse problems, what is, in fact, lost is the essential sureness and power of the "model for what science should be," and we are returned to a familiar world in which things and data are messy, disorganized, and partial, and in which theories, despite our best intentions, find themselves unable to peep over the hedge of time into the future or the past.

A familiar and by now depressing shape has revealed itself beneath the third and final simile of evolutionary psychology. It succeeds in meeting the demands of "the model for what science should be," but it succeeds in meeting those demands only at an isolated point. The rest is darkness, mystery, and magic.

IF THE CHIEF SIMILES of evolutionary psychology have not improved our understanding of the human mind in any appreciable sense, might we at least say that they have done something toward promoting the field's principal hope, namely, that the mind will in the end take its place as a material object existing in a world of other material objects?

This too is by no means clear. As Leda Cosmides has very sensibly observed, evolutionary psychology is more a research program than a body of specific results. As a program, it rather resembles a weekend athlete forever preparing to embark on a variety of strenuous exercises. In the literature of evolutionary psychology, there is thus no very de-termined effort to assess any of the classical topics in the philosophy of mind with the aim of doing more than affirming vaguely that some as-pect of the mind exists because it may well have been useful. There is, in evolutionary psychology, no account of the emotions beyond the trivial, or of the sentiments, no account of action or intention, no account of the human ability to acquire mathematical or scientific knowledge, no very direct exploration of the mind's power to act at a distance by investing things with meaning—no account, that is, of any of the features of the mind whose existence prompts a question about its origins. In its great hope as in so many other respects, evolutionary psychology has reposed its confidence on the bet that in time these things will be explained. If that is so, all that we on the outside can say is that time will tell.

Yet any essay on evolutionary psychology would be incomplete if it did not acknowledge the moving power of its chief story. For that story, involving as it does our own ancestors, suggests that the human mind

that we now occupy had its source in circumstances that, although oc-
cluded by time and damaged by distance, are nonetheless familiar.

The time is the distant past. "In Babylonia," the third-century historian Eusebius writes in recounting the lost histories of Berossos the
Chaldean, a large number of people "lived without discipline and without order, just like the animals." A frightening monster named Oannes
then appeared to the Babylonians after clambering out of the Red Sea.
"It had the whole body of a fish, but underneath and attached to the head
of the fish there was another head, human, and joined to the tail of the
fish, feet, like those of a man, and it had a human voice." The monster
"spent his days with men, never eating anything, but teaching men the
skills necessary for writing, and for doing mathematics, and for all sorts
of knowledge."

Since that time, Eusebius adds regretfully, "nothing further has
been discovered."

2005

ALL THOSE DARWINIAN DOUBTS

THE DEFENSE OF DARWIN'S THEORY OF EVOLUTION HAS NOW FALL-
en into the hands of biologists who believe in suppressing criticism
when possible and ignoring it when not. It is not a strategy calculated in
induce confidence in the scientific method. A paper published recently
in the *Proceedings of the Biological Society of Washington* concluded that
the events taking place during the Cambrian era could best be understood in terms of an intelligent design—hardly a position unknown in
the history of Western science. The paper was, of course, peer-reviewed
by three prominent evolutionary biologists. Wise men attend to the
publication of every one of the *Proceedings'* papers, but in the case of
Stephen Meyer's "The Origin of Biological Information and the Higher
Taxonomic Categories," the Board of Editors was at once given to understand that they had done a bad thing. Their indecent capitulation
followed at once.

Publication of the paper, they confessed, was a mistake. It would
never happen again. It had barely happened at all. And peer review?

The hell with it.

"If scientists do not oppose antievolutionism," Eugenie Scott, the
Executive Director of the National Center for Science Education, remarked, "it will reach more people with the mistaken idea that evolution
is scientifically weak." Scott's understanding of 'opposition' had nothing to do with reasoned discussion. It had nothing to do with reason at
all. Discussing the issue was out of the question. Her advice to her colleagues was considerably more to the point: "Avoid debates."

Everyone else had better shut up.

In this country, at least, no one is ever going to shut up, the more so since the case against Darwin's theory retains an almost lunatic vitality.

Look—The suggestion that Darwin's theory of evolution is like theories in the serious sciences—quantum electrodynamics, say—is grotesque. Quantum electrodynamics is accurate to 13 unyielding decimal places. Darwin's theory makes no tight quantitative predictions at all.

Look—Field studies attempting to measure natural selection inevitably report weak to non-existent selection effects.

Look—Darwin's theory is open at one end since there is no plausible account for the origins of life.

Look—The astonishing and irreducible complexity of various cellular structures has not yet successfully been described, let alone explained.

Look—A great many species enter the fossil record trailing no obvious ancestors and depart for Valhalla leaving no obvious descendants.

Look—Where attempts to replicate Darwinian evolution on the computer have been successful, they have not used classical Darwinian principles, and where they have used such principles, they have not been successful.

Look—Tens of thousands of fruit flies have come and gone in laboratory experiments, and every last one of them has remained a fruit fly to the end, all efforts to see the miracle of speciation unavailing.

Look—The remarkable similarity in the genome of a great many organisms suggests that there is at bottom only one living system; but how then to account for the astonishing differences between human beings and their near relatives—differences that remain obvious to anyone who has visited a zoo?

But look again—If the differences between organisms are scientifically more interesting than their genomic similarities, of what use is Darwin's theory since its otherwise mysterious operations take place by genetic variations?

These are hardly trivial questions. Each suggests a dozen others. These are hardly circumstances that do much to support the view that there are "no valid criticisms of Darwin's theory," as so many recent editorials have suggested.

Serious biologists quite understand all this. They rather regard Darwin's theory as an elderly uncle invited to a family dinner. The old boy has no hair, he has no teeth, he is hard of hearing, and he often drools. Addressing even senior members at the table as Sonny, he is inordinately eager to tell the same story over and over again.

But he's family. What can you do?

ACADEMIC EXTINCTION

W EARING PINK TASSELED SLIPPERS AND CONICAL HATS COVERED in polka dots, Darwinian biologists are persuaded that a plot is afoot to make them look silly. At Internet web sites such as The Panda's Thumb or Talk Reason, where various eminences repair to assure one another that all is well, it is considered clever beyond measure to attack critics of Darwin's theory such as William Dembski by misspelling his name as William Dumbski.

Publishing his work with the Cambridge University Press, hardly a venue known for its slack intellectual standards, Dembski has proposed that designed structures in nature might be detected by means of a rigorous analytical test. The idea of design is a staple of the social, anthropological and forensic sciences. It is the crucial metaphor in Noam Chomsky's minimalist theory. Dembski holds two PhD's, the first from the University of Chicago in mathematics, and the second from the University of Illinois in philosophy.

Dumbski indeed. Elsewhere, rhetoric is more measured, even if it conveys arguments no more compelling. After alluding to intelligent design at a faculty cocktail party—*Je m'imagine cela*—the dean of undergraduate education at the University of Calfornia at Berkeley was amazed and remarked "that colleagues indicated a great deal of sympathy for this alternative to Darwinism."

His amazement notwithstanding, the dean's defense was a model of evasive circumspection.

"Although I told them that few, if any, reputable biologists in the country subscribe to intelligent design, I could tell that they were not persuaded. Somewhat dismayed, I turned to other, more congenial issues."

Now these are remarkable words, if only because they reveal that a prominent academic regards it as quite natural to be dismayed on those occasions when his views are disputed. They are remarkable as well because they indicate that the dean is persuaded that dissent might in the case of Darwin's theory be ended by an appeal to what "reputable biologists believe."

My dear dean. Allow me to set you straight. It is precisely the reputable biologists who are under attack. For the first time, they are being asked to defend the thesis that biological design is more apparent rather than real. The effort has left them breathless. They are, of course, not about to surrender their ideological allegiances. Their rhetoric fills the op-ed columns of every liberal newspaper and is conveyed additionally by academic allies whose welfare is contingent on theirs—analytic philosophers, pop psychologists, and even newspaper columnists eager beyond measure to do anything but attentively study the evidence.

But what is at issue, of course, is not what reputable biologists believe, but whether it is true.

A great many ordinary men and women are persuaded that it is not. And even at Berkeley. Their dissatisfaction has traveled as far a field as Paris. Expertise is hardly at issue. Darwin's theory of evolution is not protected by the twelve doors mentioned in Revelation 21:21. It is right there in plain sight.

The unfathomable complexity of living systems, Darwin's theory affirms, is the result of random variation and natural selection. Is it indeed? Of these concepts, the second is hopelessly confused and the first is of no intellectual interest. Darwin's theory, when the thing is plainly considered, is no more than a form of behaviorism written on the level of the species. Like those endless psychological experiments, all of them conducted apparently at Harvard, in which some undergraduates were trained to say ouch after being stuck with a pin, and others to say ooh, species, on Darwin's view, are trained to say ouch or ooh when stuck by the environment.

B. F. Skinner is long dead, and among the dinosaurs, behaviorism in psychology has been the first to descend, honking sadly, into the tar pits.

What reputable biologists believe is one thing; what they fear is there in plain sight.

"Everyone on the Berkeley campus should be exposed to the arguments supporting real science and to the fallacies of views based on guesswork and unfounded hypotheses."

Ah, yes, Everyone should. Even at Berkeley.

THE STRENGTH OF NATURAL SELECTION IN THE WILD

LIKE HELL ITSELF, DARWIN'S THEORY OF EVOLUTION IS OFTEN SAID to be protected by walls that are at least seven miles thick, in that it is not only true, but unassailable. It is a considerable irony, therefore, that some of the most cogent criticisms of Darwin's theory are the result of work undertaken by very orthodox members of the biological establishment itself. Such criticisms are inevitably designated as calls for further research. They are, nonetheless, what they are.

A recent study by J. G. Kingsolver et al. (hereinafter Kingsolver) entitled "The Strength of Phenotypic Selection in Natural Populations," published in the March 2001 issue of *The American Naturalist*, is an interesting example. It is field studies of natural selection that is at issue in this study. Such studies are addressed to living species under natural conditions, and it comes as something of a surprise to learn that despite very long-standing claims by evolutionary biologists to have established the robust viability of natural selection as a biological force, the overwhelming number of such studies have been conducted only in the past 15 years.

Kingsolver's study is second-order in nature: It analyses and discusses 63 field studies dealing with 62 species that have been reported in the literature since the publication of J. A. Endler's well-known monograph, *Natural Selection in the Wild*, published in 1986.

The statistical methods by which Kingsolver proceeded are simple to the point of triteness. On the one hand, there are a series of quantitative biological traits, chiefly morphological in nature; and on the other hand, certain quantitative measure of fitness. Beak length in finches is a typical morphological trait, and survival, mating success or fecundity typical measure of fitness. Using the methodology first introduced by R. Lande and S. J. Arnold in their 1983 study, "The Measurement of Selection on Correlated Characteristics," published in *Evolution* 37, Kingsolver proposed to define selection in terms of the slope of the regression between a quantitative trait of interest and specific measures of fitness. This provides an estimate of the strength of selection.

Natural selection disappears as a biological force and reappears as a statistical artifact. The change is not trivial. It is one thing to say that nothing in biology makes sense except in the light of evolution; it is quite another thing to say that nothing in biology makes sense except in the light of various regression correlations between quantitative characteristics. It hardly appears obvious that if natural selection is simply a matter of correlations established between quantitative traits, that Darwin's theory has any content beyond the phenomenological, and in the most obvious sense, is no theory at all.

Be that as it may, the real burden of Kingsolver's study lies in the quantitative conclusions it reaches. Two correlations are at issue. The first is linear, and corresponds to what in population genetics is called directional selection; and the second quadratic, and corresponds either to stabilizing or disruptive selection. These are the cornerstones of the modern hill and valley model of much of mathematical population genetics.

Kingsolver reported a median absolute value of 0.16 for linear selection, and a median absolute value of 0.10 for quadratic selection. Thus an increase of one standard deviation in, say, beak finch length, could be expected to change fitness by only 16 percent in the case of linear selection, and 10 percent in the case of quadratic selection. These figures are commonly understood to represent a very weak correlation. Thus if a change in the length of a finch's beak by one standard deviation explains 16 percent of the change in the population's fitness, 84 percent of the change is not explained by selection at all.

These results, although at odds with those reported by Endler, are not in themselves astounding. It is when sample sizes pass beyond samples of 1,000 that results become far more difficult to accommodate, for under these circumstances, Kingsolver reported, both linear and quadratic selection were virtually non-existent.

The significance of these results is, of course, not entirely clear. Kingsolver goes no further than observing that "important issues about selection remain unresolved." Considering the fundamental role of both linear and quadratic selection in population genetics and in popular accounts of Darwin's theory, one of those "unresolved" issues may well be whether natural selection exists to any appreciable extent, and if it does, whether it plays any real role in biological change altogether.

These considerations may assume some importance when it comes to assessing the widely repeated claim that Darwin's theory is as well-established as physical theories such as general relativity or quantum mechanics.

AN OPEN LETTER TO
THE AMAZING RANDI

James Randi is a retired professional magician (stage name: the Amazing Randi) and self-styled skeptic on matters relating to "pseudoscience."

—EDITOR

June 13, 2005

Dear Amazing Randi:

I just read your widely publicized letter to the Smithsonian Institution about its decision to air *The Privileged Planet*, Discovery Institute's film on intelligent design. You find it "impossible to comprehend" why the Smithsonian has chosen to screen such a film. And I see that you are willing to pay the Smithsonian Institute $20,000 so that they don't do it.

I want you to know, you're doing the right thing. I figure the American people are dumb as posts. Who knows what ideas a film like that could put into their heads? You haven't seen the film either, am I right? See no evil, see no evil is what I always say.

But here's the thing, Randi. I was sort of planning to screen the film right here in my apartment in Paris. I've got a little screening room I call The Smithsonian right between the bathroom and the kitchen, I sort of figured I'd invite some friends over, open a couple cans of suds, sort of kick back and enjoy. Now you fork over $20,000 to the Smithsonian not to show the film and right away I'm showing the film here in Paris— that's just not going to work for you, if you catch my drift.

But hey, what are friends for? I mean for $20,000, I can make my screening of the *The Privileged Planet* go away too. An extra $10,000 and we spend the evening reading aloud from Daniel Dennett's autobiography. I hear it's a real snoozer, no chance at all that anyone's going to walk away from an evening like that with poor thoughts about the cosmos or anything like that. You handle the refreshments—nothing much, some cocktail franks maybe, a few kegs of French beer—and I knock 10 percent off the price. What do you say?

Now I know what you're thinking, Randi, because to tell you the truth, I've been thinking the same thing. You're thinking, hey, I'm out forty thousand seminolas to can this film in Washington and Paris, and right away, some yutz is going to figure it's show time in Oklahoma or Nebraska or even in New York, and what do I do then? I'm way ahead of you on this one. I've talked with my buddies at the Discovery Institute and for the right kind of donation, we poleax the film completely. That's right. It disappears itself, if you catch my drift. You get to keep the negatives, we keep the director's cut in our safe for insurance. Is this some sort of deal, or what? Now I know what you're thinking because I've been there myself. You're thinking, the Discovery Institute? Bunch of right-wing weirdoes, am I right? Hey, it's not like that at all, Randi, I got to tell you. We here at the Discovery Institute, we're businessmen, if you catch my drift. We want to do the right thing and we want to do it at the right price. Look at it this way. The right kind of donation gets you total peace of mind. You really can't buy that kind of protection, only in this case you can.

So give me a ring, or send me a note. I'd like to tell you we take checks, but you're a businessman, too, am I right? It's got to be cash. More than you've got lying around? Not a problem. Just give George Soros a call. Tell him it's for a friend. Do it now.

You'll sleep better at night.

Your admirer,
David Berlinski

P. S. I write a lot of stuff for *Commentary*, too. For the right price, I don't have to write anything at all. Think it over. Let me know.

Our Silent Partners

WE ARE ANIMALS IN OUR APPETITES, AND ANIMALS AGAIN IN OUR instincts and emotions. We are animals in biology. Blood is blood, tissue is tissue, and cells are cells; and when everything is stripped away, we are animals in the organization of our genes, mindful now that but for a few alterations of the human genome, we might well be slithering about in some abysmal pond or lowing in the field, or even contemplating the mysteries of life as a form of yeast. What the medievals called the great chain of being is a vibrant faith in contemporary biology, and so in evolutionary psychology as well. Everything is connected.

Human language represents an odd, irritating impediment on the otherwise smooth manifold of these reflections, for the plain fact of the matter is that while other animals hoot, chatter, grimace, moan, whistle, chirp, bang their heads upon the ground, twitter, dance, bark, meow, sing lustrously from the trees, screech in barns, or make disgusting slurping noises to attract their mates, it is only human beings who talk, and so only human beings who catch sound and meaning from the air.

During the first half of the twentieth century, American psychologists such as John Watson and B. F. Skinner assumed that language was essentially an acquired habit, one imparted in training and acquired by diligence. This view was decisively refuted by Noam Chomsky in the late 1950s. The proper object of linguistic study is not what human beings do in speaking a language, but what they must know in order to do what they do. What they do know, Chomsky demonstrated, is quite astonishing and goes beyond what they might have learned. Grammatical judgments involve both delicacy and precision. A grammar is required, a device for generating the sentences of a natural language from its words, and since there are infinitely many sentences in any human language, a grammar is inherently recursive.

These ideas constitute Chomsky's revolution in the cognitive sciences; and over 40 years they have constituted as well the strongest reason for taking seriously the view that the mind is engaged in computations. They do something more. They reveal the very deep tensions in evolutionary psychology, for the very arguments that led Chomsky to reject stimulus response theories in linguistics may with little difficulty be used to reject Darwinian theories in biology. In stimulus sampling theories, for example, subjects are presented with random stimuli; their responses follow, followed in turn by selective reinforcement. In Darwinian theories, random variations in the genome act in a parallel fashion to the psychologist's random stimuli, while natural selection provides the reinforcement. Evolution is learning undertaken by a species. The rhetoric in both disciplines has an eerie familiarity. Behavioral psychologists were quite sure that reinforcement was the only force that could account for learning, just as Darwinian biologists are quite certain that natural selection is the only force that could account for complex structures in biology.

But if behavioral theories are inadequate to explain the acquisition of language in a child, why should precisely the same theories be adequate to explain the acquisition of language in a species? Nothing beyond time is variable between the two cases, and time is of little relevance, since as Chomsky argued there are things that *no amount* of training can teach. Chomsky became known not for his disavowal of Darwinian theories, but for his discontent: "Any progress [toward the goals of linguistic theory] will deepen a problem for the biological sciences that is far from trivial: how can a system such as human language arise in the mind/brain, or for that matter, in the organic world, in which one seems not to find anything like the basic properties of human language?"

AND YET THERE IT is, that odd thing, for in a recent paper in *Science,* written jointly with Marc D. Hauser and W. Tecumseh Fitch, Chomsky has reinvented himself as an evolutionary psychologist, one prepared manfully to embrace the great apes and by placing a moist hand on their hairy shoulders welcome them as our silent partners in the great work of language. It is a remarkable effort, the more so since what the paper claims and what it concludes are utterly at odds.

Entitling their paper "The Faculty of Language: What It Is, Who Has It, and How Did It Evolve," Hauser, Fitch and Chomsky begin reasonably enough by dividing the faculty of language into three parts, each corresponding to what an organism must have in order to be able to use a language. A *sensory motor system,* to begin with. In order to say anything, an animal must be able to say something, and so requires organs of speech, articulation, hearing and comprehension. A *conceptual-intentional system* next. In order to say something, an animal must have something to say and so requires thoughts and intentions. And finally a *recursive system.* In order to spin off sentences, an animal must be able to spin them off without end, and so requires a computational system. Of these components, Hauser, Fitch and Chomsky suggest in outlining their ideas, the first two are found in other organisms, and so present evolutionary psychology with no special problems. It is only recursion that is uniquely human. But no specifically Darwinian theory is required to explain its emergence. Like a smile, it just appears.

It is by these elegant parallel motions of displacement and insouciance that Chomsky proposes to establish his credentials as an evolutionary psychologist while emptying evolutionary psychology of any conceivable intellectual interest. Whenever evidence concerning our silent partners is examined seriously, it appears that striking limitations on what they can say or think are in force.

Do our silent partners have the right kind of sensory motor system to make language possible? "The available data," Hauser, Fitch and Chomsky gamely propose, "suggest a much stronger continuity between animals and humans with respect to speech than previously believed."

At once they indicate precisely why the available data is entirely misleading. The human capacity for vocal imitation is not found in any other species to any notable extent. The average high-school student, they observe, knows as many as 60,000 words and these were acquired with little or no effort. "Herculean efforts," they admit, are required to persuade even clever chimpanzees to acquire a handful of words. "Our discussion," Hauser, Fitch and Chomsky conclude, "… is not meant to diminish the impressive achievements of monkeys and apes, but to highlight how different the mechanisms underlying the production of human and non-human primate gestures, either vocally expressed or signed, must be."

There is next the matter of what our silent partners might be thinking during the long night in which they found themselves unable to say what they had in mind—what Hauser, Fitch and Chomsky call "the mismatch between the conceptual capacities of animals and the communicative content of their vocal and visual signals." As it happens, animals do have a rich inner thought world, a conclusion that will come as no surprise to anyone who has owned an animal. But when it comes to the crucial concepts necessary for language, the point at issue, after all, animals are lacking. They do not use and cannot grasp the point of a complex, independent referential system whose elements are arbitrarily associated with things. The matter is quite beyond the reach of otherwise intelligent apes. "It appears that many of the elementary properties of words … have only weak analogs or homologs in natural animal communication systems," Hauser, Fitch and Chomsky acknowledge.

If animals can neither acquire words by imitation or use them without effort, what, one is bound to ask, remains of the thesis that there is "a much stronger continuity between animals and humans with respect to speech than previously believed"?

Undeterred by their own argument (or their own sources, for that matter), Hauser, Fitch and Chomsky conclude that "the data summarized so far, although far from complete, provide overall support for the position of continuity between humans and other animals."

There remains recursion, the jewel in the crown. And about the emergence of this key feature, Hauser, Fitch and Chomsky have nothing of interest to say, beyond speculating that it, too, might have been an earlier acquisition of our silent partners, one used, say, for navigation or foraging, and by some supremely lucky accident, breaking free, or popping out precisely when most needed.

To imagine that language arose among primates incapable of speech, and unable appropriately to think, by means of a recursive system designed for other ends, represents a conspicuous willingness to look anywhere for miracles save in the place they are generally found.

Copernicus Stages
a Comeback

More that 60 years after the famous Galileo "The Earth it Moves" trial in Rome, Copernicus is in the news again, this time in the form of a so-called theory of universal gravitation (or UG, as it has come to be known). Headquartered at the Royal Society, a think tank in London funded by well-heeled royalist donors, members of the universal gravitation movement argue that the facts of astronomy are so complicated that they require the introduction of a mysterious "universal force of gravitation." But when queried, members decline to specify the author of this force, saying only (according to a public spokesman) that, "In this, we are agnostic."

Unlike the old breed of Copernicists, universal gravitation theorists sport established degrees from well-known universities and claim to be doing cutting edge science. "Look," said one prominent member of the Royal Society, "there's plainly some sort of force at work on the surface of the earth. Unsupported objects always fall. All we're saying is that it is perfectly reasonable to follow the evidence. If the evidence leads to some sort of universal force, that's where the evidence leads."

Critics have not been impressed, pointing to the fact that members of the movement do not publish in peer-reviewed journals and instead rely on the Royal Society itself to put out their slick products.

"It's all smoke and mirrors," claimed one member of the astronomy faculty at the University of Augsburg. "It's a classic force of the gaps argument. All they're really saying is that there are some things established science can't explain yet, so there must be a mysterious force at work. If science has taught us anything over two thousand years, it's that sooner or later the gaps get filled in a perfectly natural way."

Said another critic, the professor of Ptolemaic Understanding at Oxford University: "Nothing in astronomy makes sense except in the light of Ptolemaic astronomy. There is overwhelming evidence in support of the idea that the sun revolves around the earth. It has been one of the most fruitful and productive ideas in the entire history of science."

Still another critic observed that "to claim that Ptolemy's theory is just a theory is as absurd as arguing that Galen's theory of the humors is just a theory. It betrays a fundamental ignorance of the way in which science works."

UG's most well-known figure has been Isaac Newton, a notoriously reclusive mathematician with a known taste for bizarre theology and a penchant for dangerous chemical experiments. His *Principia Mathematica* has been a surprise best-seller, one of those books, as one wag quipped, which it is "easy to get into and impossible to get out of." Real mathematicians have been almost universal in their scorn, however. "The worst sort of pretentious posturing," said the professor of Counting and Arithmetic at the University of London, adding that "for all this Newton's so-called arithmetical expertise, there is just nothing in this book that indicates that it has any relevance at all to the real data of astronomy."

Colleagues at the Department of Counting and Arithmetic agreed. "Let's face it," said one "The idea that the moon is falling is just insane. Falling? How come it never hits the earth? Where is it falling from? Who or why was it dropped? The idea that heavenly bodies are held in place by some sort of invisible force isn't even bad science. It's not science at all."

Still other mathematicians have argued that Newton's book is riddled with obvious blunders and betrays a fundamental lack of scholarly rigor. Commenting on the so-called differential calculus, the professor of Applied Numbers at the University of Manchester remarked that the "whole subject was just written in Suet," adding that "Newton seems to believe there are numbers greater than zero but less than any other number. I know of no idea likely to be less productive."

When reached in his London office, Newton declined to comment, saying only that he had "no use for little smatterers in mathematics."

"Typical," said the professor of Epicycles at the University of Canterbury. "These people can quibble about tiny details in Ptolemaic astronomy, but when anyone criticizes their own work, they start babbling about conspiracies to marginalize their views. It's all just Copernicus in a cheap tuxedo."

Nonetheless, the most recent polls indicate that over 60 percent of the English public believes that the earth revolves around the sun and that it is kept in its orbit by some sort of mysterious force.

2006

ON THE ORIGINS OF LIFE

*For those who are studying aspects of the origin of life, the question no
longer seems to be whether life could have originated by chemical processes involving non-biological components but, rather, what pathway
might have been followed.*

—NATIONAL ACADEMY OF SCIENCES (1996)

IT IS 1828, A YEAR THAT ENCOMPASSED THE DEATH OF SHAKA, THE
Zulu king, the passage in the United States of the Tariff of Abomina-
tions, and the battle of Las Piedras in South America. It is, as well, the
year in which the German chemist Friedrich Wöhler announced the
synthesis of urea from cyanic acid and ammonia.

Discovered by H. M. Roulle in 1773, urea is the chief constituent of
urine. Until 1828, chemists had assumed that urea could be produced
only by a living organism. Wöhler provided the most convincing refutation imaginable of this thesis. His synthesis of urea was noteworthy, he
observed with some understatement, because "it furnishes an example
of the artificial production of an organic, indeed a so-called animal substance, from inorganic materials."

Wöhler's work initiated a revolution in chemistry; but it also initiated a revolution in thought. To the extent that living systems are chemical
in their nature, it became possible to imagine that they might be chemical in their origin; and if chemical in their origin, then plainly physical
in their nature, and hence a part of the universe that can be explained in
terms of "the model for what science should be."[62]

---

62. I used this phrase, borrowed from the mathematicians J. H. Hubbard and B. H. West,
in "On the Origins of the Mind" (*Commentary*, November 2004). The idea that science must

In a letter written to his friend, Sir Joseph Hooker, several decades after Wöhler's announcement, Charles Darwin allowed himself to speculate. Invoking "a warm little pond" bubbling up in the dim inaccessible past, Darwin imagined that given "ammonia and phosphoric salts, light, heat, electricity, etc. present," the spontaneous generation of a "protein compound" might follow, with this compound "ready to undergo still more complex changes" and so begin Darwinian evolution itself.

Time must now be allowed to pass. Shall we say 60 years or so? Working independently, J. B. S. Haldane in England and A. I. Oparin in the Soviet Union published influential studies concerning the origin of life. Before the era of biological evolution, they conjectured, there must have been an era of *chemical* evolution taking place in something like a pre-biotic soup. A reducing atmosphere prevailed, dominated by methane and ammonia, in which hydrogen atoms, by donating their electrons (and so "reducing" their number), promoted various chemical reactions. Energy was at hand in the form of electrical discharges, and thereafter complex hydrocarbons appeared on the surface of the sea.

The publication of Stanley Miller's paper, "A Production of Amino Acids Under Possible Primitive Earth Conditions," in the May 1953 issue of *Science* completed the inferential arc initiated by Friedrich Wöhler 125 years earlier. Miller, a graduate student, did his work at the instruction of Harold Urey. Because he did not contribute directly to the experiment, Urey insisted that his name not be listed on the paper itself. But their work is now universally known as the Miller-Urey experiment, providing evidence that a good deed can be its own reward.

By drawing inferences about pre-biotic evolution from ordinary chemistry, Haldane and Oparin had opened an imaginary door. Miller

---

conform to a certain model of inquiry is familiar. Hubbard and West identify that model with differential equations, the canonical instruments throughout physics and chemistry.

But the essentials of the model, it seems to me, lie less with the particular means in which it is expressed and more with the constraints that it must meet. The idea behind the "model for what science should be" is that whatever may be a system's initial conditions, or starting point, the laws of its development must be both unique and stable. When they are, the system that results is well posed, and so a proper object of contemplation.

and Urey barged right through. Within the confines of two beakers, they re-created a simple pre-biotic environment. One beaker held water; the other, connected to the first by a closed system of glass tubes, held hydrogen cyanide, water, methane, and ammonia. The two beakers were thus assumed to simulate the prebiotic ocean and its atmosphere. Water in the first could pass by evaporation to the gases in the second, with vapor returning to the original alembic by means of condensation.

Then Miller and Urey allowed an electrical spark to pass continually through the mixture of gases in the second beaker, the gods of chemistry controlling the reactions that followed with very little or no human help. A week after they had begun their experiment, Miller and Urey discovered that in addition to a tarry residue—its most notable product—their potent little planet had yielded a number of the amino acids found in living systems.

The effect among biologists (and the public) was electrifying—all the more so because of the experiment's methodological genius. Miller and Urey had done nothing. Nature had done everything. The experiment alone had parted the cloud of unknowing.

IN APRIL 1953, JUST four weeks before Miller and Urey would report their results in *Science*, James Watson and Francis Crick published a short letter in *Nature* entitled "A Structure for Deoxyribose Nucleic Acid." The letter is now famous, if only because the exuberant Crick, at least, was persuaded that he and Watson had discovered the secret of life. In this he was mistaken: the secret of life, along with its meaning, remains hidden. But in deducing the structure of deoxyribose nucleic acid (DNA) from X-ray diffraction patterns and various chemical details, Watson and Crick *had* discovered the way in which life at the molecular level replicates itself.

Formed as a double helix, DNA, Watson and Crick argued, consists of two twisted strings facing each other and bound together by struts. Each string comprises a series of four nitrogenous bases: adenine (A), guanine (G), thymine (T), and cytosine (C). The bases are nitrogenous because their chemical activity is determined by the electrons of the nitrogen atom, and they are bases because they are one of two great chemical clans—the other being the acids, with which they combine to form salts.

Within each strand of DNA, the nitrogenous bases are bound to a sugar, deoxyribose. Sugar molecules are in turn linked to each other by a phosphate group. When nucleotides (A, G, T, or C) are connected in a sugar-phosphate chain, they form a polynucleotide. In living DNA, two such chains face each other, their bases touching fingers, A matched to T and C to G. The coincidence between bases is known now as Watson-Crick base pairing.

"It has not escaped our notice," Watson and Crick observed, "that the specific pairings we have postulated immediately suggests a possible *copying mechanism* for the genetic material" (emphasis added). Replication proceeds, that is, when a molecule of DNA is unzipped along its internal axis, dividing the hydrogen bonds between the bases. Base pairing then works to prompt both strands of a separated double helix to form a double helix anew.

So Watson and Crick conjectured, and so it has proved.

TOGETHER WITH FRANCIS CRICK and Maurice Wilkins, James Watson received the Nobel Prize for medicine in 1962. In his acceptance speech in Stockholm before the king of Sweden, Watson had occasion to explain his original research goals. The first was to account for genetic replication. This, he and Crick had done. The second was to describe the "way in which genes control protein synthesis." This, he was in the course of doing.

DNA is a large, long, and stable molecule. As molecules go, it is relatively inert. It is the proteins, rather, that handle the day-to-day affairs of the cell. Acting as enzymes, and so as agents of change, proteins make possible the rapid metabolism characteristic of modern organisms.

Proteins are formed from the alpha-amino acids, of which there are 20 in living systems. The prefix "alpha" designates the position of the crucial carbon atom in the amino acid, indicating that it lies adjacent to (and is bound up with) a carboxyl group comprising carbon, oxygen, again oxygen, and hydrogen. And the proteins are polymers: like DNA, their amino-acid constituents are formed into molecular chains.

But just how does the cell manage to link amino acids to form specific proteins? This was the problem to which Watson alluded as the king of Sweden, lost in a fog of admiration, nodded amiably.

The success of Watson-Crick base pairing had persuaded a number of molecular biologists that DNA undertook protein synthesis by the same process—the formation of symmetrical patterns or "templates"—that governed its replication. After all, molecular replication proceeded by the divinely simple separation-and-recombination of matching (or symmetrical) molecules, with each strand of DNA serving as the template for another. So it seemed altogether plausible that DNA would likewise serve a template function for the amino acids.

It was Francis Crick who in 1957 first observed that this was most unlikely. In a note circulated privately, Crick wrote that "if one considers the physico-chemical nature of the amino-acid side chains, we do not find complementary features on the nucleic acids. Where are the knobby hydrophobic... surfaces to distinguish valine from leucine and isoleucine? Where are the charged groups, in specific positions, to go with acidic and basic amino acids?"

Should anyone have missed his point, Crick made it again: "I don't think that anyone looking at DNA or RNA [ribonucleic acid] would think of them as templates for amino acids."

Had these observations been made by anyone but Francis Crick, they might have been regarded as the work of a lunatic; but in looking at any textbook in molecular biology today, it is clear that Crick was simply noticing what was under his nose. Just where *are* those "knobby hydrophobic surfaces"? To imagine that the nucleic acids form a template or pattern for the amino acids is a little like trying to imagine a glove fitting over a centipede. But if the nucleic acids did not form a template for the amino acids, then the information they contained—all of the ancient wisdom of the species, after all—could only be expressed by an indirect form of transmission: a *code* of some sort.

THE IDEA WAS HARDLY new. The physicist Erwin Schrödinger had predicted in 1945 that living systems would contain what he called a "code script"; and his short, elegant book, *What Is Life?*, had exerted a compelling influence on every molecular biologist who read it. Ten years later, the ubiquitous Crick invoked the phrase "sequence hypothesis" to characterize the double idea that DNA sequences spell a message *and* that a code is required to express it. What remained obscure was both the spelling of the message and the mechanism by which it was conveyed.

The mechanism emerged first. During the late 1950s, François Jacob and Jacques Monod advanced the thesis that RNA acts as the first in a chain of intermediates leading from DNA to the amino acids.

Single- rather than double-stranded, RNA is a nucleic acid: a chip from the original DNA block. Instead of thymine (T), it contains the base uracil (U), and the sugar that it employs along its backbone features an atom of oxygen missing from deoxyribose. But RNA, Jacob and Monod argued, was more than a mere molecule: it was a messenger, an instrument of conveyance, "transcribing" in one medium a message first expressed in another. Among the many forms of RNA loitering in the modern cell, the RNA bound for duties of transcription became known, for obvious reasons, as "messenger" RNA.

In transcription, molecular biologists had discovered a second fundamental process, a companion in arms to replication. Almost immediately thereafter, details of the code employed by the messenger appeared. In 1961, Marshall Nirenberg and J. Heinrich Matthaei announced that they had discovered a specific point of contact between RNA and the amino acids. And then, in short order, the full genetic code emerged. RNA (like DNA) is organized into triplets, so that adjacent sequences of three bases are mapped to a single amino acid. Sixty-four triplets (or codons) govern 20 amino acids. The scheme is universal, or almost so.

The elaboration of the genetic code made possible a remarkably elegant model of the modern cell as a system in which sequences of codons within the nucleic acids act at a distance to determine sequences of amino acids within the proteins: commands issued, responses undertaken. A third fundamental biological process thus acquired molecular incarnation. If replication served to divide and then to duplicate the cell's ancestral message, and transcription to re-express it in messenger RNA, "translation" acted to convey that message from messenger RNA to the amino acids.

FOR ALL THE BOLDNESS and power of this thesis, the details remained on the level of what bookkeepers call general accounting procedures. No one had established a direct—a physical—connection between RNA and the amino acids.

Having noted the problem, Crick also indicated the shape of its solution. "I therefore proposed a theory," he would write retrospectively, "in which there were 20 adaptors (one for each amino acid), together with 20 special enzymes. Each enzyme would join one particular amino acid to its own special adaptor."

In early 1969, at roughly the same time that a somber Lyndon Johnson was departing the White House to return to the Pedernales, the adaptors whose existence Crick had predicted came into view. There were twenty, just as he had suggested. They were short in length; they were specific in their action; and they were nucleic acids. Collectively, they are now designated "transfer" RNA (tRNA).

Folded like a cloverleaf, transfer RNA serves physically as a bridge between messenger RNA and an amino acid. One arm of the cloverleaf is called the anti-coding region. The three nucleotide bases that it contains are curved around the arm's bulb-end; they are matched by Watson-Crick base pairing to bases on the messenger RNA. The other end of the cloverleaf is an acceptor region. It is here that an amino acid must go, with the structure of tRNA suggesting a complicated female socket waiting to be charged by an appropriate male amino acid.

The adaptors whose existence Crick had predicted served dramatically to confirm his hypothesis that such adaptors were needed. But although they brought about a physical connection between the nucleic and the amino acids, the fact that they were themselves nucleic acids raised a question: in the unfolding molecular chain, just what acted to adapt the adaptors to the amino acids? And this, too, was a problem Crick both envisaged and solved: his original suggestion mentioned both adaptors (nucleic acids) and their enzymes (proteins).

And so again it proved. The act of matching adaptors to amino acids is carried out by a family of enzymes, and thus by a family of proteins: the aminoacyl-tRNA synthetases. There are as many such enzymes as there are adaptors. The prefix "aminoacyl" indicates a class of chemical reactions, and it is in aminoacylation that the cargo of a carboxyl group is bonded to a molecule of transfer RNA.

Collectively, the enzymes known as synthetases have the power both to recognize specific codons and to select their appropriate amino acid under the universal genetic code. Recognition and selection are ordinarily thought to be cognitive acts. In psychology, they are poorly understood, but within the cell they have been accounted for in chemical terms and so in terms of "the model for what science should be."

With tRNA appropriately charged, the molecule is conveyed to the ribosome, where the task of assembling sequences of amino acids is then undertaken by still another nucleic acid, ribosomal RNA (rRNA). By these means, the modern cell is at last subordinated to a rich narrative drama. To repeat:

- *Replication* duplicates the genetic message in DNA.

- *Transcription* copies the genetic message from DNA to RNA.

- *Translation* conveys the genetic message from RNA to the amino acids—whereupon, in a fourth and final step, the amino acids are assembled into proteins.

IT WAS ONCE AGAIN Francis Crick, with his remarkable gift for impressing his authority over an entire discipline, who elaborated these facts into what he called the central dogma of molecular biology. The cell, Crick affirmed, is a divided kingdom. Acting as the cell's administrators, the nucleic acids embody all of the requisite wisdom—where to go, what to do, how to manage—in the specific sequence of their nucleotide bases. Administration then proceeds by the transmission of information *from* the nucleic acids *to* the proteins.

The central dogma thus depicts an arrow moving one way, from the nucleic acids to the proteins, and never the other way around. But is anything ever routinely returned, arrow-like, from its target? This is not a question that Crick considered, although in one sense the answer is plainly no. Given the modern genetic code, which maps four nucleotides onto twenty amino acids, there can be no inverse code going in the opposite direction; an inverse mapping is mathematically impossible.

But there is another sense in which Crick's central dogma does engender its own reversal. If the nucleic acids are the cell's administrators, the proteins are its chemical executives: both the staff and the stuff of life. The molecular arrow goes one way with respect to information, but it goes the other way with respect to chemistry.

Replication, transcription, and translation represent the grand unfolding of the central dogma as it proceeds in one direction. The chemical activities initiated by the enzymes represent the grand unfolding of the central dogma as it goes in the other. Within the cell, the two halves of the central dogma combine to reveal a system, of coded chemistry, an exquisitely intricate but remarkably coherent temporal tableau suggesting a great army in action.

From these considerations a familiar figure now emerges: the figure of a chicken and its egg. Replication, transcription, and translation are all under the control of various enzymes. But enzymes are proteins, and these particular proteins are specified by the cell's nucleic acids. DNA requires the enzymes in order to undertake the work of replication, transcription, and translation; the enzymes require DNA in order to initiate it. The nucleic acids and the proteins are thus profoundly coordinated, each depending upon the other. Without aminoacyl-tRNA synthetase, there is no translation from RNA; but without DNA, there is no synthesis of aminoacyl-tRNA synthetase.

If the nucleic acids and their enzymes simply chased each other forever around the same cell, the result would be a vicious circle. But life has elegantly resolved the circle in the form of a spiral. The aminoacyl-tRNA synthetase that is required to complete molecular translation enters a given cell from its progenitor or "maternal" cell, where it is specified by that cell's DNA. The enzymes required to make the maternal cell's DNA do its work enter that cell from *its* maternal line. And so forth.

On the level of intuition and experience, these facts suggest nothing more mysterious than the longstanding truism that life comes only from life. *Omnia viva ex vivo*, as Latin writers said. It is only when they are embedded in various theories about the *origins* of life that the facts engender a paradox, or at least a question: in the receding molecular spiral, which came first—the chicken in the form of DNA, or its egg in the form of various proteins? And if neither came first, how could life have begun?

IT IS 1967, THE year of the Six-Day War in the Middle East, the discovery of the electroweak forces in particle physics, and the completion of a 20-year research program devoted to the effects of fluoridation on dental caries in Evanston, Illinois. It is also the year in which Carl Woese, Leslie Orgel, and Francis Crick introduced the hypothesis that "evolution based on RNA replication *preceded* the appearance of protein synthesis" (emphasis added).

By this time, it had become abundantly clear that the structure of the modern cell was not only more complex than other physical structures but complex in poorly understood ways. And yet no matter how far back biologists traveled into the tunnel of time, certain features of the modern cell were still there, a message sent into the future by the last universal common ancestor. Summarizing his own perplexity in retrospect, Crick would later observe that "an honest man, armed with all the knowledge available to us now, could only state that, in some sense, the origin of life appears at the moment to be almost a miracle." Very wisely, Crick would thereupon determine never to write another paper on the subject—although he did affirm his commitment to the theory of "directed panspermia," according to which life originated in some other portion of the universe and, for reasons that Crick could never specify, was simply sent here.

But that was later. In 1967, the argument presented by Woesel, Orgel, and Crick was simple. Given those chickens and their eggs, *something* must have come first. Two possibilities were struck off by a process of elimination. DNA? Too stable and, in some odd sense, too perfect. The proteins? Incapable of dividing themselves, and so, like molecular eunuchs, useful without being fecund. That left RNA. While it was not obviously the right choice for a primordial molecule, it was not obviously the wrong choice, either.

The hypothesis having been advanced—if with no very great sense of intellectual confidence—biologists differed in its interpretation. But they did concur on three general principles. First: that at some time in the distant past, RNA rather than DNA controlled genetic replication. Second: that Watson-Crick base pairing governed ancestral RNA. And third: that RNA once carried on chemical activities of the sort that are now entrusted to the proteins. The paradox of the chicken and the egg was thus resolved by the hypothesis that the chicken was the egg.

The independent discovery in 1981 of the ribozyme—a ribonucleic enzyme—by Thomas Cech and Sidney Altman endowed the RNA hypothesis with the force of a scientific conjecture. Studying the ciliated protozoan *Tetrahymena thermophila*, Cech discovered to his astonishment a form of RNA capable of inducing cleavage. Where an enzyme might have been busy pulling a strand of RNA apart, there was a ribozyme doing the work instead. That busy little molecule served not only to give instructions: apparently it took them as well, and in any case it did what biochemists had since the 1920s assumed could only be done by an enzyme and hence by a protein.

In 1986, the biochemist Walter Gilbert was moved to assert the existence of an entire RNA "world," an ancestral state promoted by the magic of this designation to what a great many biologists would affirm as fact. Thus, when the molecular biologist Harry Noller discovered that protein synthesis within the contemporary ribosome is catalyzed by ribosomal RNA (rRNA), and not by any of the familiar, old-fashioned enzymes, it appeared "almost certain" to Leslie Orgel that "there once *was* an RNA world" (emphasis added).

It is perfectly true that every part of the modern cell carries some faint traces of the past. But these molecular traces are only hints. By contrast, to everyone who has studied it, the ribozyme has appeared to be an authentic relic, a solid and palpable souvenir from the pre-biotic past. Its discovery prompted even Francis Crick to the admission that he, too, wished he had been clever enough to look for such relics before they became known.

Thanks to the ribozyme, a great many scientists have become convinced that the "model for what science should be" is achingly close to encompassing the origins of life itself. "My expectation," remarks David Liu, professor of chemistry and chemical biology at Harvard, "is that we will be able to reduce this to a very simple series of logical events." Although often overstated, this optimism is by no means irrational. Looking at the modern cell, biologists propose to reconstruct in time the structures that are now plainly there in space.

Research into the origins of life has thus been subordinated to a rational three-part sequence, beginning in the very distant past. First, the constituents of the cell were formed and assembled. These included the nucleotide bases, the amino acids, and the sugars. There followed next the emergence of the ribozyme, endowed somehow with powers of self-replication. With the stage set, a *system of coded chemistry* then emerged, making possible what the molecular biologist Paul Schimmel has called "the theater of the proteins." Thus did matters proceed from the pre-biotic past to the very threshold of the last universal common ancestor, whereupon, with inimitable gusto, life began to diversify itself by means of Darwinian principles.

This account is no longer fantasy. But it is not yet fact. That is one reason why retracing its steps is such an interesting exercise, to which we now turn.

IT IS PERHAPS FOUR billion years ago. The first of the great eras in the formation of life has commenced. The laws of chemistry are completely in control of things—what else is there? It is Miller Time, the period marking the transition from inorganic to organic chemistry.

According to the impression generally conveyed in both the popular and the scientific literature, the success of the original Miller-Urey experiment was both absolute and unqualified. This, however, is something of an exaggeration. Shortly after Miller and Urey published their results, a number of experienced geochemists expressed reservations. Miller and Urey had assumed that the pre-biotic atmosphere was one in which hydrogen atoms gave up (reduced) their electrons in order to promote chemical activity. Not so, the geochemists contended. The pre-biotic atmosphere was far more nearly neutral than reductive, with little or no methane and a good deal of carbon dioxide.

Nothing in the intervening years has suggested that these sour geochemists were far wrong. Writing in the 1999 issue of *Peptides*, B. M. Rode observed blandly that "modern geochemistry assumes that the secondary atmosphere of the primitive earth (i.e., after diffusion of hydrogen and helium into space)... consisted mainly of carbon dioxide, nitrogen, water, sulfur dioxide, and even small amounts of oxygen." This is not an environment calculated to induce excitement.

Until recently, the chemically unforthcoming nature of the early atmosphere remained an embarrassing secret among evolutionary biologists, like an uncle known privately to dress in women's underwear; if biologists were disposed in public to acknowledge the facts, they did so by remarking that every family has one. This has now changed. The issue has come to seem troubling. A recent paper in *Science* has suggested that previous conjectures about the pre-biotic atmosphere were seriously in error. A few researchers have argued that a reducing atmosphere is not, after all, quite so important to pre-biotic synthesis as previously imagined.

In all this, Miller himself has maintained a far more unyielding and honest perspective. "Either you have a reducing atmosphere," he has written bluntly, "or you're not going to have the organic compounds required for life."

IF THE COMPOSITION OF the pre-biotic atmosphere remains a matter of controversy, this can hardly be considered surprising: geochemists are attempting to revisit an era that lies four billion years in the past. The synthesis of pre-biotic chemicals is another matter. Questions about them come under the discipline of laboratory experiments.

Among the questions is one concerning the nitrogenous base cytosine (C). Not a trace of the stuff has been found in any meteor. Nothing in comets, either, so far as anyone can tell. It is not buried in the Antarctic. Nor can it be produced by any of the common experiments in pre-biotic chemistry. Beyond the living cell, it has not been found at all.

When, therefore, M. P. Robertson and Stanley Miller announced in *Nature* in 1995 that they had specified a plausible route for the pre-biotic synthesis of cytosine from cyanoacetaldehyde and urea, the feeling of gratification was very considerable. But it has also been short-lived. In a lengthy and influential review published in the 1999 *Proceedings of the National Academy of Science*, the New York University chemist Robert Shapiro observed that the reaction on which Robertson and Miller had pinned their hopes, although active enough, ultimately went nowhere. All too quickly, the cytosine that they had synthesized transformed itself into the RNA base uracil (U) by a chemical reaction known as deamination, which is nothing more mysterious than the process of getting rid of one molecule by sending it somewhere else.

The difficulty, as Shapiro wrote, was that "the formation of cytosine and the subsequent deamination of the product to uracil occur[ed] at about the same rate." Robertson and Miller had themselves reported that after 120 hours, half of their precious cytosine was gone—and it went faster when their reactions took place in saturated urea. In Shapiro's words, "It is clear that the yield of cytosine would fall to 0 percent if the reaction were extended."

If the central chemical reaction favored by Robertson and Miller was self-defeating, it was also contingent on circumstances that were unlikely. Concentrated urea was needed to prompt their reaction; an outhouse whiff would not do. For this same reason, however, the pre-biotic sea, where concentrates disappear too quickly, was hardly the place to begin—as anyone who has safely relieved himself in a swimming pool might confirm with guilty satisfaction. Aware of this, Robertson and Miller posited a different set of circumstances: in place of the pre-biotic soup, drying lagoons. In a fine polemical passage, their critic Shapiro stipulated what would thereby be required:

> An isolated lagoon or other body of seawater would have to undergo extreme concentration. . . .
> It would further be necessary that the residual liquid be held in an impermeable vessel [in order to prevent cross-reactions].
> The concentration process would have to be interrupted for some decades… to allow the reaction to occur.
> At this point, the reaction would require quenching (perhaps by evaporation to dryness) to prevent loss by deamination.
> At the end, one would have a batch of urea in solid form, containing some cytosine (and urea).

Such a scenario, Shapiro remarked, "cannot be excluded as a rare event on early earth, but it cannot be termed plausible."

LIKE CYTOSINE, SUGAR MUST also make an appearance in Miller Time, and, like cytosine, it too is difficult to synthesize under plausible pre-biotic conditions.

In 1861, the Russian chemist Aleksandr Butlerov created a sugar-like substance from a mixture of formaldehyde and lime. Subsequently refined by a long line of organic chemists, Butlerov's so-called formose reaction has been an inspiration to origins-of-life researchers ever since.

The reaction is today initiated by an alkalizing agent, such as thallium or lead hydroxide. There follows a long induction period, with a number of intermediates bubbling up. The formose reaction is autocatalytic in the sense that it keeps on going: the carbohydrates that it generates serve to prime the reaction in an exponentially growing feedback loop until the initial stock of formaldehyde is exhausted. With the induction over, the formose reaction yields a number of complex sugars.

Nonetheless, it is not sugars in general that are wanted from Miller Time but a particular form of sugar, namely, ribose—and not simply ribose but dextro ribose. Compounds of carbon are naturally right-handed or left-handed, depending on how they polarize light. The ribose in living systems is right-handed, hence the prefix "dextro." But the sugars exiting the formose reaction are racemic, that is, both left- and right-handed, and the yield of usable ribose is negligible.

While nothing has as yet changed the fundamental fact that it is very hard to get the right kind of sugar from any sort of experiment, in 1990 the Swiss chemist Albert Eschenmoser was able to change substantially the way in which the sugars appeared. Reaching with the hand of a master into the formose reaction itself, Eschenmoser altered two molecules by adding a phosphate group to them. This slight change prevented the formation of the alien sugars that cluttered the classical formose reaction. The products, Eschenmoser reported, included among other things a mixture of ribose-2,4,-diphosphate. Although the mixture was racemic, it did contain a molecule close to the ribose needed by living systems. With a few chemical adjustments, Eschenmoser could plausibly claim, the pre-biotic route to the synthesis of sugar would lie open.

It remained for skeptics to observe that Eschenmoser's ribose reactions were critically contingent on Eschenmoser himself, and at two points: the first when he attached phosphate groups to a number of intermediates in the formose reaction, and the second when he removed them.

What had given the original Miller-Urey experiment its power to excite the imagination was the sense that, having set the stage, Miller and Urey exited the theater. By contrast, Eschenmoser remained at center stage, giving directions and in general proving himself indispensable to the whole scene.

Events occurring in Miller Time would thus appear to depend on the large assumption, still unproved, that the early atmosphere was reductive, while two of the era's chemical triumphs, cytosine and sugar, remain for the moment beyond the powers of contemporary pre-biotic chemistry.

IN THE GRAND PROGRESSION by which life arose from inorganic matter, Miller Time has been concluded. It is now 3.8 billion years ago. The chemical precursors to life have been formed. A limpid pool of nucleotides is somewhere in existence. A new era is about to commence.

The historical task assigned to this era is a double one: forming chains of nucleic acids from nucleotides, and discovering among them those capable of reproducing themselves. Without the first, there is no RNA; and without the second, there is no life.

In living systems, polymerization or chain-formation proceeds by means of the cell's invaluable enzymes. But in the grim inhospitable pre-biotic, no enzymes were available. And so chemists have assigned their task to various inorganic catalysts. J. P. Ferris and G. Ertem, for instance, have reported that activated nucleotides bond covalently when embedded on the surface of montmorillonite, a kind of clay. This example, combining technical complexity with general inconclusiveness, may stand for many others.

In any event, polymerization having been concluded—by whatever means—the result was (in the words of Gerald Joyce and Leslie Orgel) "a random ensemble of polynucleotide sequences": long molecules emerging from short ones, like fronds on the surface of a pond. Among these fronds, nature is said to have discovered a self-replicating molecule. But how?

Darwinian evolution is plainly unavailing in this exercise or that era, since Darwinian evolution begins with self-replication, and self-replication is precisely what needs to be explained. But if Darwinian evolution is unavailing, so, too, is chemistry. The fronds comprise "a *random* ensemble of polynucleotide sequences" (emphasis added); but no principle of organic chemistry suggests that aimless encounters among nucleic acids must lead to a chain capable of self-replication.

If chemistry is unavailing and Darwin indisposed, what is left as a mechanism? The evolutionary biologist's finest friend: sheer dumb luck.

Was nature lucky? It depends on the payoff and the odds. The payoff is clear: an ancestral form of RNA capable of replication. Without that payoff, there is no life, and obviously, at some point, the payoff paid off. The question is the odds.

For the moment, no one knows how precisely to compute those odds, if only because within the laboratory, no one has conducted an experiment leading to a self-replicating ribozyme. But the minimum length or "sequence" that is needed for a contemporary ribozyme to undertake what the distinguished geochemist Gustaf Arrhenius calls "demonstrated ligase activity" is known. It is roughly 100 nucleotides.

Whereupon, just as one might expect, things blow up very quickly. As Arrhenius notes, there are 4^{100} or roughly 10^{60} nucleotide sequences that are 100 nucleotides in length. This is an unfathomably large number. It exceeds the number of atoms contained in the universe, as well as the age of the universe in seconds. If the odds in favor of self-replication are 1 in 10^{60}, no betting man would take them, no matter how attractive the payoff, and neither presumably would nature.

"Solace from the tyranny of nucleotide combinatorials," Arrhenius remarks in discussing this very point, "is sought in the feeling that strict sequence specificity may not be required through all the domains of a functional oligmer, thus making a large number of library items eligible for participation in the construction of the ultimate functional entity." Allow me to translate: why assume that self-replicating sequences are apt to be rare just because they are long? They might have been quite common.

They might well have been. And yet all experience is against it. Why should self-replicating RNA molecules have been common 3.6 billion years ago when they are impossible to discern under laboratory conditions today? No one, for that matter, has ever seen a ribozyme capable of *any* form of catalytic action that is not very specific in its sequence and thus unlike even closely related sequences. No one has ever seen a ribozyme able to undertake chemical action without a suite of enzymes in attendance. No one has ever seen anything like it.

The odds, then, are daunting; and when considered realistically, they are even worse than this already alarming account might suggest. The discovery of a single molecule with the power to initiate replication would hardly be sufficient to establish replication. What template would it replicate *against*? We need, in other words, at least two, causing the odds of their joint discovery to increase from 1 in 10^{60} to 1 in 10^{120}. Those two sequences would have been needed in roughly the same place. And at the same time. And organized in such a way as to favor base pairing. And somehow held in place. And buffered against competing reactions. And productive enough so that their duplicates would not at once vanish in the soundless sea.

In contemplating the discovery by chance of two RNA sequences a mere 40 nucleotides in length, Joyce and Orgel concluded that the requisite "library" would require 10^{48} possible sequences. Given the weight of RNA, they observed gloomily, the relevant sample space would exceed the mass of the earth. And this is the same Leslie Orgel, it will be remembered, who observed that "it was almost certain that there once was an RNA world."

To the accumulating agenda of assumptions, then, let us add two more: that without enzymes, nucleotides were somehow formed into chains, and that by means we cannot duplicate in the laboratory, a pre-biotic molecule discovered how to reproduce itself.

A NEW ERA IS now in prospect, one that begins with a self-replicating form of RNA and ends with the system of coded chemistry character-istic of the modern cell. The *modern* cell—meaning one that divides its labors by assigning to the nucleic acids the management of information and to the proteins the execution of chemical activity. It is 3.6 billion years ago.

It is with the advent of this era that distinctively conceptual prob-lems emerge. The gods of chemistry may now be seen receding into the distance. The cell's system of coded chemistry is determined by two dis-crete combinatorial objects: the nucleic acids and the amino acids. These objects are discrete because, just as there are no fractional sentences containing three-and-a-half words, there are no fractional nucleotide sequences containing three-and-a-half nucleotides, or fractional pro-teins containing three-and-a-half amino acids. They are combinatorial because both the nucleic acids and the amino acids are combined by the cell into larger structures.

But if information management and its administration within the modern cell are determined by a discrete combinatorial system, the *work* of the cell is part of a markedly different enterprise. The periodic table notwithstanding, chemical reactions are not combinatorial, and they are not discrete. The chemical bond, as Linus Pauling demonstrated in the 1930s, is based squarely on quantum mechanics. And to the extent that chemistry is explained in terms of physics, it is encompassed not only by "the model for what science should be" but by the system of differential equations that play so conspicuous a role in every one of the great theo-ries of mathematical physics.

What serves to coordinate the cell's two big shots of information management and chemical activity, and so to coordinate two fundamen-tally different structures, is the universal genetic code. To capture the remarkable nature of the facts in play here, it is useful to stress the word *code*.

By itself, a code is familiar enough: an arbitrary mapping or a system of linkages between two discrete combinatorial objects. The Morse code, to take a familiar example, coordinates dashes and dots with letters of the alphabet. To note that codes are arbitrary is to note the distinction between a code and a purely physical connection between two objects. To note that codes embody mappings is to embed the concept of a code in mathematical language. To note that codes reflect a linkage of some sort is to return the concept of a code to its human uses.

In every normal circumstance, the linkage comes first and represents a human achievement, something arising from a point beyond the coding system. (The coordination of dot-dot-dot-dash-dash-dash-dot-dot-dot with the distress signal S-O-S is again a familiar example.) Just as no word explains its own meaning, no code establishes its own nature.

The conceptual question now follows. Can the origins of a system of coded chemistry be explained in a way that makes no appeal whatsoever to the kinds of facts that we otherwise invoke to explain codes and languages, systems of communication, the impress of ordinary words on the world of matter?

In this regard, it is worth recalling that, as Hubert Yockey observes in *Information Theory, Evolution, and the Origin of Life* (2005), "there is no trace in physics or chemistry of the control of chemical reactions by a sequence of any sort or of a code between sequences."

WRITING IN THE 2001 issue of the journal *RNA*, the microbiologist Carl Woese referred ominously to the "dark side of molecular biology." DNA replication, Woese wrote, is the extraordinarily elegant expression of the structural properties of a single molecule: zip down, divide, zip up. The transcription into RNA follows suit: copy and conserve. In each of these two cases, structure leads to function. But where is the coordinating link between the chemical structure of DNA and the third step, namely, translation? When it comes to translation, the apparatus is baroque: it is incredibly elaborate, and it does not reflect the structure of any molecule.

These reflections prompted Woese to a somber conclusion: if "the nucleic acids cannot in any way recognize the amino acids," then there is no "fundamental *physical* principle" at work in translation (emphasis added).

But Woese's diagnosis of disorder is far too partial; the symptoms he regards as singular are in fact widespread. What holds for translation holds as well for replication and transcription. The nucleic acids cannot *directly* recognize the amino acids (and vice versa), but they cannot directly replicate or transcribe themselves, either. Both replication and translation are enzymatically driven, and without those enzymes, a molecule of DNA or RNA would do nothing whatsoever. Contrary to what Woese imagines, no fundamental physical principles appear directly at work *anywhere* in the modern cell.

The most difficult and challenging problem associated with the origins of life is now in view. One half of the modern system of coded chemistry—the genetic code and the sequences it conveys—is, from a chemical perspective, arbitrary. The other half of the system of coded chemistry—the activity of the proteins—is, from a chemical perspective, necessary. In life, the two halves are coordinated. The problem follows: how did *that*—the whole system—get here?

THE PREVAILING OPINION AMONG molecular biologists is that questions about molecular-biological systems can only be answered by molecular-biological *experiments*. The distinguished molecular biologist Horoaki Suga has recently demonstrated the strengths and the limitations of the experimental method when confronted by difficult conceptual questions like the one I have just posed.

The goal of Suga's experiment was to show that a set of RNA catalysts (or ribozymes) *could* well have played the role now played in the modern cell by the protein family of aminoacyl synthetases. Until his work, Suga reports, there had been no convincing demonstration that a ribozyme was able to perform the double function of a synthetase—that is, recognizing both a form of transfer RNA and an amino acid. But in Suga's laboratory, just such a molecule made a now-celebrated appearance. With an amino acid attached to its tail, the ribozyme managed to cleave itself and, like a snake, affix its amino-acid cargo onto its head. What is more, it could conduct this exercise backward, shifting the amino acid from its head to its tail again. The chemical reactions involved acylation: precisely the reactions undertaken by synthetases in the modern cell.

Horoaki Suga's experiment was both interesting and ingenious, prompting a reaction perhaps best expressed as, "Well, would you look at that!" It has altered the terms of debate by placing a number of new facts on the table. And yet, as so often happens in experimental prebiotic chemistry, it is by no means clear what interpretation the facts will sustain.

Do Suga's results really establish the existence of a primitive form of coded chemistry? Although unexpected in context, the coordination he achieved between an amino acid and a form of transfer RNA was never at issue in principle. The question is whether what was accomplished in establishing a chemical connection between these two molecules was anything like establishing the existence of a *code*. If so, then organic chemistry itself could properly be described as the study of codes, thereby erasing the meaning of a code as an arbitrary mapping between discrete combinatorial objects.

Suga, in summarizing the results of his research, captures rhetorically the inconclusiveness of his achievement. "Our demonstration indicates," he writes, "that catalytic precursor tRNA's *could have provided* the foundation of the genetic coding system." But if the association at issue is not a code, however primitive, it could no more be the "foundation" of a code than a feather could be the foundation of a building. And if it is the foundation of a code, then what has been accomplished has been accomplished by the wrong agent.

In Suga's experiment, there was no sign that the execution of chemical routines fell under the control of a molecular administration, and no sign, either, that the missing molecular administration had anything to do with executive chemical routines. The missing molecular administrator was, in fact, Suga himself, as his own account reveals. The relevant features of the experiment, he writes, "allow[ed] *us* to select active RNA molecules with selectivity toward a *desired* amino acid" (emphasis added). Thereafter, it was Suga and his collaborators who "applied *stringent* conditions" to the experiment, undertook "*selective amplification* of the self-modifying RNA molecules," and "*screened*" vigorously for "self-aminoacylation activity" (emphasis added throughout).

IF NOTHING ELSE, THE advent of a system of coded chemistry satisfied the most urgent of imperatives: it was needed and it was found. It was needed because once a system of chemical reactions reaches a certain threshold of complexity, nothing less than a system of coded chemistry can possibly master the ensuing chaos. It was found because, after all, we are here.

Precisely these circumstances have persuaded many molecular biologists that the explanation for the emergence of a system of coded chemistry must in the end lie with Darwin's theory of evolution. As one critic has observed in commenting on Suga's experiments, "If a certain result can be achieved by direction in a laboratory by a Suga, surely it can also be achieved by chance in a vast universe."

A self-replicating ribozyme meets the first condition required for Darwinian evolution to gain purchase. It is by definition capable of replication. And it meets the second condition as well, for, by means of mistakes in replication, it introduces the possibility of variety into the biological world. On the assumption that subsequent changes to the system follow a law of increasing marginal utility, one can then envisage the eventual emergence of a system of coded chemistry—a system that can be explained in terms of "the model for what science should be."

It was no doubt out of considerations like these that, in coming up against what he called the "dark side of molecular biology," Carl Woese was concerned to urge upon the biological community the benefits of "an all-out Darwinian perspective." But the difficulty with "an all-out Darwinian perspective" is that it entails an all-out Darwinian impediment: notably, the assignment of a degree of foresight to a Darwinian process that the process could not possibly possess.

The hypothesis of an RNA world trades brilliantly on the idea that a divided modern system had its roots in some form of molecular symmetry that was then broken by the contingencies of life. At some point in the transition to the modern system, an ancestral form of RNA must have assigned some of its catalytic properties to an emerging family of proteins. This would have taken place at a given historical moment; it is not an artifact of the imagination. Similarly, at some point in the transition to a modern system, an ancestral form of RNA must have acquired the ability to code for the catalytic powers it was discarding. And this, too, must have taken place at a particular historical moment.

The question, of course, is which of the two steps came first. Without life acquiring some degree of foresight, neither step can be plausibly fixed in place by means of any schedule of selective advantages. How could an ancestral form of RNA have acquired the ability to code for various amino acids before coding was useful? But then again, why should "ribozymes in an RNA world," as the molecular biologists Paul Schimmel and Shana O. Kelley ask, "have expedited their own obsolescence?"

Could the two steps have taken place simultaneously? If so, there would appear to be very little difference between a Darwinian explanation and the frank admission that a miracle was at work. If no miracles are at work, we are returned to the place from which we started, with the chicken-and-egg pattern that is visible when life is traced backward now appearing when it is traced forward.

It is thus unsurprising that writings embodying Woese's "all-out Darwinian perspective" are dominated by references to a number of unspecified but mysteriously potent forces and obscure conditional circumstances. I quote without attribution because the citations are almost generic (emphasis added throughout):

+ The aminoacylation of RNA initially *must* have provided some selective advantage.

+ The products of this reaction *must* have conferred some selective advantage.

+ However, the development of a crude mechanism for controlling the diversity of possible peptides *would* have been advantageous.

+ [P]rogressive refinement of that mechanism *would* have provided *further* selective advantage.

And so forth—ending, one imagines, in reduction to the all-purpose imperative of Darwinian theory, which is simply that what was must have been.

At the conclusion of a long essay, it is customary to summarize what has been learned. In the present case, I suspect it would be more prudent to recall how much has been *assumed*:

First, that the pre-biotic atmosphere was chemically reductive; second, that nature found a way to synthesize cytosine; third, that nature also found a way to synthesize ribose; fourth, that nature found the means to assemble nucleotides into polynucleotides; fifth, that nature discovered a self-replicating molecule; and sixth, that having done all that, nature promoted a self-replicating molecule into a full system of coded chemistry.

These assumptions are not only vexing but progressively so, ending in a serious impediment to thought. That, indeed, may be why a number of biologists have lately reported a weakening of their commitment to the RNA world altogether, and a desire to look elsewhere for an explanation of the emergence of life on earth. "It's part of a quiet paradigm revolution going on in biology," the biophysicist Harold Morowitz put it in an interview in *New Scientist*, "in which the radical randomness of Darwinism is being replaced by a much more scientific law-regulated emergence of life."

Morowitz is not a man inclined to wait for the details to accumulate before reorganizing the vista of modern biology. In a series of articles, he has argued for a global vision based on the biochemistry of living systems rather than on their molecular biology or on Darwinian adaptations. His vision treats the living system as more fundamental than its particular species, claiming to represent the "universal and deterministic features of *any* system of chemical interactions based on a water-covered but rocky planet such as ours."

This view of things—metabolism first, as it is often called—is not only intriguing in itself but is enhanced by a firm commitment to chemistry and to "the model for what science should be." It has been argued with great vigor by Morowitz and others. It represents an alternative to the RNA world. It is a work in progress, and it may well be right. Nonetheless, it suffers from one outstanding defect. There is as yet no evidence that it is true.

It is now more than 175 years since Friedrich Wöhler announced the synthesis of urea. It would be the height of folly to doubt that our understanding of life's origins has been immeasurably improved. But whether it has been immeasurably improved in a way that vigorously confirms the daring idea that living systems are chemical in their origin and so physical in their nature—that is another question entirely.

In "On the Origins of the Mind," I tried to show that much can be learned by studying the issue from a computational perspective. Analogously, in contemplating the origins of life, much—in fact, more—can be learned by studying the issue from the perspective of coded chemistry. In both cases, however, what seems to lie beyond the reach of "the model for what science should be" is any success beyond the local. All questions about the *global* origins of these strange and baffling systems seem to demand answers that the model itself cannot by its nature provide.

It goes without saying that this is a tentative judgment, perhaps only a hunch. But let us suppose that questions about the origins of the mind and the origins of life do lie beyond the grasp of "the model for what science should be." In that case, we must either content ourselves with its limitations or revise the model. If a revision also lies beyond our powers, then we may well have to say that the mind and life have appeared in the universe for no very good reason that we can discern.

Worse things have happened. In the end, these are matters that can only be resolved in the way that all such questions are resolved. We must wait and see.

Reprinted from *Commentary*, February 2006, by permission;
copyright © 2006 by Commentary, Inc.

2007

INSIDE THE
MATHEMATICAL MIND

WHEN PHYSICISTS WRITE BOOKS FOR THE GENERAL PUBLIC, THEY write about black holes, dark matter, or strings that wriggle like mad. The universe is their subject. Mathematicians write about mathematics and what it all means. Their subject is their subject.

The mathematician David Ruelle is well known for his work on nonlinear dynamics and turbulence, and his new book, *The Mathematician's Brain* (Princeton University Press, 2007), is a book about mathematics and what it all means.

If the entomologist studies bugs, and the linguist languages, just what is it the mathematician studies? Sets, numbers, equations—that much is clear. Thereafter, everything solid dissolves into thin air. What is a number? Or a set? Or a shape, for that matter?

If mathematicians cannot say what mathematics is about, neither can they say why its conclusions are certain. "The instability of human knowledge is one of our few certainties," the journalist Janet Malcolm remarked recently. Yet mathematicians believe that their conclusions are forever.

In answering his own questions, Mr. Ruelle advances two general theses. The first is that "the structure of human science is largely dependent on the special nature and organization of the human brain." The second is that "the scientific method is a different thing in different disciplines."

The first claim is empty. We do not know how the brain generates its thoughts. If the brain is simply a physical organ, there is no reason to suppose that it has access to any form of certainty beyond the calculations needed to climb the greasy pole of life. If the brain does have such access, then the structure of human science cannot be largely dependent on its physical organization.

It is quite true that there is no such thing as the scientific method. There are many methods. What counts in mathematics is proof, a systematic way of deriving conclusions from assumptions. Under ordinary circumstances, a mathematical proof is written in the mathematician's vernacular. Precision is demanded—physicists need not apply—but not logical obsession. Yet mathematicians have in the twentieth century learned just how an informal proof may be expressed as the driest of dry structures, a system in which meaning has been stripped from symbols and their manipulation governed by precisely stated rules. The result resembles a text written in an alien language, or a program written in assembly code. No mathematician would dream of presenting his proofs in this way. They would take too long to write and once written, they would never be read.

While Mr. Ruelle is an excellent mathematician, he is no logician. And it shows. "We can in principle," Mr. Ruelle writes, "give a completely formalized presentation of mathematics." This is quite true. It hardly follows, as Mr. Ruelle concludes, that "mathematics is the unique human endeavor where the use of a human language is, in principle, not necessary." If proofs are stripped to their syntactic shell, they have no interest. The human language that Mr. Ruelle is eager to dismiss reasserts itself the moment the mathematician asks about the meaning of the proof.

If *The Mathematician's Brain* does not answer the questions it poses, this is because no other book has answered these questions either. The book's value lies in Mr. Ruelle's description of the curious inner life of mathematicians. Their subject is very difficult. It requires unusual gifts. Physicists may disguise the triviality of their results by bustling about in large research groups. Mathematicians work alone. They are professionally naked.

As a result, many mathematicians have unstable personalities. Alexandre Grothendieck is an extreme example. His is hardly a household name, especially in the English-speaking world. Yet for the 15 years between 1958 and 1973, Mr. Grothendieck dominated the field of algebraic geometry and ruled like a prince over a court comprising some of the most talented mathematicians in the world. His immense treatise on algebraic geometry is, as Mr. Ruelle observes, the last great mathematical *oeuvre* written in the French language.

To algebraic geometry, Mr. Grothendieck brought an entirely new level of power and abstraction, so much so that his colleague René Thom—a Field medalist and a great mathematician—acknowledged that he left pure mathematics because he was oppressed by Mr. Grothendieck's "crushing technical superiority." His technique was only a part of his genius. Mr. Grothendieck was a great mathematical visionary. Like mystics searching for the face of God, he was passionately concerned to see the unity of form behind various mathematical experiences. He did not simply solve isolated problems but, as Mr. Ruelle writes, enveloped them "in a rising tide of very general theories."

If Mr. Grothendieck was a magnificent mathematician, he was also a political simpleton. Departing the Institut des Hautes Études Scientifiques after a pointless dispute about the Institut's military funding, he found it impossible to obtain a position in France commensurate with his stature. For a time, he taught undergraduate mathematics at a provincial French university. Mr. Ruelle considers Mr. Grothendieck's internal exile a great injustice. It was nothing of the sort. Just as he was a political simpleton, Mr. Grothendieck was a personal pest, sending endless letters to leading mathematicians accusing them of insufficiently appreciating his genius or stealing his ideas.

Mr. Grothendieck is now said to be living in a shepherd's hut in the Pyrénées, where he lives on a vegetarian diet and is sustained by meditation. During the 1980s, he wrote an immensely long autobiography entitled *Récoltes et Semailles* (Harvest and Sowings). As Mr. Grothendieck devoted a large part of the book to the denunciation of his colleagues, no French publisher accepted it for publication. Sections have appeared on the Internet. They are often interesting in the extent to which they express the nature of mathematical longing.

Mr. Ruelle is far too discreet to tell Mr. Grothendieck's story well, but it is important that he has told it at all. Of all the arts, mathematics expresses most completely the peculiar human desire for perfect understanding. It is a desire that when fulfilled evokes a form of gratitude, but the terms of the mathematician's contract are severe, and if gratitude is the mathematician's reward, a certain personal deformation is the price that all too often he must pay.

2008

CONNECTING
HITLER AND DARWIN

ONE MAN—CHARLES DARWIN—SAYS: "IN THE STRUGGLE FOR survival, the fittest win out at the expense of their rivals...."

Another man—Adolf Hitler—says: Let us kill all the Jews of Europe.

Is there a connection?

Yes obviously is the answer of the historical record and common sense.

Published in 1859, Darwin's *On the Origin of Species* said nothing of substance about the origin of species. Or anything else, for that matter. It nonetheless persuaded scientists in England, Germany and the United States that human beings were accidents of creation. Where Darwin had seen species struggling for survival, German physicians, biologists, and professors of hygiene saw races.

They drew the obvious conclusion, the one that Darwin had already drawn. In the struggle for survival, the fittest win out at the expense of their rivals. German scientists took the word "expense" to mean what it meant: The annihilation of less fit races.

The point is made with abysmal clarity in the 2008 documentary *Expelled*. Visiting the site at which those judged defective were killed—a hospital, of course—the narrator, Ben Stein, asks the curator what most influenced the doctors doing the killing.

"Darwinism," she replies wanly.

It is perfectly true that prominent Nazis were hardly systematic thinkers. They said whatever came into their heads and since their heads were empty, ideas tended to ricochet. Heinrich Himmler proclaimed himself offended by the idea that he might been descended from the apes.

If Himmler was offended, the apes were appalled.

Nonetheless, even stupid men reach their conclusions because they have been influenced in certain ways. At Hitler's death in May of 1945, the point was clear enough to the editorial writers of the *New York Times*. "Long before he had dreamed of achieving power," they wrote, "[Hitler] had developed the principles that nations were destined to hate, oppose and destroy one another; [and] that the law of history was the struggle for survival between peoples..."

Where, one might ask, had Hitler heard those ideas before? We may strike the Gospels from possible answers to this question. Nonetheless, the thesis that there is a connection between Darwin and Hitler is widely considered a profanation. A professor of theology at Iowa State University, Hector Avalos is persuaded that Martin Luther, of all people, must be considered Adolf Hitler's spiritual advisor. Luther, after all, liked Jews as little as Hitler did, and both men suffered, apparently, from hemorrhoids. Having matured his opinion by means of an indifference to the facts, Roger Friedman, writing on the Fox News website, considers the connection between Darwin and Hitler and in an access of analytical insight thinks only to remark, "Urgggh."

The view that we may consider the sources of Nazi ideology in every context except those most relevant to its formation is rich, fruity, stupid and preposterous. It is for this reason repeated with solemn incomprehension at the website Expelled Exposed: "Anti-Semitic violence against Jews," the authors write with a pleased sense of discovery, "can be traced as far back as the Middle Ages, at least seven centuries before Darwin."

LET ME IMPART A secret. It can be traced even further. "Oh that mine head were waters and mine eyes a fountain of tears," runs the lamentation in Jeremiah 9:1, "that I might weep day and night for the slain daughters of my people."

And yet if anti-Semitism has been the white noise of European history, to assign it causal powers over the Holocaust is simply to ignore very specific ideas that emerged in the nineteenth century, and that at once seized the imagination of scientists throughout the world.

What is often called social Darwinism was a malignant force in Germany, England and the United States from the moment that social thinkers forged the obvious connection between what Darwin said and what his ideas implied. Justifying involuntary sterilization, Justice Oliver Wendell Holmes argued that "three generations of imbeciles is enough." He was not, it is understood, appealing to Lutheran ideas. Germany reached a moral abyss before any other state quite understood that the abyss was there to be reached because Germans have always had a congenital weakness for abysses and seem unwilling, when one is in sight, to avoid toppling into it.

These historical connections are so plain that from time to time, those most committed to Darwin's theory of evolution are moved to acknowledge them. Having dismissed a connection between Darwin and Hitler with florid indignation, the authors of Expelled Exposed at once proceed to acknowledge it: "The Nazis appropriated language and concepts from evolution," they write, "as well as from genetics, medicine (especially the germ theory of disease), and anthropology as propaganda tools to promote their perverted ideology of 'racial purity.'"

Just so.

Would he care to live in a society shaped by Darwinian principles? The question was asked of Richard Dawkins.

Not at all, he at once responded.

And why not?

Because the result would be fascism.

In this, Richard Dawkins was entirely correct; and it is entirely to his credit that he said so.

THE SCIENTIFIC
EMBRACE OF ATHEISM

AT SOMETIME AFTER THE RUSSIAN COSMONAUT YURI GAGARIN
first entered space, stories began to circulate that he had been given
secret instructions by the Politburo. Have a look around, they told him.
Suitably instructed, Gagarin looked around. When he returned without
having seen the face of God, satisfaction in high circles was considerable.

The commissars having vacated the scene, it is the scientific community that has acquired their authority. Richard Dawkins, Daniel Dennett, Steven Weinberg, Victor Stenger, Sam Harris, and most recently
the mathematician John Paulos, have had a look around: They haven't
seen a thing. No one could have seen less.

It is curious that so many scientists should have recently embraced
atheism. The great physical scientists—Copernicus, Kepler, Galileo,
Newton, Clerk Maxwell, Albert Einstein—were either men of religious
commitment or religious sensibility.

The distinguished physicist Steven Weinberg has acknowledged
that this is what the great scientists believed: But we know better, he has
insisted, because we know more.

This prompts the obvious question: Just what have scientists learned
that might persuade the rest of us that they know better? It is not, presumably, the chemistry of boron salts that has done the heavy lifting.

There is quantum cosmology, I suppose, a discipline in which the
mysteries of quantum mechanics are devoted to the question of how the
universe arose or whether it arose at all. This is the subject made popular
in Stephen Hawking's *A Brief History of Time*. It is an undertaking radi-

ant in its incoherence. Given the account of creation offered in Genesis and the account offered in *A Brief History of Time*, I know of no sane man who would hesitate between the two.

And there is Darwin's theory of evolution. It has been Darwin, Richard Dawkins remarked, that has made it possible to be an intellectually fulfilled atheist.

A much better case might be made in the other direction. It is atheism that makes it possible for a man to be an intellectually fulfilled Darwinist. In the documentary *Expelled*, one of those curious exercises in which some scientists, at least, say what they really think, Ben Stein interviews a number of Darwinian biologists eager to evade the evidence whenever possible or to ignore it when not. Rich in self-satisfaction, Dawkins appears at the film's end.

How did life on earth arise?

The question, Dawkins acknowledges, is very difficult.

Perhaps the seeds of life were sent here from outer space?

It could well be.

Or by a vastly superior intelligence?

Well, yes.

Questions and their answers follow one another, but in the end Stein says nothing. There is no absurdity Dawkins is not prepared to embrace so long as he can avoid a transcendental inference.

Beyond quantum cosmology and Darwinian biology—the halt and the lame—there is the solemn metaphysical aura of science itself. It is precisely the aura to which so many scientists reverently appeal. The philosopher John Searle has seen the aura. The "universe," he has written, "consists of matter, and systems defined by causal relations."

Does it indeed? If so, then God must be nothing more than another material object, a class that includes stars, starlets and solitons. If not, what reason do we have to suppose that God might not exist?

We have no reason whatsoever. If neither the sciences nor its aura have demonstrated any conclusion of interest about the existence of God, why then is atheism valued among scientists?

It takes no very refined analytic effort to determine why Soviet Commissars should have regarded themselves as atheists. They were unwilling to countenance a power higher than their own. Who knows what mischief Soviet citizens might have conceived had they imagined that the Politburo was not, after all, infallible?

By the same token, it requires no very great analytic effort to understand why the scientific community should find atheism so attractive a doctrine. At a time when otherwise sober individuals are inclined to believe that too much of science is too much like a racket, it is only sensible for scientists to suggest aggressively that no power exceeds their own.

2009

THE STATE OF THE MATTER

"Nature allows us to see the lion only by its tail."

—ALBERT EINSTEIN

IN THE SUMMER OF 1996, THE PHYSICIST ALAN SOKAL PUBLISHED A paper entitled "Transgressing the Boundaries: Toward a Transformative Hermeneutics of Quantum Gravity," in the journal *Social Text*. The essay was a hoax, one designed to ridicule post-modern thought. Satisfaction among physicists was considerable. Post-modern thought never quite recovered. Only Stanley Fish remained faithful, a one man multitude. Yet curiously enough, physicists have since then grown no easier in their minds. A fear of contamination remains widespread. The department of physics at Harvard is now often referred to as the department of *post-modern* physics by its critics.

How on earth did *that* happen?

It happened because senior members of the physics establishment went over to the other side at precisely the moment the other side went home. After warmly congratulating Sokal in the *New York Review of Books*, Steven Weinberg determined shortly thereafter that "we may be at a new turning point, a radical change in what we accept as a legitimate foundation for a physical theory." The changes that Weinberg had in mind were, indeed, very radical, suggesting as they did precisely the doctrines that previously he was prepared to reject. Addressing a conference on string theory held in 2005, Weinberg indicated that he was prepared, after all, to welcome his new insect overlords.

An informal poll indicated that the audience of physicists rejected his views by a margin of four to one.

"We win some and we lose some," Weinberg remarked equably.

QUESTIONS ABOUT THE MATERIAL world have a particular urgency in twenty-first century thought. It is an appeal to matter that acts as the anchor of a grand narrative, one that explains the origins of mind in terms of the biology and evolution of living systems, and the origins of living systems in terms of the physics of material objects. There is no counter-narrative and certainly none that has captured the allegiance of the scientific community. I have called this narrative a *narrative* because it is by no means obvious that it is true; and I have called it *grand* because like the Colossus at Rhodes, it is big without being awesome. Psychologists, biologists and biochemists must now give way. The mathematical physicists are in charge. A *ta dum* or even a *te Deum* is appropriate. The material world is their business, and if we could not trust them to tell us what's up and so what matters, whom could we trust?

Questions about matter and its origins are in their largest sense metaphysical; and when it comes to metaphysics, a great many physicists accept what Alan Sokal has called "materialist monism" as a default position of their faith. It is a position severe in its demands. The monist dismisses anything beyond matter. The materialist embraces nothing but matter. Everything, it would seem, must end with matter. What else is there? Questions will inevitably remain about the origins of matter, but they are not questions it is possible rationally to answer. To explain the origins of matter in terms that are not material is to embrace a contradiction, and to explain the origins of matter in material terms is to invite a regress. The matter may be left to theologians and quantum cosmologists. Miracles are their business. The rest of us must accept that in going downward in our explanations we must at some place stop.

If so, what cause for complaint? Explanations *do* come to an end.

The question remains, of course, whether they have come to an end in the right place?

Or whether they have come to an end at all?

WHATEVER THE PROSPECTS FOR a revolution in our time, it is worthwhile to recall that quantum mechanics remains the last revolution to have left blood on the streets. When in 1930, Paul Dirac published his treatise, *Quantum Mechanics*, it was commonly assumed that while the new theory was to no one's liking, it was to everyone's benefit. Benefits have since accrued steadily. Quantum mechanics is still no better liked.

The era that came to an end in 1930 began 25 years earlier with the submission of Albert Einstein's paper "On the Electrodynamics of Moving Bodies" to the influential German periodical *Zeitschrift für Physik*. It was read by its editor, Max Planck, who at once understood its importance and assured its publication. It is in this extraordinary essay that Einstein presented his special theory of relativity. Two assumptions governed the flow of his thoughts. The first reprises a claim made by Galileo in the seventeenth century. Imagining himself aboard a smoothly sailing ship, Galileo conjectured that under ideal conditions no shipboard experiment could reveal that the ship was in motion. The principle is known as Galilean relativity.

In special relativity, Galilean relativity undergoes a far-reaching promotion. The laws of "electrodynamics and optics," Einstein declared, must be "valid for all frames of reference for which the equations of mechanics hold good." Since the equations of mechanics already hold in every frame of reference by Galilean relativity, the laws of physics must all be invariant with respect to uniform motion. As far as they are concerned, the ship can be at rest or sailing placidly. It is all the same to them. Or to us.

Einstein's second postulate affirms "that light is always propagated in empty space with a definite velocity c which is independent of the state of motion of the emitting body." An observer chasing a beam of light at half its speed and an observer at rest will coincide in their judgment about the speed of light, agreeing not only that it is going very fast but how fast it is going.

Einstein's second postulate seemed radically at odds with common sense when it was first advanced. It remains radically at odds with common sense today. Velocities had since the time of Newton obeyed a simple rule: They added up. Two automobiles traveling at fifty miles an hour approach one another at a closing speed of one hundred miles an hour. This rule lapses in special relativity. Two beams of light approaching one another from opposite sides of the galaxy close at the same speed. This postulate has the most dramatic consequences. The speed of light is, after all, the ratio of the distance light travels divided by the time it has taken to travel that distance. In order for the speed of light to be the same to all observers, they must revise their estimates of how fast things are going and thus how far they have gone in going fast. Before Einstein, no one imagined that space and time were revisable at all.

The postulates of special relativity are strange in any number of ways, but they are most obviously strange in the way they divide their reference. Einstein's second postulate makes a claim about the material world. Einstein's first postulate places a condition on physical theories and their laws. These are human artifacts and like epic poems, dry point etchings, or contracts at law, they are things made up. The goal of the physical sciences, Einstein wrote later in life, is a description of the "external world independent of the perceiving subject." How *any* description of the external world could be independent of the perceiving subject, he did not say. This view, and the first postulate of special relativity are not necessarily in contradiction. They are certainly in conflict. Einstein's theory of special relativity places the physicist and his instruments at the very center of the material world. Once he is there, as one might expect, there is no getting rid of him.

It is this that is strange.

ISAAC NEWTON COMPLETED THE *Principia Mathematica* in 1685, and for 220 years thereafter material objects throughout the cosmos attracted one another with a force that is proportional to their mass and inversely proportional to the square of the distance between them. It is a force that acts at a distance through the intervening modality of nothing whatsoever, and it acts everywhere at once.

Having established special relativity to his own satisfaction in 1905, Einstein quickly came to understand that the second postulate of his theory and Newton's theory of gravitational attraction could not both be true. If nothing can travel faster than the speed of light, then neither can the force of gravity. The general theory of relativity followed as Einstein's resolution of this conflict. It was an undertaking that as Einstein later recounted required "years of anxious searching in the dark."

For all that, the theory had its origins in two remarkably simple thought-experiments.

In the first, Einstein considered an object falling freely. Although plunging toward the earth, an observer in free-fall does not experience the force of gravity. His usual heavy dragging sense of being weighted down vanishes. Objects that he may have dropped linger at his side. Illusions about weightlessness come to an end only after he ceases to fall freely after an encounter with the surface of the earth.

In the second, Einstein imagined an elevator being steadily accelerated upward in space. Standing within, an observer might well conclude that even though the earth is far away, he has been reacquainted with all its old familiar properties. He feels his weight; his feet remain rooted to the ground; objects fall downward as they should, and a beam of light passing through the elevator crosses the intervening space in a parabolic arc.

Einstein's first thought experiment persuaded him that objects in free fall do not experience the force of gravity; his second, that gravitation and acceleration are locally the same.

There remained the question why objects in free fall are *falling?*

Einstein's essay on special relativity contains no mathematics beyond the level of sophisticated finger counting. Within two years of its publication, the mathematician Herman Minkowski undertook its rehabilitation by describing the structure in which the postulates of special relativity are satisfied. Four dimensions are required to identify a point in this universe: Three of space, and one of time. The geometry that results is not difficult, but it is strange. In tracing a circle with a stiffened index finger, it is natural to think that the circle exhausts the shape. With four dimensions at work, the circle gives way to a spiral as time is allowed to separate the circle into stages. "Henceforth," Minkowski affirmed, "space by itself, and time by itself, are doomed to fade away into mere shadows, and only a kind of union of the two will preserve an independent reality." This is a conclusion, curiously enough, that Edgar Allen Poe reached some years before in his cosmological treatise, *Eureka*. If either space *or* time have recently faded away into mere shadows, the news has been withheld.

A Minkowski universe—a Minkowski *manifold*, as it is known in trade—is flat in the sense that objects following their natural or inertial path travel in straight lines. This is an assumption of Newtonian mechanics. Force is required to change either the direction of a moving object or its velocity. Now a straight line, as geometers say, is the shortest distance between two points. But a straight line may also be defined as the line traversed by a traveler who does not allow his nose to move to the left or to the right. Over short distances, the judgment of the man of mathematics and the judgment of the man of action coincide. Short is short: Straight is straight.

But in some settings, while short remains short, straight does not necessarily remain straight. The shortest distance between Paris and New York is a great *circle*. What counts as a straight line depends on the space in which it is embedded and the way that space is described.

After allowing himself a brief moment in which he observed with evident satisfaction that now that the mathematicians had gotten their hands on his theory, *he* could no longer grasp it, Einstein came to understand that by changing the geometry of space and time, he might displace gravity as a force of nature and re-describe it as a contingency of curvature. The mathematical theories that he required had been discovered by the German mathematician Bernhardt Riemann more than

40 years before. The details are by no means simple. Einstein struggled to master them. By 1913, he felt ready to confide the details of his new theory to Max Planck, who unhesitatingly affirmed that it could not be true and that it would not be believed.

Within general relativity, the distinction between what is straight and what is bent persists, as it so often does. It is now a distinction contingent on the presence of matter. Light crosses an empty universe in a line that is both short and straight. The introduction of matter—a sun, a starlet, a solar system—changes the geometry of space and time. Short stays short: Straight gets bent. A beam of light traveling past the sun describes a curve in space because it is traveling through a space that curves.

This extraordinary assumption allowed Einstein to eliminate the distinction between a falling object and an object following its natural inertial path. Both are doing what comes naturally. An observer falling freely—the earth far below, a failed parachute, a forthcoming plop—follows the shortest path between two points in space and time. If he seems to be accelerating, this is only because he is following a curved path. To say that the path is curved, Einstein argued, is to assign to the nature of space and time all of the properties of gravity that Newton assigned to a mysterious universal force.

This is one half of the conceptual structure that in 1915 Einstein introduced into physical thought. A fused form of space and time determines how material objects move. The other half is reciprocal to the first. Material objects determine how space and time are curved. Matter has an effect on the space and time in which it is embedded, bunching it up here, squeezing it there, and in general compressing it whenever it gets the chance. The more matter, the greater the effect. That matter has this power to command space and time is apparently one of its intrinsic properties. A deeper understanding is not possible.

The field equation that Einstein introduced in 1915 is a majestic identity in which curvature on the one side, and mass on the other, are placed in the balance and found equal.

In the early years of the nineteenth century, Thomas Young demonstrated that light behaves like a wave. After shining a beam of light through two slits, he observed interference patterns forming on a screen placed behind them. Wave crests met wave crests to form bigger crests; wave troughs met wave troughs to form deeper troughs; and when crests and troughs were not meeting companionably, they interfered with one another in order to extinguish themselves.

What could be simpler? Light is a wave.

Ah, but on the other hand, Einstein demonstrated in 1905 that light is comprised of particles. Send a beam of light toward a metal surface, and electrons pop out. Plainly they pop out because they have been knocked off. To accommodate both popping out and knocking off, Einstein found it necessary to think of light as if it were composed of discrete packets of energy or photons.

What could be simpler? Light is a particle.

It is not entirely clear how in the matter of *Young v. Einstein*, both men could have been right.

By the third decade of the twentieth century, it had become plain that light is both a wave and a particle, and what is more, it is a wave and a particle on the level of the individual photon.

Against every reasonable expectation, *Young v. Einstein* turned out a draw.

Quantum mechanics was created in the third decade of the twentieth century by a consortium of physicists: Niels Bohr, Werner Heisenberg, Erwin Schrödinger, Max Born.

There are two equivalent ways in which to formulate the mathematical details, one due to Heisenberg, the other to Schrödinger. Having disappeared into the Austrian Alps with another man's wife, Schrödinger achieved his insights during what Hermann Weyl described as a "late-in-life erotic outburst." He has been an inspiration to physicists ever since, and if not to physicists, then certainly to me.

A quantum particle—an electron or a photon, say—is here and somewhat later, it is there. The old here and there, Schrödinger's equation specifies in terms of the properties of a wave. It is here where the wave mounts and there where it dips. The introduction of this metaphor seems at first to suggest how photons might interfere with themselves. They do so as waves. Passing through two slits, their wave peaks at the left and peaks as well at the right.

It is easy to persuade oneself that something familiar is at work. If a particle is a wave, then this is so because it is part of a team, as when particles of mud form a wave during a landslide. Whatever particles of mud may be doing, photons are not doing it. The idea that in a photon stream, photons might be interfering with one another, Dirac rebutted by observing that the result would violate the conservation of energy. The inescapable conclusion, as Dirac noted, is that "each photon interferes *only* with itself." And, indeed, experiments indicate that even when photons are sent toward a screen *one by one*, interference patterns form nonetheless.

The formalism of quantum mechanics ratifies this conclusion by committing the physicist to a form of legerdemain that has to this day resisted all attempts at explication. When physicists talk of quantum mechanical states, they mean to designate the position, momentum, or even the spin of quantum particles. Quantum theory enforces a remarkable principle of superposition on quantum states. If being here *or* being there are two separate quantum states, then so is the combined state of being *here and there*. The two original states are superimposed. They form a packet. They exist together.

It is the wave packet as a whole that is assigned to quantum systems; and it is the wave packet as a whole that enjoys all of what is left of reality when it comes to such systems. Schrödinger's equation describes their undulations.

IT WAS MAX BORN who in 1926 provided the standard interpretation of the quantum mechanical formalism. The details are complex, but in a rough and ready way, Born suggested that waves in quantum mechanics must be understood in terms of the probabilities they reveal. The amplitude of a wave is a sign that quite likely there is a particle there, and the distance between its peaks, a sign that quite likely the particle is traveling with a particular momentum. A wave with two peaks rising like the devil's horns might represent a particle dividing its allegiances equally between two slits.

Under Born's interpretation of quantum mechanics, the identity of a particle undergoes further deconstruction. The old here or there having long passed to the new here *and* there, what is here and there is now a matter of chance. Having impossibly divided itself between two slits by adopting an incarnation as a wave, a single photon undergoes further demotion to appear in quantum mechanics as the ghost of its position. It *could* be here; it *could* be there; and somehow it *could* be at both places at once.

These divided allegiances come to an end abruptly when an observer, paddling in from beyond the quantum system, undertakes a measurement. All at once, the particle that could be here *and* there becomes here *or* there all over again. The wave packet collapses into just one of its possibilities. The wave itself disappears. The other quantum states that it embodies vanish, and they vanish instantaneously. No one knows why.

Widely considered to be inscrutable in conversation owing to the particular flavor of his Danish Grope and Mumble, Bohr embraced this interpretation of quantum mechanics, whence its designation as the Copenhagen interpretation.

"So long as the wave packet reduction is an essential component," John Bell has written, "and so long as we do not know when and how it takes over from the Schrödinger equation, we do not have an exact and unambiguous formulation of our most fundamental physical theory."

If this is so, why is our most fundamental physical theory *fundamental?*

I'm just asking.

QUANTUM MECHANICS IS A theory about the quantum world. Special relativity is a theory about theories and the line they must all toe. Whatever the line, Schrödinger's equation fails to toe it. From the first, physicists understood that major renovations were required to accommodate the plush of one theory in the frame of the other. Having been at once undertaken, renovations proved workable, but they did not appear natural, and the result to this day has some of the unfortunate aspect of home plumbing repairs. The sink works and so, too, the theory.

Classical physicists had always assumed that a particle's position and its momentum could be measured with ever increasing precision. Not so. In 1927, Heisenberg demonstrated that the greater the certainty about a particle's position, the less the certainty about its momentum. The uncertainty is fundamental. It has nothing to do with the limitations of any measuring apparatus. There is no further level of reality in which it is eliminated.

What holds for a particle's position and momentum holds again when its energy and time are measured. If time is short, uncertainties about energy begin ominously to accumulate. But energy and matter are by special relativity deep down the same. A disturbing profligacy now enters into the material world. Staring into empty space, the classical physicist might have said, *look, there is nothing there*, and shrugged. This cannot be true in a quantum system subject to the constraints of special relativity. Since the uncertainty principle holds dominion over the quantum world, uncertainties about energy *are* uncertainties about matter. New pairs of particles are forever lurking in the ontological background and threatening to erupt into existence.

This contingency classical quantum mechanics could not accommodate. It takes its particles neat—one at a time. With a troubled sense that they were being driven by inferences they could barely control, physicists concluded that the very idea of a particle of matter was an artifact of theory, and so an illusion. On a still deeper level of reality, other structures prevailed and other equations were needed to describe them.

It is the quantum field that physicists invoked to explain the jiggling fundaments of the quantum world. A field is something displayed over the whole of space and time. To each point in the field, the physicist associates a set of numbers. The field characterizing the electron is thus marked by numbers representing its mass, momentum, electric charge, as well as its helicity and isospin. A field may be blank and inscrutable as the sun, its structure for a moment indicating nothing that is vibrating on its surface. Under such circumstances, the field is in its vacuum state. But under other circumstances, the field may be excited, quantum numbers, and the properties that they signify, bubbling into agitated life. A compact and excited region of space and time is one in which particles appear like sprightly ghosts. These field-like particles may themselves interact, the natural vocabulary describing particles in motion carried over to fields in action. The great merit of a field-theoretic perspective— indeed, its only merit—is that a field is intrinsically capable of infinite particle creation and annihilation.

First introduced in the late 1920s by Pascual Jordan, Max Born and Paul Dirac, quantum field theory represented an enterprise of enviable mathematical obscurity. This is an advantage it yet possesses. Almost at once, quantum field theory proved mathematically inconsistent, predicting exuberantly that certain finite physical magnitudes were infinite. In the late 1940s, quantum field theory was restored to health by the miracle of a technique known as renormalization. By substituting the measured properties of certain physical parameters (such as charge) for their theoretical value, physicists got infinities to cancel one another. It was not an undertaking calculated to appeal to those unwilling to see how sausages are made. As renormalization achieved its triumphs, Paul Dirac and Richard Feynman both expressed doubts with respect to its methods.

Other physicists were prepared to tuck right in.

At the beginning of the 1960s, physicists understood that there were four forces in play in the material world: The force of gravitation, the electromagnetic force, and the weak and strong nuclear forces. There were no others. Nonetheless, the existence of four fundamental forces seemed as burdensome to physicists as the four wives permitted in the Koran. Why not one?

Within 15 years, physicists had decided the issue in favor of monotheism (or monogamy), their theories converging to a structure known as the Standard Model of particle physics. Within the Standard Model, the strong, the weak and the electromagnetic force have adopted a dignified incarnation as aspects of a single force. "As all suns shrivel in a single sun," Chesterton remarked, "the word is many but the Word is one." The Standard Model is the great triumph of physical thought in the twentieth century. It is as well the last.

The Standard Model comprises three theories. The first is Quantum Electrodynamics or QED. Completed in the late 1940s by Richard Feynman, Julian Schwinger and Sin-Itiro Tomonaga, QED unifies light, electricity, and magnetism, and for this reason retains a direct contact with the world of experience, from computer chips to electric toasters. Without it we would be lost, or at best, inconvenienced.

In 1954, C. N. Yang and Robert Mills, writing with enviable brevity, outlined a daring generalization of QED. Their paper described a new physical structure, but since it predicted particles that no one had seen obeying symmetries that no one could find, their work was studied by a few attentive professionals and then allowed to lapse.

Just a few years later, Murray Gell-Mann and Yuval Ne'eman managed in their theory of the eight-fold way to bring a certain organization to the hadrons by conjecturing that they were all so many varieties of quark. With the proliferation of quarks and their varieties, new symmetries emerged, and they proved to be precisely the symmetries that Yang and Mills had predicted.

Thereafter, the development of the Standard Model acquired an aspect of intellectual inevitability. In the late 1960s, Steven Weinberg, Sheldon Glashow and Abdus Salam created in their Electroweak Theory a structure in which the weak nuclear force and the electromagnetic force were assimilated. Unification is predicated on a form of symmetry conjectured to obtain shortly after the Big Bang. To account for the fact that in the world as it is observed, the weak and electromagnetic forces are not unified, Weinberg, Glashow and Salam appealed to the audacious idea that what physicists could today see of the weak and electromagnetic forces represented a form of broken symmetry, as when couples remember how happy they were amidst the shambles of their discontent.

The strong nuclear force yet remained as a profound puzzle in physical thought. Experiments had indicated that particles bound by the strong nuclear force behaved in ways quite unlike particles governed by the weak nuclear force, or any other force, for that matter. Their interactions seemed to grow stronger as the distance between them increased.

And then in the early 1970s, David Gross, David Politzer and Frank Wilczek discovered in their theory of asymptotic freedom that this was an expected consequence of a Yang-Mills theory of the strong nuclear force. The road to Quantum Chromodynamics—the theory of the strong nuclear force—lay open.

The Standard Model was complete.

IF THE STANDARD MODEL represents a very considerable summing up, just what does the sum tell us about the material world? If nothing else, the Standard Model suggests that the God of the underworld is not much interested in economy of effect. Elementary particles appear either as bosons or fermions. The bosons *act*; the fermions are acted *on*. It is the photon—a boson to the core—that conveys the electromagnetic force in QED. The fermions are, in turn, divided into quarks and leptons. Quarks come in six varieties; and although they come in those six varieties, they are never seen, confined as they are within the hadrons by a force that grows weaker at short distances and stronger at distances that are long. There are equally six leptons. Depending on just how things are counted, matter has as its fundamental constituents 24 elementary particles; it is dominated by four forces; it is described by any number of quantum fields; and the fields in turn obey powerful special-interest symmetries, some of them long broken, and others hustling in the here and now.

It is strange to think that any stable macroscopic order—let alone human life—could have emerged from this seething pre-historic world.

ALTHOUGH A VERY GREAT achievement, the Standard Model proved in some respects unsatisfying. No physicist has ever suggested otherwise. For one thing, it required that a great many numerical parameters be set by hand because they cannot be derived from any underlying theory. For another thing, even within its proper domain of application, the Standard Model leaves certain questions unanswered. Just why *does* the neutrino have mass? No one knows. And finally, the Standard Model does not account for bound states of matter among the elementary particles. It remains a solitary operator.

If there were questions that the Standard Model did not answer, physicists assumed, this indicated only that the Standard Model was a work in process and so a work in progress. There yet remained one respect in which the Standard Model was *radically* unsatisfying: It did not incorporate the force of gravity.

Twentieth-century physics has been divided into alien kingdoms, the one dominated by Einstein's theory of general relativity, the other by quantum mechanics. Since general relativity was designed to satisfy the assumptions of special relativity, tensions between theories run very deep. The Standard Model represents the apotheosis of quantum mechanical ideas, bringing to its appointed conclusion one revolution in thought. But general relativity is unlike the theories that comprise the Standard Model. It is severely non-linear. It is not amenable to the miracle of renormalization. It gives out to another universe. Three of the four forces have been unified with one another, but none of them has been unified with the force of gravity. General relativity and quantum mechanics resemble two aging matadors facing a bull, the both of them retiring flustered after a number of half-hearted veronicas and ineffective passes.

The bull is still there, snorting through velvet nostrils. He does not seem the least bit fatigued.

"THOUSANDS HAVE LIVED WITHOUT love," W. H. Auden observed, "not one without water." Love is important: Water is necessary. If water is necessary, so, too, a great many other things. In a paper entitled "Large Number Coincidences and the Anthropic Principle in Cosmology," and published just as the Standard Model was nearing completion, the physicist Brandon Carter observed that many physical properties of the universe appeared finely tuned to permit the appearance of living systems.

What a lucky break—*things have just worked out.*

What an odd turn of phrase—*finely tuned.*

What an unexpected word—*permit.*

Whether lucky, odd or unexpected, the facts are clear or, at least, suggestive. The cosmological constant is a number controlling the expansion of the universe. If it were larger than it is, the universe would have expanded too quickly to allow the formation of living systems; and if smaller, it would have collapsed too early. Very similar observations have been made with respect to the fine structure constant, the ratio of neutrons to protons, the ratio of the electromagnetic force to the gravitational force, even the speed of light.

Why stop? The second law of thermodynamics affirms that in a general way, things are running down. Except for new-born infants, who would doubt it? The entropy of the universe is everywhere increasing. Time, Death and Longing have by this means entered the world. But if things are running down, what are they running down *from*? This is the question that the distinguished mathematician Roger Penrose has asked. And considering the run-down, he could only conclude that the run-up was an initial state of the universe whose entropy was very, very low, and so very finely tuned.

Who ordered *that*, one might ask?

"Scientists," the physicist Paul Davies has observed, "are slowly waking up to an inconvenient truth—the universe looks suspiciously like a fix."

QUESTIONS THAT PHYSICISTS HAVE asked about the parameters of their theories, they have also asked about their laws. In the end, they are the same question. The fact that the cosmological constant is tuned to an order of 120 decimal places is vexing because it is arbitrary. The same might be said of Newton's law of gravitational attraction or Schrödinger's equation in quantum mechanics.

Why are they true?

Why, indeed, *are* they true?

An appeal to logic is unavailing. The laws of nature are not logical truths. "Blind metaphysical necessity," as Newton observed, "which is certainly the same always and everywhere, could produce no variety of things." An appeal to still another theory is unenlightening. It makes no sense to clamber up an infinite ladder.

If the obvious answers to the question why a final theory is true are either unavailing or unenlightening, in what sense could a final theory be final?

And if not final, how could it anchor a narrative, whether grand or otherwise?

AS HISTORY FLOWS TOWARD certain dates, so it recedes from them. Within the austere confines of mathematical physics, 1973 appears now as a turning point. It marks on the one hand the triumph of the Standard Model; and it marks on the other, the appearance in physical thought of certain troubling questions about the nature of physical thought itself. Such questions have traditionally occupied philosophers. They are, Richard Feynman observed, *always* barging into the conference hall of physics "asking stupid questions." The physicists themselves expected that the serene progression of their discipline would proceed as it had always proceeded: From triumph to triumph. Making every allowance for the inevitable limitations of any system of thought, Steven Weinberg argued in the 1980s that a final theory, although not in hand, was nonetheless within reach. A theory embracing both quantum mechanics, as it had come to be expressed in the Standard Model, and general relativity, would complete the arch of thought.

It was not to be, and if it were to be, it would be on terms physicists had never imagined.

For almost 35 years, a very substantial portion of the community of particle physicists have been studying a subject known as string theory. The effort has consumed the best minds of two generations.

Whereupon there is the inevitable, wait a minute: *Strings?*

Yes, strings. A string is just what its name suggest. It is a wiggling one dimensional object, something like a garden hose although somewhat smaller (roughly the Planck length of 10^{-33} cm.), and extended in length but not width.

Strings can be straight; they can be curved; they can join with themselves to form loops; and what is more, since they are strings, they can vibrate under tension. If nothing else, the idea that nature was deep down a matter of how things were strung along or strung out had one wonderfully invigorating consequence. It made possible in theory the recovery of the elementary particles from *one* fundamental object vibrating in various ways. In place of the complicated system of precisely adjusted forces and parameters characteristic of the Standard Model, string theory pointed to two, and only two, fundamental constraints: The first reflected the string's tension, and so served as the key to its powers of creation; and the second, its coupling constant, the measure of how likely it was to break into two.

Nothing more was needed.

If nothing more was needed of strings, a great deal more was demanded of the universe in which they had taken occupancy. The first quantum theory of strings demanded 26 dimensions, 22 more than are commonly observed. This is an idea with a long history. Very soon after Einstein's theory of general relativity was published, Oskar Klein and

Theodor Kaluza both proposed that gravitational and electromagnetic forces might be unified if only the universe contained a fifth spatial dimension, one hidden under ordinary circumstances. Although absurd, this idea is not *obviously* absurd. An ordinary pane of window glass appears to have only two spatial dimensions: The third dimension is in plane sight but hidden from view. Theories in which extra spatial dimensions figure are known as Kaluza-Klein theories. Einstein was initially charmed by them and quickly grew skeptical.

In addition to those 26 dimensions, bosonic string theory—the first of many—suffered an additional liability. It seemed to sanction tachyons, particles that enjoyed the power to travel backwards in time.

Those extra dimensions were an inconvenience; tachyons were an embarrassment.

Bosonic string theory became a path to glory owing to the discovery of super-symmetry by Pierre Raymond, Julius Weiss and Bruno Zumino in the early 1970s. Super-symmetry allows bosons and fermions to be shuffled around freely. A bosonic string theory facing a mirror saw reflected a theory in which fermions appeared.

Symmetries do not come more super than this, the more so since super-symmetry allowed physicists to reduce the requisite number of dimensions from 26 to 10. If this still left physicists with at least six dimensions more than anyone might wish, at least it allowed them to say to one another that they were moving in the right direction.

On the other hand, super-symmetry did not come cheap. The super-symmetry between fermions and bosons demanded that fundamental fermions be matched to a bosonic *super*-partner and so too the other way around. No one had ever seen such things. The infernal engine of particle multiplication that physicists hoped would stop turning in the night at once coughed into sleep-disturbing life. It is still coughing up super-partners to all the elementary particles: Sleptons, squarks, selectrons—these are the bosonic partners to fermions; and then winos, zinos, photinos, higgsinos and gravitinos—the fermionic partners to bosons.

These bizarre conjectures and their consequences might well have suggested to physicists that they were hopelessly lost were it not for an illumination they received, one that lit up all of particle physics. Working very much in isolation during the late 1970s, Joel Scherk and John Schwarz observed that string theory, no matter how manipulated, seemed to predict the existence of a new massless particle. Almost against their will, Scherk and Schwarz realized that the particle making an appearance amidst all those twitching strings could be interpreted as a graviton, the boson conveying the force of gravity. For the first time, a fundamental theory in particle physics had incorporated a force long missing from any calculations whatsoever.

From that moment on, a number of physicists had the rarest of all experiences: They came to believe that they could hear nature herself knocking at their door.

As FAR AS ANYONE knew in the late 1970s and early 1980s, string theory could well have involved something akin to division by zero. When in 1985, the physicists John Schwartz and Michael Green published a fundamental paper entitled "Anomaly Cancellations in Super-symmetric $D = 10$ Gauge Theory and Superstring Theory," this system of anxieties was eased. The anomalies at issue, as the word might suggest, were mathematically undesirable: By the time-tested method of allowing the left hand to withdraw what the right hand had placed in various equations, Schwartz and Green demonstrated that they cancelled one another. Very shortly thereafter, similar results were achieved for a new class of theories called heterotic string theories by David Gross, Jeff Harvey, Emil Martinec and Ryan Rohm. In an excess of whimsy, the group is known now as the Princeton String Quartet.

These papers initiated what is commonly referred to as the first superstring revolution. In 1995 Edward Witten, the most considerable mathematical physicist of his generation, brought to a close one revolution by inflating the second. Various forms of string theory, Witten observed, were profoundly connected by a quivering network of relationships, strange dualities, and bizarre coincidences. Writing in the notices of the *American Mathematical Society*, he presented a popular account of his thoughts under the title, "Magic, Mystery, Matrix," the alliterations in M serving to draw attention to the new theory that Witten was proposing, which he called M theory. "The five string theories traditionally studied," Witten argued, "are limiting cases of one richer and still little understood theory." In its low energy limit in which nothing much is happening, M theory recovered an old bruiser from the 1970s known as super-gravity. To accommodate super-gravity, *eleven* dimensions were required.

If the various string theories had taxed the mathematical physicists, this new theory would tax them even more. It was very complicated. If no physicist alive could say quite what M theory was, a good many of them working in other fields had rather the reaction to the thing that one might have during a horror movie. They averted their eyes.

There followed a most unusual event in the history of mathematical physics. The theory was criticized in the popular press by a distinguished theoretical physicist and a mathematician. *The Trouble with Physics*, written by Lee Smolin, and *Not Even Wrong*, by Peter Woit, examined string theory with some sympathy and found it wanting. Neither author could find a theory where theoreticians said a theory should be; and both authors noted with some asperity that string theory had no apparent connections to experiment and that none were in prospect. Woit went so far as to observe that the mathematical structure on which string theory rested, far from being a thing of great beauty, was the most horrible thing he had ever seen.

ALL STRING THEORIES ARE characterized by a dimensional overload: There are 26 dimensions if physicists are inclined to welcome tachyons; ten in classical string theories, eleven in M theory. Our own universe contains only three or four dimensions, but in any case, no more than a handful. It is one thing to consider higher dimensions as mathematical artifacts. Mathematicians have no difficulty in dealing with infinite dimensional spaces. They do it all the time. They can take 26 dimensions in their stride. But the extra dimensions of string theory are not purely mathematical. They are within string theory quite real, if only because they have useful work they must do. If real, those extra dimensions are nonetheless invisible. As one might imagine—as one might *easily* imagine—the conflict between the demands of theory—*get me those extra dimensions*—and the constraints of common sense—*no extra dimensions here, Boss, and we looked*—was not easily resolved.

In 1985, Philip Candelas, Gary Horowitz, Andrew Strominger and Edward Witten made the argument that by choosing the right kind of mathematical structures in which to house the dimensional overflow, super-symmetric string theory could very successfully keep one hand in the world of four dimensions (three of space and one of time), while keeping the other in the world of seven (the overflow). It was hardly an idea calculated to appeal to a man with a taste for desert landscapes. The requisite structures were available to string theorists in the form of geometrical objects known as Calabi-Yau manifolds. At each point in space and time, string theorists conjectured, one would find a tiny Calabi-Yau manifold, and curled up within, the extra dimensions of string theory itself.

Having identified the geometrical structures that might absorb string theory's dimensional overflow, physicists were confronted by an embarrassment of riches. There are thousands of Calabi-Yau manifolds and for a time, it appeared as if string theories could be arranged continuously so that between any two theories, another might be found. This was a prospect that elicited little enthusiasm. If extra dimensions seemed a luxury of thought, burying them all somewhere in the wilderness of geometrical shapes seemed perversely to solve one problem—*how to find the extra dimensions?*—only at the expense of uncovering another—*how to hide them all over again?* "It is quite disgusting," the Japanese physicist Noboru Nakanishi wrote, "to revive the old Kaluza-Klein theory without any essentially new idea for resolving the fundamental problem of how to expel the extra dimensions from the physical world."

Writing in 2000, Rafael Bousso and Joseph Polchinski argued by means of an elaborate construction that it was possible to elicit discrete solutions from the machinery of string theory after all. The extra dimensions stayed where they were. They were simply encouraged to better behavior. Shortly thereafter, Renata Kallosh, Andrei Linde, Shamit Kachru and Sandip Trivedi enlarged on these results by means of what Leonard Susskind called a "Rube Goldberg contraption." The number of distinct theories that stood ready to emerge from the contraption varied as a function of the cosmological constant. If the constant is negative or zero, there are infinitely many solutions. No one thought this a good thing. If the cosmological constant is greater than zero, as it is in our universe, there are more than 10^{500} versions of string theory lounging indolently about, and so at least 10^{500} mathematical structures making claims to theoretical domination, each leading to a different version of the theory, a point in an enormous space of possibilities, a landscape of a sort never seen before, a place where each point seemed to embody a different scheme of physical thought, and so a different universe governed by the scheme.

In its appearance in various popular journals the mutant thing was depicted as a gigantic set of bubbles floating in space, our own universe a dimpled dot, lost somewhere amidst that infernally expanding froth.

Having survived any number of near-death experiences, physicists have again found themselves at death's door.

STRING THEORY CONFRONTED THE community of particle physicists with an exquisite dilemma. A theory that initially seemed too *good* to be true had by the late 1990s seemed too good to be *true*. This was widely considered monstrously unjust.

If string theory did not uniquely describe one universe, physicists reasoned, the fault lay with our universe. One universe having proved inadequate, more would be required. Endeavoring to unify the forces of nature, physicists determined to multiply the universes in which they were satisfied. Very few physicists appreciated the irony involved in pursuing the first ambition by embracing the second. The physicist Leonard Susskind thus claimed that "the narrow twentieth-century view of a unique universe, about ten billion years old and ten billion light years across with a unique set of physical laws, is giving way to something far bigger and pregnant with new possibilities."

In service to this idea, Susskind wrote that "physicists and cosmologists are coming to see our ten billion light years as an infinitesimal pocket of a stupendous megaverse." On reflection, Susskind came to understand that the word "megaverse" carried negative class associations, as in mega-blockbuster (a movie no one wishes to see) or mega-mall (a place no one wishes to go), whereupon he re-named the megaverse, the Landscape.

The Landscape at once suggested the radical changes to come. "Theoretical physicists," Susskind wrote, "are proposing theories which demote our ordinary laws of nature to a tiny corner of a gigantic landscape of mathematical possibilities."

Each of the versions of string theory is thus free to find its home in some particular universe. Like Odysseus worshiping in alien temples, there is a universe in which a very large cosmological constant is made to feel right at home. The MIT physicist Max Tegmark is persuaded that this is so, and if in some universe he is persuaded that it is not so, he has learned to accept the emotional incoherence that would trouble others with equanimity.

THE LANDSCAPE IS A new idea in physical thought, but it is not a new idea. Philosophers have long found a single universe burdensome. In the late 1960s, David Lewis assigned possible worlds ontological benefits previously reserved to worlds that are real. In some possible world, Lewis argued, Julius Caesar is very much alive. He is endeavoring to cross the Hudson instead of the Rubicon, and fuming, no doubt, at the delays before the toll both on the George Washington Bridge. It is just as parochial to reject this world as unreal, Lewis argued, as it would be to reject Chicago as unreal because it cannot be seen from New York. Lewis argued brilliantly for modal realism. The absurdity of the resulting view was not an impediment to his satisfaction. Or to mine, needless to say.

Quantum mechanics has also invited the promotion of possible worlds to the ontological Big Time. Under the Copenhagen interpretation of quantum mechanics, a quantum system exists in a superimposed state of all of its possibilities.

In 1957, Hugh Everett III suggested that the superimposed states of a quantum system are all quite real so that a physicist on making a measurement liberates one state to take up residence in one world and another to take up residence in a second. Some physicists were charmed by this idea; and others warmed to it during the subsequent decades.

During the 1980s, the physicist Alan Guth suggested that the early universe was characterized by a period of exponential inflation. Very soon after it blew up in the first place, it blew up again. When suitably blown up, it stopped blowing up. The Stanford physicist André Linde carried this idea a step further in his theory of eternal chaotic inflation. Universes are blowing up all over the place. They cannot stop themselves.

When string theorists talk about the Landscape, they are among friends. If their friends were willing to believe in anything, string theorists, having so lately consorted with 26 dimensions, are hardly in a position to complain.

There is no need to turn to such esoteric doctrines to capture the underlying current of thought that animates the Landscape. It is simply the claim that given sufficiently many universes, what is true in one need not be true in another. This thesis has been current in every college classroom for at least 50 years. It arises spontaneously in discussion, like soap bubbles in water. It is expressed in the same way and often by the same stolid, heavy-thighed undergraduate—a Mr. Waldburg, in my case and class. After raising his hand with the air of a man compelled to observe the obvious, this is what he has to say:

There are no absolute truths.

Waldburg meet Weinberg.

ALTHOUGH INITIATED AS A whim, the Landscape has been welcomed by string theorists as a deliverance. Whether string theory *is* rescued by the Landscape is relatively a trivial matter. Theories come and go and if this one goes, another is sure to come. The Landscape has acquired a life of its own because it is addressed to issues that arise *whatever* the theory *whenever* it arrives. If science, as the French mathematician René Thom once remarked, is an attempt to reduce the arbitrariness of our descriptions, then every theory short of one that is logically necessary must in the end provoke the same two questions: Why are its numerical parameters *as* they are and why are its assumptions *what* they are?

The Landscape provides a generic answer. It is all-purpose in its intent. It works no matter the theory. And it works by means of the simple principle that by multiplying universes the Landscape dissolves improbabilities. To the question *what are the odds?* the Landscape provides the invigorating answer that it hardly matters. If the fine structure constant has in our universe the value of one over one hundred and thirty seven, in another universe it has another value. Given sufficiently many universes, things improbable in one must from the perspective of them all appear certain.

The same reasoning applies to questions about the laws of nature. Why is Newton's universal law of gravitation true?

No need to ask. In another universe, it is not.

The Big Fix has by this maneuver been supplanted by the Sure Thing.

AS ONE HALF OF the radical turn, the Landscape does what it can, and what it does, it does very well. It dilutes the acid of improbability. But as philosophers and physicists at once observed, the Landscape offers a general solution to what is, in fact, a particular problem. The multiplication of universes serves to establish that in some universe, the fine structure constant will take the value of one over one hundred and thirty seven. It is a Sure Thing. Nonetheless, the Sure Thing establishes only that life's lucky numbers will sooner or later turn up somewhere or other.

And yet *they* have turned up *here*, just where *we* need them the most.

Requiring certain amenities, we find ourselves in a universe in which they have been liberally supplied. This may not be a paradox in thought, but surely it seems a very good deal. We might well have found ourselves in a far less agreeable universe, one in which none of life's lucky numbers were tuned to their sweet spot.

And where would we have been then?

The Landscape now works hand in glove with a second radical idea in physical thought. In the same paper in which he drew attention to the question of fine tuning, Brandon Carter observed that "the universe must be such as to admit the creation of observers within it at some stage." Such is the Anthropic Principle, or, at least, one of them, since the Principle now comes in a variety of forms and flavors. It consists, when analyzed, of two quite separate claims. The first is a matter of common sense. If the universe had not admitted the creation of observers at some stage, why, then, we would not be here.

The second is a claim about the facts of life. If surprised by a universe in which we have been given what we need, some part of that surprise, Carter suggested, represents a form of bad faith. If the necessities of life are necessary, they must be inevitable. And if inevitable, whence the surprise?

The simple fact that we are *where* we are is sufficient to explain *why* we have *what* we have.

What more could anyone ask?

The question why the ultimate laws of nature are true, and why its numerical parameters have the value that they do, now admits of a two-part response.

The first is provided by the Landscape. Neither the numbers nor the laws represent anything improbable.

And the second by the Anthropic Principle: If they were false, or if they had different values, where would you be?

Nowhere, right?

And yet here you are.

What *did* you expect?

THE LANDSCAPE AND THE Anthropic Principle together comprise the radical turn favored by Steven Weinberg and Leonard Susskind. Although many distinguished physicists (such as David Gross) have joined them in turning, they have not turned in any great spirit of intellectual optimism. If physicists had hoped to achieve a description of the external world "independent of the perceiving subject," allowing the perceiving subject access to the microphone did not seem obviously a helpful maneuver. Nor has the Landscape afforded physicists concerned to champion materialist monism a desired sense of tranquility. In its attitude toward ontology, the Landscape is, after all, exuberantly indiscriminate—anything goes on because everyone gets in. Yet if the radical turn represents a retreat, it is hardly a retreat that is recent. The Landscape and the Anthropic Principle express what psychopathologists call latent tendencies. They have been there all along.

In the middle of the seventeenth century, Galileo *observed* that sailors aboard smoothly sailing ships could not determine that they were in motion; at the beginning of the twentieth, Einstein *demanded* that the laws of nature be the same in every inertial frame of reference.

Einstein was led to this demand because Clerk Maxwell's theory of the electromagnetic field did not conform to Galilean invariance. "It is known," as Einstein wrote, "that Maxwell's electrodynamics... when applied to moving bodies, leads to asymmetries which do not appear to be inherent in the phenomenon." So it does. A magnet rotating around a conductor produces an electrical field. A conductor rotating around a magnet does not. But inertial frames, Einstein believed, exhibit a natural symmetry. No matter their relative motion (one to the other), deep down, they stay the same. What stays the same when they stay the same are the laws of physics. It is the second postulate of special relativity that preserves Galilean invariance, and that allows Maxwell's equations to be brought panting and to heel. A symmetry in nature whose existence no one suspected had been ratified by a constraint whose power no one had recognized. The result was a great success, with only a few senior physicists, such as Hendrik Lorenz, wondering forlornly what it was all about.

In the reasoning that he employed, and in the conclusions that he reached, Einstein had demonstrated that it is possible to investigate the physical world by placing formal restrictions on the theories that describe it. It is the essence of the revolution that he initiated. Henceforth, physicists would find this style of argument inescapable on those occasions when they found a resolution of their difficulties otherwise impossible.

Brought into existence by assumption, the Landscape serves the purpose of eliminating "certain asymmetries that do not appear to be inherent in the phenomenon." What could be more asymmetrical than a theory that is all brain (too many solutions) and no brawn (too few universes)? Looking at string theory, narrow-minded physicists could see only its brains. Physicists with a broader vision understood that by multiplying universes, brains could be brought into symmetrical alignment with brawn. What might have seemed a deficiency in string theory was by this means revealed as a virtue. If to many philosophers the reasoning employed seemed obviously circular inasmuch as the Landscape, unlike the sun, existed only because string theory would collapse without it, physicists assured themselves that circular reasoning had a place in physical thought after all.

The radical turn has a number of ancestors, and if special relativity is one, quantum theory is another. "Although our situation is not necessarily central," Brandon Carter wrote in explaining the Anthropic Principle, "it is inevitably privileged to some extent." This is a remarkably frank affirmation because it violates the Copernican principle that "our situation" has nothing to do with anything. But although frank, Carter's admission was also obvious. It is the essence of the Copenhagen interpretation of quantum mechanics. The quantum observer stands outside the quantum system, gratefully occupying a classical world in which quantum rules play no part. To make sense of particles that do not respect the common decencies of position and momentum, the classical observer makes use of probabilities: It is the only way in which he can make sense of the quantum world at all.

If the radical turn has been prefigured by latent tendencies within the history of twentieth century physics, it has been prefigured as well by rather *louche* tendencies in post-modern thought. This is not a family connection either side is much minded to advertise—the physicists because, God forbid, they might be associated with an intellectual movement whose members—Jacques Lacan, Jacques Derrida, Julia Kristeva, even Richard Rorty—were of a demonstrable degree of scientific ineptitude, and the post-modernists because of late they have either died or left off being post-modernists. Stanley Grenz, writing in *A Primer on Post-Modernism*, nonetheless expressed his views in a way apt to make physicists wonder whether they and their enemies might not have been friends after all? "Post-modernism affirms that whatever we accept as truth and even the way we envision truth are dependent on the community in which we participate…. There is no absolute truth, rather truth is relative to the community in which we participate." If the odious word "community" is in this quotation replaced by the word "universe," many physicists would be pleased to embrace the result.

Why *did* we ever fight?

THE GREAT DIFFICULTY WITH the Landscape and the Anthropic Principle is that physicists prepared to welcome these ideas had no way in which to control them, while physicists prepared to reject them had no way in which to avoid them. In a stimulating paper entitled "Multiverses and Physical Cosmology," the distinguished cosmologists G. F. R. Ellis, U. Kirchner, and W. R. Stoeger considered the idea that in the Landscape anything goes because everything is possible. "In some universes," they write, "there will be a fundamental unification of physics expressible in a basic 'theory of everything,' in others this will not be so."

But having advanced this opinion, Ellis, Kirchner and Stoeger have neglected to tell us whether *it* is true across the Landscape? If so, then not everything goes; and if not, how could it be of interest?

This is, to be sure, something that Ellis, Kirchner and Stoeger recognize. At the beginning of their essay, they observe that "the very existence of [the Landscape] is based on an assumed set of laws... which all universes... have in common." It is only later in their essay that they forget what they have written.

I know just how it is. I can *never* remember where I left my keys.

The speed with which a commitment to the Landscape ends in incoherence, while it is alarming, is not unexpected. Nothing exceeds like excess, most especially revolutions. "Any scientist," Weinberg writes in defending his endorsement of anthropic reasoning, "must live in a part of the landscape where physical parameters take values suitable for the appearance of life and its evolution into scientists." To say that portions of the Landscape are "suitable" for the appearance of life, is to say that it is *there* that life is *possible*. But if life is possible there, it is not possible elsewhere. Otherwise life would be possible everywhere. Human beings could not, presumably, investigate the universe from the interior of the sun. If life is not possible elsewhere, then it is *necessarily* impossible elsewhere. But what might justify this powerful claim if not some physical principle true *everywhere*? If a principle about life itself is general throughout the Landscape, this would seem to make purely local matters of biology supreme matters of physical thought. This assigns to living systems a degree of cosmic importance that only theologians suspected they possessed.

And if there is at least one such principle, why not others?

And if others, what on earth is the usefulness of either the Landscape or the anthropic principle in the first place?

If no use at all, we are returned to the question with which we began: Why are the laws of Nature true, and why do its numerical parameters have the value that they do?

THESE ARGUMENTS GO BACKWARDS as well as forwards. The mathematical physicist Noboru Nakanishi has been a critic of string theory from the beginning. He has never compromised his conviction that mathematical physics has as its aim a description of an external world "independent of the perceiving subject." If the description has not yet been achieved, that is because it has been pursued with an insufficient sense of how austere a structure it really is. The constraints that a proper theory of everything must meet are very stringent. Quantum theory does not meet them. It does not comes close. And yet, Nakanishi writes, "almost all physicists believe [that] the fundamental laws of physics governing nature [are those] of quantum theory." The conclusion of Nakanishi's argument follows as a prescription: It is to get rid of a good deal. The good deal comprises what Max Tegmark calls "baggage," and what the rest of us might regard as the unavoidable human aspect of any physical theory. "We do not want," Nakanishi writes, "to incorporate the theory of the observation into the framework of the ultimate theory." Such a theory must be free of the human stain, and so disinfected. "It is quite natural to expect that quantum theory is closed within itself without aid of classical theory, because it is impossible to believe that nature has utilized the notion invented by human beings in order to control itself."

With intellectual austerity in mind, Nakanishi proposes the following "principle of quantum priority." In the ultimate theory, "any concept of classical physics must not appear logically prior to its quantum-theoretical construction."

If Nakanishi is strong for restrictive entrance policies when it comes to classical concepts, the mathematician Alexey Kryukov is disposed to draw the line when it comes to space and time. He agrees with Nakanishi about quantum mechanics. He goes further. String theory, although it suggests a unification of forces, does so only because it retains an old-fashioned view of space as the place in which something happens, and time as the arena when it happens. This will not do, if only because it is in conflict with general relativity. He is eager to provide a "model of emergence" in which space and time are derived from a structure in which nothing ever happens and it does not happen at any place either.

If notable in the severity of mind they express, these views are defective in the limitations they impose. Concepts excised from one theory tend inexorably to reappear in another. Having undergone liposuction, dieters will know precisely what I mean.

Whatever else it may be, a physical theory is something written down. It is a *theory*, after all, and not a planet or a beam of light. Like any claim made against the world, theories in physics have an intended interpretation. It is this interpretation that specifies what they mean and where they mean it. The worlds in which theories are satisfied, logicians call models. Theories and their models are the subject of a branch of mathematical logic known as model theory.

A theory having been put on the table, the logician's task is to endow it with meaning. Post-modernists may now be observed rising from their tables in a wave. *Endow it with meaning?* But from what perspective? It is not at all a foolish question. Of what use is a final theory stripped entirely of its human baggage if in explaining it, that human baggage simply reappears? That it must follows directly from mathematical logic itself—a note of caution, a long-distance rebuke.

A theory is an artifact: It comprises a set of sentences. A model is a world. It is out there where worlds generally are. An interpretation is the logician's effort to fix the theory in the model. The logician's perspective—his platform of declaration—encompasses both the theory and its models. Working from a Chinese text, an English translator must see the Chinese for what it means, and see the what it means for what is meant.

The repressed now makes its return, as it always does. The logician's professional perspective is in effect a meta-theory, one piggy-backed on the theory at hand. Model theory establishes that no matter the theory, it is contained within its meta-theory, and so for this reason cannot be logically prior to its meta-theory. A meta-theory *is* a better theory, just as Einstein suspected. This is the content of both Gödel's incompleteness theorem and Tarski's theorem on the indefinability of truth. Both are theorems that apply only to severely formalized languages. No theory in mathematical physics conforms to the pattern that these theorems suggest. It would be reckless to suggest otherwise. But what physicist determined to purge his theory of its human baggage would be inclined to defend the result on the grounds that it is logically defective?

Common sense suggests as well that whatever the meta-theory, it is unavoidably classical. How are we to understand the numbers without the riotous and humanly stained apparatus of classical thinking—no, of *ordinary* thinking? The numbers may well exist beyond space and time. We do not. In order to grasp the simple fact that two follows one, we must understand what it means for one thing to follow another, and this is a concept that makes use of time. In order to get at the numbers, we must write down their numerals, and confronted by a world of shapes, accept assumptions that are in force throughout daily life: That shapes may be recognized; that they are likely to stay the same from one moment to the next; that we can agree on their meaning; that the operations we undertake we undertake in a certain order; and that throughout, we pretty much know where we are and when we are there, the world that we invoke in contemplating even the most abstract of mathematical structures, a familiar one in which orders are given and understood, things undertaken and accomplished, demonstrations offered and approved, certainty suggested and then ratified.

Is this a postmodern conclusion? I suppose it is. I am as surprised as anyone else.

Why did we *ever* fight?

AS HISTORY TENDS TOWARD certain dates, so it recedes from them. Coming in toward 1973—Edward Wilson's *Sociobiology*; the RNA world; the Standard Model. And going out—string theory, fine tuning, evolutionary psychology, post-modernism, the Landscape and the Anthropic Principle. If the mood going in was optimistic, going out it has become fractious—hardly conditions making for smooth sailing. If we are not disposed to accept the Landscape, Leonard Susskind has warned, we shall be "hard pressed" to answer critics prepared to welcome theories of intelligent design. Implacably opposed to the Landscape *and* the Anthropic Principle, the experimental physicist (and Nobel Laureate) Burton Richter has expressed the same morbid concern, but on markedly different grounds. Those who accept the Anthropic Principle, he has argued, are "creationists." No rebuke could be sterner. Both Susskind and Richter agree that were views they condemn to prevail, the result would be a very great misfortune, Susskind because he believes that unless his views are accepted, the result would be some form of intelligent design, and Richter because he believes that unless those views are rejected, the result would be some form of creationism. Since one view or the other must be right, the only certainty to emerge from this exchange is one that neither man intended.

Having entered this essay with its origins unknown, the physical world leaves this essay with its nature undetermined. The Standard Model remains the last unalloyed triumph of the physical sciences. Like a great stone Buddha, it remains squatting on the roads of time. Another great stone Buddha embodying general relativity is at a distance. The two Buddhas do not talk to one another. Their continued commitment to peaceful coexistence is widely regarded as a miracle. The theories that we possess are "magnificent, profound, difficult, sometimes phenomenally accurate," as the distinguished mathematician Roger Penrose has observed, but, as he at once adds, they also comprise a "tantalizingly *inconsistent* scheme of things." The theories that we might wish, such as string theory, do achieve a form of unity, but at such a cost in plausibility as to place unity itself in doubt as a goal. If one universe will not do, why not settle for *none* instead? What is left when this renunciation is enforced are various physical *theories*—general relativity and quantum mechanics, most obviously; but all the others as well. Each describes what it must. The universe remains what it was.

For more than three hundred years, the physical sciences have exploited a single method. The method comprises the "model for what science should be." The model is so simple and it has led to so many triumphs, that it is easy to conclude that it has some aspect of intellectual inevitability. The simplicity of the model is derived from the obvious way in which it divides our intellectual experience. To account for change in nature, it is necessary to specify *how* things evolve and the conditions *from which* they evolved. Western science is coextensive with this method. There is no escaping it.

One might at least wonder whether the method might finally have exhausted itself? What can be discovered by its employment has been discovered; what can be known by these means is known. Although much has since 1973 been proposed, much less has been accomplished, and for obvious reasons. The imperative that theories in physics be judged by experiment is becoming more and more difficult to respect. Getting to the rock face of data now requires a massive apparatus, such as the hadron collider under construction at CERN. Experiments may reveal clues about the elementary particles, but physicists have not over the course of three hundred years *ever* been satisfied with clues.

It is for this reason that string theory is so conspicuously dominated by the mathematics involved in its construction. And why not? "Mathematics," the mathematician Vladimir Arnold has argued, "is the part of physics where experiments are cheap." This view has also been defended by the American logician Gregory Chaitin. Max Tegmark has argued the reverse. It is physics that is a part of mathematics. The universe is entirely a mathematical structure. If it is not clear whether mathematics is a part of physics, or whether physics is a part of mathematics, perhaps the only conclusion safely to be drawn is that something is not clear.

What implications in all this for the grand narrative of our times? Where do the arrows of explanation in the end point?

The plain truth—no trivial thing, of course—is that no one knows. It is odd and remarkable that in the face of theories that have proven inconclusive such as string theory, physicists have concluded that they must at once change the standards by which their theories are judged. When it is not possible to argue the facts, lawyers quite understand, then it is necessary to argue the law. In this the physicists have unwittingly drawn close to doctrines that previously they had rejected as frivolous. But neither physicists disposed radically to change the law, nor physicists disposed radically to reject the change, have made arguments that have persuaded the other side. And if they cannot persuade one another, surely it is unreasonable for either side to expect that they have persuaded *us*.

What we *can* say is that something is changing and that something has changed. The idea that the goal of the physical sciences is to reach a perspective "independent of the perceiving subject" has acquired the unfortunate aspect of a principle that while it commands assent no longer commands respect. The principles that have been proposed as its replacement, while they command respect, have not yet commanded assent. "I know well what I am fleeing from," Montaigne observed, "but not what I am in search of."

Why *are* we surprised?

INDEX

Breinigsville, PA USA
04 October 2009

225198BV00002B/2/P